ROBUSTNESS OF STATISTICAL METHODS

AND NONPARAMETRIC STATISTICS

THEORY AND DECISION LIBRARY

General Editors: W. Leinfellner and G. Eberlein

Series A: Philosophy and Methodology of the Social Sci

Editors: W. Leinfellner (University of Nebraska)
G. Eberlein (University of Technology, Munich)

Series B: Mathematical and Statistical Methods

Editor: H. Skala (University of Paderborn)

Series C: Game Theory and Decision Making

Editor: S. Tijs (University of Nijmegen)

Series D: Organization and Systems Theory

Editor: W. Janko (University of Karlsruhe)

SERIES B: MATHEMATICAL AND STATISTICAL METHODS

Scope

The series focuses on the application of methods and ideas of logic, m and statistics to the social sciences. In particular, formal treatment phenomena, the analysis of decision making, information theory and pr inference will be central themes of this part of the library. Besides results, empirical investigations and the testing of theoretical models of problems will be subjects of interest. In addition to emphasizing interd communication, the series will seek to support the rapid dissemination results.

ROBUSTNESS OF STATISTICAL METHODS AND NONPARAMETRIC STATISTICS

Edited by
Dieter Rasch and
Moti Lal Tiku

D. REIDEL PUBLISHING COMPANY
A MEMBER OF THE KLUWER ACADEMIC PUBLISHERS GROUP
DORDRECHT / BOSTON / LANCASTER / TOKYO

Library of Congress Cataloging-in-Publication Data

CIP

Main entry under title:

Robustness of statistical methods and nonparametric statistics.

1. Robust statistics — Addresses, essays, lectures.
2. Nonparametric statistics — Addresses, essays, lectures.
I. Rasch, Dieter. II. Tiku, Moti Lal.
QA276.16.R63 1985 519.5 85-18286
ISBN-13: 978-94-009-6530-0 e-ISBN-13: 978-94-009-6528-7
DOI:10.1007/978-94-009-6528-7

Distributors for the Socialist countries
VEB Deutscher Verlag der Wissenschaften, Berlin

Distributors for the U.S.A. and Canada
Kluwer Academic Publishers,
190 Old Derby Street, Hingham, MA 02043, U.S.A.

Distributors for all remaining countries
Kluwer Academic Publishers Group,
P.O. Box 322, 3300 AH Dordrecht, Holland.

Published by VEB Deutscher Verlag der Wissenschaften, Berlin
in co-edition with
D. Reidel Publishing Company, Dordrecht, Holland.

Preface

This volume contains most of the invited and contributed papers presented at the Conference on Robustness of Statistical Methods and Nonparametric Statistics held in the castle of Schwerin, Mai 29 — June 4 1983. This conference was organized by the Mathematical Society of the GDR in cooperation with the Society of Physical and Mathematical Biology of the GDR, the GDR-Region of the International Biometric Society and the Academy of Agricultural Sciences of the GDR. All papers included were thoroughly reviewed by scientist listed under the heading "Editorial Collaboratories". Some contributions, we are sorry to report, were not recommended for publication by the reviewers and do not appear in these proceedings. The editors thank the reviewers for their valuable comments and suggestions.

The conference was organized by a Programme Committee, its chairman was Prof. Dr. Dieter Rasch (Research Centre of Animal Production, Dummerstorf-Rostock). The members of the Programme Committee were

Prof. Dr. Johannes Adam (Martin-Luther-University Halle)
Prof. Dr. Heinz Ahrens (Academy of Sciences of the GDR, Berlin)
Doz. Dr. Jana Jurečkova (Charles University Praha)
Prof. Dr. Moti Lal Tiku (McMaster University, Hamilton, Ontario)

The aim of the conference was to discuss several aspects of robustness but mainly to present new results regarding the robustness of classical statistical methods especially tests, confidence estimations, and selection procedures, and to compare their performance with nonparametric procedures. Robustness in this sense is understood as intensivity against violation of the normal assumption. Three approaches can be found

— analytical approach for continuous distributions
— combinatorial approach for k-point distributions
— simulations in a system of distributions (Pearson system, Fleishmann system)

The simulation studies were well designed and some papers deal with testing the pseudo-random number generators used. Most of the results have not been published elsewhere and appear in these proceedings for the first time. Some papers deal with the robustness issues on the lines of Huber although the main emphasis of the conference was to study robustness in the Pearson framework. Some papers deal with robust experimental designs and some with nonparametric methods (classification, estimation, etc.).

We thank the members of the programme committee, the assistant editor, the reviewers and the contributors of papers for their cooperation and assistance in making these proceedings possible.

Dieter Rasch and Moti Lal Tiku
Rostock and Hamilton
February 1984

ROBUSTNESS OF STATISTICAL METHODS AND NONPARAMETRIC STATISTICS

Edited by Dieter Rasch and Moti Lal Tiku

Proceedings of a Conference held on these topics at Schwerin
May 29 – June 4 1983
Organized by the Mathematical Society of the GDR

Editors:

Dieter Rasch
Research Centre of Animal Production Dummerstorf-Rostock of the Academy of Agricultural Sciences of the GDR and
Mathematical Department of Wilhelm Pieck University Rostock, GDR
Moti Lal Tiku
Department of Statistics McMaster University, Hamilton (Ontario), Canada

Assistent Editor:

Günter Herrendörfer
Research Centre of Animal Production Dummerstorf-Rostock of the Academy of Agricultural Sciences of the GDR

Editorial Collaboratories:

Ahrens, H. (Berlin); Atkinson, A. (London); Bellach, J. (Berlin); Bunke, H. (Berlin); Dijkstra, J. B. (Eindhoven); Enderlein, G. (Berlin); Enke, H. (Halle); Field, Ch. (Halifax); Gaffke, N. (Aachen); Heckendorff, H. (Karl-Marx-Stadt); Herrendörfer, G. (Rostock); Hischer, K. (Rostock); Horn, M. (Jena); Jurečkova, J. (Prag); Laan, P. van der (Wageningen); Liese, F. (Jena); Neumann, P. (Dresden); Pilz, J. (Freiberg); Posten, H. O. (Connecticut); Rasch, D. (Rostock); Rudolph, E. (Rostock); Schmidt, W. (Berlin); Tiku, M. L. (Hamilton); Toutenburg, H. (Berlin); Verdooren, P. (Wageningen); Zielinski, R. (Warschau); Zwanzig, S. (Berlin)

CONTENTS Page

Department of Statistics, Charles University, Prague

Behaviour of L-Estimators of Location from the Point of View of Large Deviations

JAROMÍR ANTOCH

Abstract

Let $X_1, \ldots, X_n$ be a random sample from a population with density $f(x-\Theta)$ such that $f(x)$ is symetric and positive. It is proved that the tails of the logarithmic derivative of the density of L-estimators of Θ converge at most n-times faster than the tails of the logarithmic derivative of the basic density and, on the other hand, there are estimators which behave from this point of view in the same way as one single observation. It is shown that both extreme cases may happen for the sample mean. Moreover, behaviour of some typical L-estimators of Θ is studied from this point of view.

1. Introduction

Let $X_1, \ldots, X_n$ be a sequence of iid rv's distributed according to an absolutely continuous and symetric density $f(x-\Theta)$, $x \in R_1$, $\Theta \in R_1$. For each fixed n let $T_n = T_n(X_1, \ldots, X_n)$ be an equivariant estimator of Θ based on observations $X_1, \ldots, X_n$. Different measures of performance of T_n have been suggested and investigated. Our approach differs from more customary methods of investigation of the behaviour of T_n mainly in two points:

I. We consider rather the tail-behaviour than the local behaviour of the distribution of T_n ; the sample size is fixed.

II. The behaviour of T_n is studied through the logarithmic derivative of its density rather than through its distribution function or its density itself; this is done due to the following reasons:

As it was pointed out by Hampel (1973) and Field and Hampel (1982), $q(x-\Theta) = f'(x-\Theta) / f(x-\Theta)$ provides more basic description of a probability distribution than density or the cumulative distribution. This function describes well the behaviour of the distribution at the fixed point and, morover, it is often of a simple form. The fact that it is linear for the normal distribution reflects the important position of the normal distribution among the other distributions; the approximation of $q(x-\Theta)$ locally by a linear function corresponds to the approximation by the normal distribution.

Let us denote density of T_n by means $f_n(x-\Theta)$ and put $q_n(x-\Theta) = f'_n(x-\Theta) / f_n(x-\Theta)$. Some of upper arguments can be used when taking $q_n(x-\Theta)$ as basic description of distribution of T_n. Moreover, the paper of Field and Hampel (1982) gives asymptotic approximations of $q_n(x-\Theta)$ very precisely for very small sample sizes ($n = 3, 4$) even in the extreme tails. Methods of this paper can be very effectively used in robust testing the hypotheses as was shown by Ronchetti (1982).

These were the main reasons why we decided to describe the behaviour of T_n with the aid of $q_n(x-\Theta)$ for $x \to +\infty$. More precisely, our criterion is

$$\lim_{x \to +\infty} \left(-q_n(x-\Theta)\right) / \left(-q(x-\Theta)\right), \tag{1.1}$$

if this limit exists.

We shall show that the rate of convergence of $q_n(x-\Theta)$ cannot be more than n-times faster than that of $q(x-\Theta)$, while the rate of $q_n(x-\Theta)$ can be also as slow as the rate of $q(x-\Theta)$. Both extreme cases may happen even for the sample mean. The upper bound is attained, e. g., for a sample from the normal distribution while the lower bound for a sample from the Cauchy distribution.

If we trimm-off some extreme observations, then the rate of convergence cannot attain the upper bound. The same result holds for lower bound and samples from distributions with exponential tails while, surprisingly, this is not the case for lower bound and samples from distribution with heavy tails, see Theorem 3.2.

2. Model

In this section three basic models will be introduced and some properties of them discussed. Before doing it we shall give some necessary notions.

ASSUMPTION A: The random variable X has absolutely continuous distribution function $F(x)$, $F(x) + F(-x) = 1$ for all $x \in R_1$ and absolutely continuous density $f(x)$, $f(x) > 0$ for all $x \in R_1$ and has finite and positive Fisher information.

Let us denote:

$$r(x) = f(x) / \left(1 - F(x)\right), \qquad q(x) = f'(x) / f(x) \tag{2.1}$$

and

$$B(x) = \left(f(x)/F(x)\right) / q(x), \quad A(x) = r(x)/q(x). \tag{2.2}$$

Now we can introduce the models we shall be interested in.

M I. Let F(E) be a class of random variables fulfilling the assumption A and such that $f'(x) < 0$, $f''(x)$ exists and $r(x)$ is increasing for all $x \geq K_1(f) > 0$, $\lim_{x \to +\infty} A(x) = 1$.

M II. Let F(P) be a class of random variables fulfilling the assumption A and such that $f'(x) < 0$, $f''(x)$ exists and $r(x)$ is decreasing for all $x \geq K_2(f) > 0$.

M III. Let F(H) be a class of random variables fulfilling the assumption A and such that $f'(x) < 0$, $f''(x)$ exists and $r(x)$ is constant for all $x \geq K_3(f) > 0$.

Remark 2.1. (i) The class F(E) is usually called that of densities of exponential type for large values of x, see e. g. Gumbel (1956). This is due to the fact that the tails of densities of random variables from F(E) decrease at least exponentially fast. On the other hand, not all densities with exponentially decreasing tails belong to F(E). This is, e. g., the case of class F(H), family of symetric lognormal distribution etc. Nevertheless, F(E) covers most commonly used densities of exponential type like normal, logistic, symetric Gamma, symetric Weibull distributions as well as some classes of Pearson's, Burr's and Johnson's curves etc.

(ii) The basic representants of F(P) are random variables with densities having the tails $\sim A \cdot x^s$, for all $x \gg 0$, $s < -1$ and $A > 0$ (with Cauchy's distribution beeing the most typical representant. Nevertheless, there exist families of random variables with the tails of exponential type decreasing more slowly than the tails of Laplace's distribution, what is the case of the family of symetric lognormal distribution etc. Moreover, some classes of Pearson's, Burr's and Johnson's curves belong to F(P), too.

(iii) If we compare M I., M II. and M III., we can see that F(H) create the natural bound between F(P) and F(E). To F(H) belong all densities satisfying condition A with the tails $\sim M \cdot \exp(-cx)$ for all $x \geq K_3(f)$, $c > 0$, $0 < M < +\infty$; with the family of Laplace's densities being the most typical representant.

(iv) Let us denote $F_c = F(E) \cup F(P) \cup F(H)$.

3. Behaviour of L-Estimators

The following theorem describes the behaviour of order statistics.

Theorem 3.1. Let random variables $X_1, \ldots, X_n$ be independent copies of $X \in F_c$ with common density $f(x-\Theta)$, $\Theta \in R_1$ and $X_{(1)} < \ldots < X_{(n)}$ be their respective order statistics. Let $f_{(k)}(x-\Theta)$ be the density of $X_{(k)}$ and

$$U_{(k)}(x-\Theta) = \frac{f'_{(k)}(x-\Theta)}{f_{(k)}(x-\Theta)} \bigg/ \frac{f'(x-\Theta)}{f(x-\Theta)}, \quad 1 \leq k \leq n.$$

It holds:

(i) if $X \in F(E)$ or $F(H)$, then $\lim\limits_{x \to +\infty} U_{(k)}(x-\Theta) = n-k+1$, $1 \leq k \leq n$;

(ii) if $X \in F(P)$ then

$$1 \leq \underline{\lim}_{x \to +\infty} U_{(k)}(x-\Theta) \leq \overline{\lim}_{x \to +\infty} U_{(k)}(x-\Theta) \leq n-k+1, \quad 1 \leq k \leq n.$$

Proof: Without loss of generality we can put $\Theta = 0$. Using (2.2.) we can rewrite $U_{(k)}(x)$ in the form

$$\begin{aligned} U_{(k)}(x) &= 1-(k-1)B(x)+(n-k)A(x) = \\ &= 1-(k-1)A(x)\big((1-F(x))/F(x)\big)+(n-k)A(x). \end{aligned}$$

Part (i) now follows immediately from M I., resp. M III. and remark 2.1. (iii), because $(1-F(x))/F(x) \to 0$ for $x \to +\infty$. It was proved in Barlow and Proschan (1966), Theorem 4.1., that for every $X \in F(P)$ exists positive and finite constant $K_2(f)$ such that $0 < A(x) < 1$ for all $x \geq K_2(f)$ and this implies (ii).

qed

Corolary: Under the assumptions of Theorem 3.1.

$$1 \leq \underline{\lim}_{x \to +\infty} U_{(k)}(x-\Theta) \leq \overline{\lim}_{x \to +\infty} U_{(k)}(x-\Theta) \leq n, \quad 1 \leq k \leq n.$$

The following theorem gives upper and lower bounds for the rate of convergence of the tails of the logaritmic derivative of density of L-estimators of location.

Theorem 3.2.: Let random variables $X_1, \ldots, X_n$ be independent copies of $X \in F_c$ with common density $f(x-\Theta)$, $\Theta \in R_1$ and $X_{(1)} < \ldots < X_{(n)}$ be their respective order statistics. Let $c_1, \ldots, c_n$ be nonnegative constants such that $\sum\limits_{k=1}^{n} c_k = 1$ and T_n be an L-estimator of Θ of the form

Then

$$T_n = \sum_{i=1}^{n} c_i X_{(i)}.$$

$$1 \leq \underline{\lim}_{x \to +\infty} H_n(x-\Theta) \leq \overline{\lim}_{x \to +\infty} H_n(x-\Theta) \leq n, \tag{3.1}$$

where $g_n(x-\Theta)$ is the density of T_n and

$$H_n(x-\Theta) = \frac{g'_n(x-\Theta)}{g_n(x-\Theta)} \bigg/ \frac{f'(x-\Theta)}{f(x-\Theta)}.$$

Proof: Without loss of generality we can put $\Theta = 0$. The existence of $q_n(x-\Theta) = g'_n(x-\Theta) / g_n(x-\Theta)$ was proved, e. g. in Klaassen (1981). Let $f_{(k)}(x)$ denote the density of $X_{(k)}$, $1 \leq k \leq n$. Let us show that there exists a finite constant L_1 such that

$$f_{(1)}(x) \leq g_n(x) \leq f_{(n)}(x) \quad \text{for all } x > L_1. \tag{3.2}$$

We shall show in details only lower inequality in (3.2.) for the case $c_1 > 0$, $c_n > 0$, $n > 2$, because in all other cases the proof follows analogously.

Using the well known density of the vector of order statistics $(X_{(1)}, \ldots, X_{(n)})$ and transformations

$$Z_1 = \sum_{i=1}^{n} c_i X_{(i)}, \; Y_2 = X_{(2)}, \ldots, Y_n = X_{(n)}.$$

resp.

$$Z_2 = X_{(1)}, \; Y_2 = X_{(2)}, \ldots, Y_n = X_{(n)},$$

we can $g_n(z)$ and $f_{(1)}(z)$ express in the form

$$g_n(z) = n! \int \ldots \int \prod_{i=2}^{n} f(y_i) \cdot f\left(\frac{z - \sum_{i=2}^{n} c_i y_i}{c_1}\right) \cdot I(A_z) \cdot c_1^{-1} \, dy_2 \ldots dy_n$$

resp.

$$f_{(1)}(z) = n! \int \ldots \int \prod_{i=2}^{n} f(y_i) \cdot f(z) I(B_z) \, dy_2 \ldots dy_n,$$

where

$$A_z = \left\{ y = (y_2, \ldots, y_n) \;\middle|\; \frac{z - \sum_{i=2}^{n} c_i y_i}{c_1} < y_2 < \ldots < y_n \right\}$$

and

$$B_z = \{ y = (y_2, \ldots, y_n) \mid z < y_2 < \ldots < y_n \}.$$

It is easy to show that $A_z \supset B_z$, $A_z - B_z \neq \emptyset$ and for each $y \in B_z$

$$z > c_1^{-1} \cdot \left(z - \sum_{i=2}^{n} c_i y_i \right).$$

From assumptions M I.–M. III. we know that there exists a constant $K(f)$, $0 < K(f) < +\infty$, such that $f(x)$ is decreasing for all $x > K(f)$, so that for all $z > K(f)$ and all $y \in B_z$

$$f(z) < f\left(c_1^{-1} \cdot \left(z - \sum_{i=2}^{n} c_i y_i\right)\right) < c_1^{-1} \cdot f\left(c_1^{-1} \cdot \left(z - \sum_{i=1}^{n} c_i y_i\right)\right).$$

Lower inequality in (3.2.) now follows immediately becouse

$$g_n(z)-f_{(1)}(z)=n!\int\dots\int\prod_{i=2}^{n} f(y_i)\cdot I(B_z)\cdot$$

$$\cdot\left(c_1^{-1}\cdot f\left(c_1^{-1}\cdot\left(z-\sum_{i=2}^{n} c_i y_i\right)\right)-f(z)\right)dy_2\dots dy_n+$$

$$+\frac{n!}{c_1}\int\dots\int\prod_{i=2}^{n} f(y_i)\cdot f\left(c_1^{-1}\cdot\left(z-\sum_{i=2}^{n} c_i y_i\right)\right)\cdot I(A_z-B_z)$$

$$dy_2\dots dy_n.$$

Starting from (3.2.) and regarding that $f(x)\downarrow 0$ as $x\to+\infty$, there exists finite constant $L_2>L_1$ such that $-\ln f(x)>0$ for all $x\geq L_2$, hence

$$\frac{-\ln f_{(n)}(x)}{-\ln f(x)}\leq\frac{-\ln g_n(x)}{-\ln f(x)}\leq\frac{-\ln f_{(1)}(x)}{-\ln f(x)}\quad\text{for all } x\geq L_2$$

and

$$\overline{\lim_{x\to+\infty}}\frac{-q_n(x)}{-q(x)}=\overline{\lim_{x\to+\infty}}\frac{-\ln g_n(x)}{-\ln f(x)}\leq\overline{\lim_{x\to+\infty}}\frac{-\ln f_{(1)}(x)}{-\ln f(x)}=$$

$$=\overline{\lim_{x\to+\infty}}\left(-\frac{f'_{(1)}(x)}{f_{(1)}(x)}\right)\Big/\left(-\frac{f'(x)}{f(x)}\right)\leq n$$

according theorem 3.1.

Analogously

$$\underline{\lim_{x\to+\infty}}\frac{-q_n(x)}{-q(x)}=\underline{\lim_{x\to+\infty}}\frac{-\ln g_n(x)}{-\ln f(x)}\geq\underline{\lim_{x\to+\infty}}\frac{-\ln f_{(n)}(x)}{-\ln f(x)}=$$

$$=\underline{\lim_{x\to+\infty}}\left(-\frac{f'_{(n)}(x)}{f_{(n)}(x)}\right)\Big/\left(-\frac{f'(x)}{f(x)}\right)=1.$$

qed

The following theorem shows the effect of trimming off some extreme order statistics on the tail behaviour of T_n.

Theorem 3.3.: Let random variables $X_1,\dots,X_n$ be independent copies of $X\in F_c$ with common density $f(x-\Theta)$, $\Theta\in R_1$ and let $X_{(1)}<\dots<X_{(n)}$ be their respective order statistics. Let $c_1,\dots,c_n$ be nonnegative constants such that $\sum_{k=1}^{n}c_k=1$. Put $c_0=c_{n+1}=0$ and assume that $c_s=0$ for $0\leq s\leq i$ and $n-j+1\leq s\leq n+1$, $0\leqq i+j<n$, $c_{i+1}>0$ and $c_{n-j}>0$. Denote $g_n(x-\Theta)$ density of the statistic

$$T_n=\sum_{i=1}^{n}c_i X_{(i)}$$

and put

$$H_n(x-\Theta)=\frac{g_n'(x-\Theta)}{g_n(x-\Theta)}\Big/\frac{f'(x-\Theta)}{f(x-\Theta)}$$

It hold:

(i) If $X\in F(P)$ then

$$1\leq\underline{\lim_{x\to+\infty}}H_n(x-\Theta)\leq\overline{\lim_{x\to+\infty}}H_n(x-\Theta)\leq n-i;\tag{3.3}$$

(ii) if $X\in F(E)$ or $F(H)$ then

$$j+1\leq\underline{\lim_{x\to+\infty}}H_n(x-\Theta)\leq\overline{\lim_{x\to+\infty}}H_n(x-\Theta)\leq n-i.\tag{3.4}$$

Proof: We can put $\Theta=0$ without loss of generality. It follows from the assumptions that

$$X_{(i+1)}\leq T_n\leq X_{(n-j)}.$$

Proceeding in the same way as in the proof of the theorem 3.2., we can show that exists finite constant L_3 such that

$$f_{(i+1)}(x)\leq g_n(x)\leq f_{(n-j)}(x)\quad\text{for all } x\geq L_3,$$

hence

$$\frac{-\ln f_{(n-j)}(x)}{-\ln f(x)}\leq\frac{-\ln g_n(x)}{-\ln f(x)}\leq\frac{-\ln f_{(i+1)}(x)}{-\ln f(x)}$$

for all $x\geq L_4>L_3$; that gives the desired conclusions.

qed

4. Examples

Let us illustrate the results on the behaviour of the sample mean and sample median.

Theorem 4.1.: Let $X_1,\dots,X_n$ be independent copies of random variable $X\in F_c$ with the common density $f(x-\Theta)$, $\Theta\in R_1$. Let $g_n(x-\Theta)$ denote the density of the sample mean T_n and $H_n(x-\Theta)$ be defined as above. Then

$$1\leq\underline{\lim_{x\to+\infty}}H_n(x-\Theta)\leq\overline{\lim_{x\to+\infty}}H_n(x-\Theta)\leq n.\tag{4.1}$$

Proof: Follows immediately from theorem 3.2.

The example demonstrates that both bounds in (4.1.) are attainable.

Example 4.1.: Let $\mathcal{L}(X_i)\sim N(\Theta,\sigma^2)$, $i=1,\dots,n$,

Then $\mathcal{L}(T_n)\sim N(\Theta,\sigma^2/n)$ and

$$H_n(x-\Theta)=(n(x-\Theta)/2)/((x-\Theta)/2)=n\quad\text{for all } x\in R_1,$$

so that the upper bound is attained not only for $x\to+\infty$, but for all real x.

(b) Let $f(x-\Theta)=\pi^{-1}\cdot(1+(x-\Theta)^2)^{-1}$ for all $x\in R_1$, $\Theta\in R_1$. Then T_n is distributed according to the same Cauchy distribution and

$$H_n(x-\Theta)=1\quad\text{for all real } x,$$

so that the lower bound is attained not only for $x\to+\infty$, but for all real x.

Theorem 4.2.: Let $X_1,\dots,X_n$ be independent copies of random variable $X\in F_c$ with common density $f(x-\Theta)$, $\Theta\in R_1$. Let $g_n(x-\Theta)$ denote the density of the sample median T_n and $H_n(x-\Theta)$ be defined as above. It holds:

(i) if $X\in F(P)$ then

$$1\leq\underline{\lim_{x\to+\infty}}H_n(x-\Theta)\leq\overline{\lim_{x\to+\infty}}H_n(x-\Theta)\leq\left[\frac{n+2}{2}\right]\tag{4.2}$$

(ii) if $X\in F(E)$ or $F(H)$ then

$$\left[\frac{n+1}{2}\right]\leq\underline{\lim_{x\to+\infty}}H_n(x-\Theta)\leq\overline{\lim_{x\to+\infty}}H_n(x-\Theta)\leq\left[\frac{n+2}{2}\right];\tag{4.3}$$

where $[\cdot]$ means the function integral part.

Proof: Follows immediately from theorems 3.1. and 3.3.

Remark: If we compare these results with those of Jurečková (1981) we can see that it is the lower bound in (4.2.) which is surprising. Nevertheless, it is easy to show that it is attainable in such a way that for every ε, $0<\varepsilon<1$, exists $X\in F(P)$ such that

$$1<\lim_{x\to+\infty}H_n(x-\Theta)<1+\varepsilon.$$

Actually, let, say, $n=2k+1$, $\Theta=0$ and

$$\begin{aligned}f(x)&=K &&\text{for all } x\ |x|<A,\\ &=K\cdot|x|^{-s-1} &&\text{for all } x\ |x|\geq A,\end{aligned}$$

where $0 < A < +\infty$ and K is normalizing constant. Then (4.4.) is true if

$$0 < s < \frac{\varepsilon}{K-\varepsilon}.$$

This phenomenon is typical for densities with extremely heavy tails and one can see that every shift of some mass from the tails to the center substantially improves the behaviour of the sample median from the point of view of the chosen criterion.

6. Refrences

BARLOW, R. E., PROSCHAN, F.
Mathematical theory of reliability.
J. Wiley and Sons., New York, 1966.

FIELD, Ch. A., HAMPEL, F. R.
Small sample asymptotic distribution of M-estimators of location.
Biometrika 69, (1973), 29–46.

GUMBEL, E. J.
Statistics of extremes.
Columbia University Press, New York, 1962.

HAMPEL, F. R.
Some small sample asymptotics.
Proceedings of the 1st Prague Conference on Asymptotic Statistics, Vol. II, (1973), 109–126.

JUREČKOVÁ, J.
Tail behaviour of location estimators.
AS 9, (1981), 578–585.

KLAASSEN, C. A. J.
Statistical performance of location estimators.
Mathematical Centre Tract 133, Amsterdam 1981.

RONCHETTI, E.
Robust testing in linear models – infinitesimal approach.
Dissertation ETH No. 7084, Zürich, 1982.

Imperial College, London

Simulation in Research on Linear Models

ANTHONY C. ATKINSON

Abstract

Sketches are given of six applications of simulation in research on linear models. Fairly full references are given to more extended treatments of the topics, as they are to recent developments in the generation of pseudo-random numbers and variables.

1. Introduction

Although the mathematics of least squares has been understood for over 150 years, research into statistical methods for the linear model can still yield problems to which an analytical solution is not possible. This is clearly frequently the case in recent developments, such as the generalized linear model (Nelder and Wedderburn, 1972; McCullagh and Nelder, 1983) where, usually, only asymptotic results are available. To answer many inferential questions recourse has then to be made to simulation. The purpose of the talk, on which this paper is based, was to describe examples of the use of simulation. These, it is hoped, both exemplify some recent advances in statistical techniques and also illustrate general principles in the design and analysis of simulation experiments.

2. Robust Regression

The estimating equation for the location parameter of a simple sample can be written as

$$\sum_{i=1}^{n} (y_i - \hat{\mu}) = 0. \tag{1}$$

which is the least squares solution yielding the sample mean. In robust estimation using M-estimates (1) is replaced by

$$\sum_{i=1}^{n} \psi\left\{(y_i - \tilde{\mu})/\sigma\right\} = 0. \tag{2}$$

One example is Huber's 'Proposal 2' in which

$$\psi(z) = \begin{cases} -c & (z \leq -c) \\ z & (-c \leq z \leq c) \\ c & (z \geq c). \end{cases}$$

Details of this and other methods of robust estimation are given in Huber (1981).

For least squares regression the analogue of (1) is the set of p equations

$$X^T\left(y - X\hat{\beta}\right) = X^T r = 0. \tag{3}$$

where r is the vector of n least squares residuals. The robustified version of (3), analogous to (2) is

$$X^T\psi\left\{\left(y - X\tilde{\beta}\right)/\sigma\right\} = X^T\psi(r) = 0. \tag{4}$$

In which the ψ function acts on the robust scaled residuals. See, for example, equation (6.11) of Bock (1982), who also discusses estimation of the scale parameter σ.

The numerical solution of (4) usually starts from the least squares estimates satisfying (3). A difficulty is that, due to their position in X space, some observations have small residuals, irrespective of the value of the response. To see this consider the least squares residuals

$$r = y - X\hat{\beta} = \left\{I - X\left(X^TX\right)^{-1}X^T\right\}y = (I - H)y. \tag{5}$$

The idempotent matrix H, often called the 'hat' matrix (Hoaglin and Welsch, 1978), has diagonal elements h_i. The variance of the ith residual is given by

$$\operatorname{var}(r_i) = \sigma^2(1 - h_i). \tag{6}$$

For remote points in X space, so-called 'leverage' points, $h_i \rightarrow 1$ and, from (6), $\operatorname{var}(r_i) \rightarrow 0$ as the prediction at x_i comes increasingly to depend only on y_i. Thus observations with large values of h_i, which can be caused by erroneous values of the carriers x_i, will have small residuals and will not be down-weighted by the M-estimate (4). This form of robust regression therefore does not protect either against erroneous leverage points, nor against leverage points with outlying responses.

Alternative methods of robust regression which are intended to protect against these departures are described by Huber (1981, Cap. 7), Krasker and Welsch (1982) and by Huber (1983). In an investigation by simulation of the properties of estimates of location given by (1), Andrews et al. (1972) used conditional Monte-Carlo methods which reduced the computation involved by factors of powers of ten. Can such efficient methods be developed to aid our understanding of robust regression?

3. Regression Diagnostics

In diagnostic regression analysis the aim is the identification of features of the data, often groups of one or a few observations, which either have an appreciable effect on the fitted model or which indicate ways in which the model is systematically inadequate. The aim of identification can be contrasted with that of robust analysis where the aim is accomodation, that is inference when a small, but unidentified, set of observations is allowed to come from some other process (Cook and Weisberg, 1983). Diagnostic regression analysis is the subject of the books by Belsley, Kuh and Welsch (1980), Cook and Weisberg (1982) and Atkinson (1985). In all three books graphical methods play an important part.

To detect observations which have an appreciable effect on conclusions drawn from the data Cook (1977) suggested the measure

$$D_i = \left(\hat{\beta}_{(i)} - \hat{\beta}\right)^T X^T X \left(\hat{\beta}_{(i)} - \hat{\beta}\right)/ps^2 \tag{7}$$

where the p elements of $\beta_{(i)}$ are the least squares estimates of the parameters when observation i is deleted and s^2 is the residual mean square estimate of σ^2. The motivation for (7) is inspection of the distance $\hat{\beta}_{(i)} - \hat{\beta}$ relative to the confidence region for β.

An expression for (7) which is both computationally more convenient and also more revealing is found by rewriting D_i in terms of the residuals r_i. A further development (Atkinson, 1981) is the modified Cook statistic

$$C_i = \left(\frac{n-p}{p}\right)^{1/2}\left(\frac{h_i}{1-h_i}\right)^{1/2}|r_i^*|, \tag{8}$$

which is proportional to the square root of (7) when s^2 is replaced by the deletion estimate $s^2_{(i)}$. In (8)

$$r_i^* = (y_i - \hat{y}_i)\Big/\sqrt{s^2_{(i)}(1-h_i)} \tag{9}$$

is often called a deletion residual.

If there are no outliers the deletion residuals follow a t-distribution. Normal, or half normal, plots of the r_i^* therefore can provide a diagnostic plot for the presence of outliers, which will be detected by departure of the plot from linearity. But, except for a balanced design when all h_i are equal, there is no reason why a half normal plot of the modified Cook statistics C_i should be straight. To aid interpretation of such half normal plots Atkinson (1981) suggests use of a simulation envelope found by ordering the values of the C_i calculated from 19 samples. Several examples of the use of these plots are given in Chapter 6 of Atkinson (1985).

The simulation can be made quite straightforward. Because the r_i^* are residuals, the parameters of the linear model, and also the scale parameter σ are not important. The deletion residuals are calculated from (9) using values from a standard normal sample fitted to the linear model with the same matrix of carriers X as that observed. Calculation of the residuals is simple. If n is sufficiently small, (5) shows that premultiplication of y by the $n \times n$ matrix I–H yields the residuals. Alternatively $X^T y$ can be found and premultiplied by the stored value of $V = (X^T X)^{-1}$ to yield the parameter estimates. In neither case is a matrix inversion required for each sample.

4. Selection of Regression Models

In the regression models of Sections 2 and 3 it was assumed that the carriers in the matrix X were known. Choice of these carriers from a set of explanatory variables is usually by a rather ad hoc process of hypothesis testing and inspection of residuals. More formal and algorithmic methods for the choice of a 'best' model include several information criteria.

The residual sum of squares from the least squares parameter estimates $\hat{\beta}$ defined in (3) is

$$R = r^T r = y^T(I-H)y. \tag{10}$$

For the jth model let this sum of squares be R_j. As terms are added to the model the residual sum of squares will decrease. But this desirable progression is offset by an increase in the variance of the parameter estimates and in the mean squared error of predictions based on the model. These effects can be balanced by use of Mallows' C_p (Mallows, 1973) or Akaike's AIC (Akaike, 1973) to select the regression equation. For both criteria the model is chosen for which $R_j + 2p_j\,\sigma^2$ is a minimum, where p_j is the number of parameters in the jth model. If the value of σ^2 is not known, a suitable estimate is employed. This criterion can be extended to that of finding the model for which the generalized information criterion

$$AIC(\alpha) = R_j + \alpha p_j \sigma^2, \tag{11}$$

is a minimum. The value of α is at our disposal. Consistent choice of the true model is found by replacing α by an $\alpha(n)$, an increasing function of n such that $\alpha(n)/n \to 0$ as $n \to \infty$. The conditions are discussed by Hannan and Quinn (1979).

Understanding the choice of α for finite samples requires simulation. A review and examples are given by Atkinson (1980a). Unlike the investigation of regression diagnostics in Section 3, the simulation here does not require individual observations and their residuals. If the errors are assumed to be normally distributed, all that is required are the values of the sufficient statistics. When the observations are normally distributed, $X^T y$ has a multivariate normal distribution with mean $X^T X\beta$ and variance $\sigma^2 X^T X$. Independently of $X^T y$, the residual sum of squares has a distribution which is $2\sigma^2$ times a gamma distribution with index $(n-p)/2$. The residual sums of squares for models with less than p parameters are readily calculated from these quantities, without the need to re-sample for a variety of β values. The speed and efficiency of the calculations are further increaced by applying all selection rules to each sample.

5. Generalized Linear Models

Many of the techniques of the linear model can be extended to the anlysis of non-normal sets of data with structure in the means by use of the generalized linear model (McCullagh and Nelder, 1983). The goodness of fit of such models is ascertained by the deviance which asymptotically has a χ^2 distribution. But if the number of observations is not large relative to the number of parameters, the distribution may be far from its asymptotic form. The simulation technique of Monte-Carlo testing, discussed by Marriott (1979), can be used to give an idea of the significance of observed results.

An example is given by Williams (1982) on 'passive smoking', which is the name given to the apparent effect on the death rate from cancer of non-smoking wives of husbands who smoke. Part of Williams' results are reproduced in Table 1, which shows an observed effect with a χ^2 value of 8.7 on 2 degrees of freedom, seemingly highly significant. The strange feature of the data is that the simple model, which this χ^2 value rejects, has a deviance of 91.7 on 96 degrees of freedom. If this model really were inadequate the deviance should be appreciably greater than its expectation, rather than slightly less.

Table 1 also shows the results of 19 simulations when the simple model is assumed true. All 19 simulated values of the deviance for the simple model are less than the observed value. The simple model does not include a smoking effect. The 19 χ^2 values for this estimate obtained from the simulation in the absence of a real effect range from 0.1 to 2.9, well in line with expectation, as against 8.7 for the observed effect. Williams' conclusion, supported by further simulations of the model with a smoking effect, is that the simple model is unacceptable. This powerful procedure is a slightly elaborated version of the straightforward Monte-Carlo test. In the basic version the observed value of a test statistic is ordered amongst values simulated under the null hypothesis. The rank, rather than any distributional form, is often used to determine significance.

6. Tests of Transformations

In the parametric family of power transformations analysed by Box and Cox (1964), the loglikelihood is shown to be proportional to the residual sum of squares of the observations after the normalized transformation

$$z(\lambda) = \frac{y^{\lambda} - 1}{\lambda \dot{y}^{\lambda - 1}}, \qquad (12)$$

where $\dot{y}$ is the geometric mean of the observations. To test hypotheses about the value of λ in (12) requires maximization of the likelihood over λ. An advantage of the approximate score test introduced by Atkinson (1973) is that maximization is not required.

The model leading to this test is that, for some λ and to a sufficient degree of approximation, the transformed observations satisfy the linear model

$$z(\lambda) = X\beta + \varepsilon.$$

Expansion of this model about the hypothesiszed value λ_0 yields the linearized model

$$z(\lambda_0) = X\beta - (\lambda - \lambda_0) w(\lambda_0) + \varepsilon. \qquad (13)$$

Box (1980) calls the derivative $w(\lambda_0) = \delta z(\lambda)/\delta\lambda \mid \lambda = \lambda_0$ a constructed variable. Often in diagnostic work $\lambda_0 = 1$, corresponding to the hypothesis of no transformation.

The test of the significance of regression on $w(\lambda_0)$ in (13) is localy equivalent to testing the null hypothesis $\lambda = \lambda_0$. The expression for this t test is obtained by analogy with expressions from the analysis of covariance (Cox and McCullagh, 1982).

Other constructed variables have been suggested leading to the exact test of Andrews (1971) and to Tukey's celebrated one degree of freedom for non-additivity (Tukey, 1949). The relationship between these tests is developed by Atkinson (1982). To examine the comparative behaviour of two of the tests and the likelihood ratio test, Atkinson (1973) simulated the power of tests for the hypothesis of the inverse transformation in the survival time data presented by Box and Cox (1964). As with the other simulations mentioned in this paper, all tests were applied to every simulated set of observations. This is equivalent to use of a randomized block design to increase efficiency. More importantly, the results were presented graphically as a normal plot of the proportion of tests which were significant. Not only is the impact of such a plot greater than that of the corresponding table, but interpretation is facilitated. The power of the tests is indicated by the slope of the plots, differences in the size of the tests causing a change in intercept. Such plots are highly commended for the presentation of simulation results.

7. Tests of Separate Families of Hypotheses

A test for the choice between a gamma model with log link and a log-normal model is an example of a test of separate families of hypotheses of the kind introduced by Cox (1961, 1962). The test statistic is the log likelihood ratio for the separate models, from which is subtracted the expected value of the ratio under the null hypothesis. This corrected ratio is then divided by the square root of the asymptotic variance of the ratio, again calculated under the null hypothesis. Calculation of the expectation and variance is usually complicated and, as the examples in Atkinson (1970) show, the resulting test statistic usually has a distribution which is far from the asymptotic limit of normality.

Under such conditions the Monte-Carlo procedure of Section 5 provides a comparatively easy way of assessing the significance of an observed test statistic. Rather than calculate the complete statistic, including expectation and variances, it is enough, and much simpler, to simulate the distribution of the log-likelihood ratio. For the example of the gamma and log-normal distributions Atkinson (1982) plotted simulated values of the residual sum of squares for the log-normal model against the deviance for the gamma model, which are the two components of the ratio. The simulations were performed with each distribution as the null hypothesis. In addition to the relative simplicity of the procedure, an advantage is that the results can be simply presented as a plot.

8. Generation and Testing of Pseudo-Random Numbers

The simulation methods outlined in this paper rely heavily on the availability of a supply of pseudo-random numbers which can be converted into pseudo-random variables. The most important property in determining the quality of pseudo-random numbers from a linear congruential generator, is that overlapping or successive k-tuples of numbers fall on a k-dimensional lattice. Examples of the structure and properties of the generators are given by Atkinson (1980b) who also demonstrates the damage that a poor lattice structure can do to a generator of normal variables. A description of tests for the lattice structure is given by Knuth (1981, Section 3.3.4). Recent advances in the theory of lattice tests are due to Ripley (1983b).

9. Computer Generation of Random Variables

As a result of the continual increase in speed and power of computers, fast algorithms for the generation of pseudo-random variables are becoming less important. The spread of micro-computers, many with inadequate software, has directed attention more to the provision of portable algorithms which are easy to program. The most important recent general algorithm for continuous random variables which meets these requirements whilst being relatively fast is the ratio of uniforms method (Kinderman and Monahan, 1977). An example of the resulting algorithm for the normal distribution is given by Knuth (1981, Section 3.4.1) and by Ripley (1983b) who also provides a survey of recent work in the area.

Table 1
Analysis of Deviance Table for Hirayamas' "passive smoking" data, from Williams (1982)
The Monte-Carlo test clearly shows the significance of the smoking effect, despite the low deviance (91.7) for the simple model.

	Observed deviance	Degrees of freedom	Results of 19 simulations	Comment
Ignore husband's smoking	91.7	96	59 –90	all < observed
Include husband's smoking	83.0	94	57 –89	13 < observed
Smoking effect	8.7	2	0.1– 2.9	all < observed

10. References

ANDREWS, D. F.
A note on the selection of data transformations.
Biometrika 58 (1971), 249–254.

ANDREWS, D. F., BICKEL, P., HAMPEL, F., HUBER, P., ROGERS, W. H. and TUKEY, J.
"Robust Estimates of Location".
Princeton, N. J.: Princeton (1972).

AKAIKE, H.
Information theory and an extension of the maximum likelihood principle.
Pp. 267–281 of "Proceedings of the Second International Symposium on Information Theory" (ed B. N. Petrov and F. Csaki). Budapest: Akademie Kiado. (1973).

ATKINSON, A. C.
A method for discriminating between models (with discussion).
J. R. Statist. Soc. B 32 (1970), 323–353.

ATKINSON, A. C.
Testing transformations to normality.
J. R. Statist. Soc. B 35 (1973), 473–479.

ATKINSON, A. C.
A note on the generalized information criterion for choice of a model.
Biometrika 67 (1980a), 413–418.

ATKINSON, A. C.
Tests of pseudo-random numbers.
Appl. Statist. 29 (1980b), 164–171.

ATKINSON, A. C.
Two graphical displays for outlying and influential observations in regression.
Biometrika 68 (1981), 13–20.

ATKINSON, A. C.
Regression diagnostics, transformations and constructed variables (with discussion).
J. R. Statist. Soc. B 44 (1982), 1–36.

ATKINSON, A. C.
"Plots, Transformations and Regression".
Oxford: Oxford University Press. (1985).

BELSLEY, D. A., KUH, E. and WELSCH, R. E.
"Regression Diagnostics: Identifying Influential Data and Souuces of Collinearity. New York: Wiley.

BOCK, J.
Robuste Regression.
Pp. 123–42 of "Probleme der angewandten Statistik Heft 7: Robustheit III (Autorenkollektiv unter Leitung von D. Rasch und G. Herrendörfer). Rostock: Forschungszentrum für Tierproduktion Dummerstorf-Rostock.

BOX, G. E. P.
Sampling and Bayes inference in scientific model building and robustness (with discussion).
J. R. Statist. Soc. A 143 (1980), 383–430.

BOX, G. E. P. and COX, D. R.
An analysis of transformations (with discussion).
J. R. Statist. Soc. B 26 (1964), 211–246.

COOK, R. D.
Detection of influential observations in linear regression.
Technometrics 19 (1977), 15–18.

COOK, R. D. and WEISBERG, S.
"Residuals and Influence in Regression".
New York and London: Chapman and Hall (1982).

COOK, R. D.
Comment on Huber.
J. Amer. Statist. Assoc. 78 (1983), 74–75.

COX, D. R.
Tests of separate families of hypotheses.
Proc. 4th Berkeley Symposium 1 (1961), 105–123.

COX, D. R.
Further results on tests of separate families of hypotheses.
J. R. Statist. Soc. B 24 (1962), 406–424.

COX, D. R. and McCULLAGH, P.
Some aspects of analysis of covariance.
Biometrics 38 (1982), 541–561.

HANNAN, E. J. and QUINN, B. G.
The determination of the order of an autoregression.
J. R. Statist. Soc. B 41 (1979), 190–195.

HOAGLIN, D. C. and WELSCH, R. E.
The hat matrix in regression and ANOVA.
Amer. Statist. 32 (1978) 17–22.

HUBER, P.
"Robust Statistics".
New York: Wiley (1981).

HUBER, P.
Minimax aspects of bounded-influence regression.
J. Amer. Statist. Assoc. 78 (1983) 66–72.

KINDERMAN, A. J. and MONAHAN, J. F.
Computer generation of random variables using the ratio of uniform deviates.
A. C. M. Trans. Math. Soft. 3 (1977), 257–260.

KNUTH, D. E.
"The Art of Computer Programming.
Volume 2 Seminumerical Algorithms".
Second Edition. Reading, Mass: Addison-Wesley.

KRASKER, W. S. and WELSCH, R. E.
Efficient bounded influence regression estimation.
J. Amer. Statist. Assoc. 77 (1982), 595–604.

MALLOWS, C. L.
Some comments on Cp.
Technometrics 15 (1973), 661–675.

MARRIOTT, F. H. C.
Barnard's Monte-Carlo test: how many simulations?
Appl. Statist. 28 (1979), 75–77.

McCULLAGH, P. and NELDER, J. A.
"Generalized Linear Models".
London: Chapman and Hall.

NELDER, J. A. and WEDDERBURN, R.
Generalized linear models.
J. R. Statist. Soc. A 135 (1972) 370–384.

RIPLEY, B. D.
Multidimensional randomness.
Pp 241–247 of "Probability, Statistics and Analysis" (ed. J. F. C. Kingman and G. E. H. Reuter).
Cambridge: Cambridge University Press. (1983a).

RIPLEY, B. D.
Computer generation of random variables – a tutorial.
Int. Statist. Rev. 51 (1983b) 301–319.

TUKEY, J. W.
One degree of freedom for nonadditivity.
Biometrics 5 (1949) 232–242.

WILLIAMS, D. A.
GLIM and Hirayama's data.
RSS News and Notes, 9 (4) (1982) 7.

Institute of Mathematics, Polish Ac. of Sc., Wrocław Branch

Note About Solutions of Asymptotic Minimax Test Problems for Special Capacities

TADEUSZ BEDNARSKI

Abstract

Explicit formulas for the optimal test statistics are given in the case of an asymptotic minimax test problem with neighbourhoods generated by a class of special capacities.

1. Introduction

The first contribution to local asymptotic robust test problem for parametric models contaminated in term of capacities, was done by Huber-Carol (1970), then generalized and considerably developed by Rieder (1978). The results presented here constitute further extension of these papers.

Let $\{P_\vartheta : |\vartheta| \leq \tau\}$ be a family of probability measures indexed by real parameters. Let n be the number of independent observations with distributions that come either from $\mathfrak{P}_{0n}$ or $\mathfrak{P}_{1n}$. It is assumed that $P_{-\tau/\sqrt{n}} \in \mathfrak{P}_{0n}$. $P_{\tau/\sqrt{n}} \in \mathfrak{P}_{1n}$ and the sets $\mathfrak{P}_{in}$ are viewed as possible departures from the distributions in the parametric family. The statistical inference in such situations is formalized as a sequence of test problems for the product sets $\mathfrak{P}_{0n}^{\oplus n}$ and $\mathfrak{P}_{0n}^{\oplus n}$. It is intuitively clear that an asymptotic minimax sequence of tests, say $\{\varphi_n\}$, can be understood as a robust solution for the sequence of hypotheses $P^{\oplus n}_{-\tau/\sqrt{n}}$ against $P^{\oplus n}_{\tau/\sqrt{n}}$. It has been proved by Rieder (1978) that, under suitable regularity conditions, when $\mathfrak{P}_{in}$ are ε-contamination and total variation neighbourhoods of $P_{-\tau/\sqrt{n}}$ and $P_{\tau/\sqrt{n}}$ respectively, then the optimal solution is given by the sequence of statistics

$$T_n = \left(1/\sqrt{n}\right) \sum_{i=1}^{n} IC_i ,$$

where IC is a truncation of the logarithmic derivative of dP_Θ/dP_0 at $\vartheta = 0$. The tests $\{\varphi_n\}$ are then indicators of $\{T_n > t\}$ for some fixed t depending on the asymptotic significance level.

Rieder (1980) has further developed his results to some estimation problems preserving the same type of contaminating neighbourhoods. In Bednarski (1983) the problem of asymptotic minimax testing is studied, however the neighbourhoods are allowed to be generated by a class of special capacities, see also Bednarski (1981). The object of the study was to determine the dependence between the optimal IC and the employed contamination. Here we show that under suitable regularity conditions one can give an "almost" explicit formula for IC and see that in some circumstances it may differ considerably from a simple truncation of the logarithmic derivative of the likehood ratio.

2. Basic Notions and Assumptions

Let Ω be a polish space with Borel σ-field $\mathfrak{B}$ and let $\mathfrak{M}$ be the set of all probability measures on $\mathfrak{B}$. Let $\mathfrak{F}$ be a class of concave functions from [0,1] to [0,1] such that for every $f \in \mathfrak{F}$ $f(1) > 0$. For every $f \in \mathfrak{F}$ and $P \in \mathfrak{M}$ we define a special capacity $v_{f,P}$ as a set function from $\mathfrak{B}$ to [0, 1] such that $v_{f,P}(\emptyset) = 0$ and for all $A \neq \emptyset$, $A \in \mathfrak{B}$

$$v_{f,P}(A) = \left[P(A) + f \circ P(A)\right] \wedge 1 .$$

The symbols $\wedge$ and $\vee$ will denote minimum and maximum respectively. Each capacity $v_{f,P}$ generates a set $\mathfrak{P}_{f,P} = \{H \in \mathfrak{M}: H(A) \leq v_{f,P}(A) \text{ for all } A \in \mathfrak{B}\}$. This set can be viewed as a contamination of P. It is convex and if $f(0) = 0$ and f is continuous in 0, then $\mathfrak{P}_{f,P}$ is weakly compact. For further information about the capacities see Bednarski (1981), Buja (1980) and Huber and Strassen (1973).

Let now $\{P_\Theta \in \mathfrak{M}: |\vartheta| \leq \tau\}$ be the given parametric family and let $f_0, f_1 \in \mathfrak{F}$. The sets $\mathfrak{P}_{0n}$ and $\mathfrak{P}_{1n}$ will be here generated by the capacities

$$\left[P_{-\tau/\sqrt{n}} + \left(1/\sqrt{n}\right) f_0 \circ P_{-\tau/\sqrt{n}}\right] \wedge 1$$

$$\text{and } \left[P_{\tau/\sqrt{n}} + \left(1/\sqrt{n}\right) f_1 \circ P_{\tau/\sqrt{n}}\right] \wedge 1$$

respectively. The following regularity conditions are assumed to hold through the paper. Compare Bednarski (1983).

A 1: There exists an exponential family $\{Q_\Theta \in \mathfrak{M}: |\Theta| \leq \tau\}$ so that $Q_\Theta \sim c(\Theta) \exp(\Theta\Delta)\, dP_0$ for some random variable Δ and $\limsup nH^2(P_{\Theta/\sqrt{n}}, Q_{\Theta/\sqrt{n}}) = 0$ where H stands for the Hellinger n $|\Theta| \leq \tau$ distance.

A 2. The distribution F of Δ under P_0 has a density with respect to Lebesque measure and the distribution has a convex support.

A 3. There exists $A \in \mathfrak{B}$ so that

$$-2\tau \int_A \Delta\, dP_0 + f_0 \circ P_0(A) + f_1 \circ P_0(A^c) < 0 .$$

A 4. The functions f_0 and f_1 are differantiable on (0,1).

3. The Result and Examples

Theorem. Under Conditions A 1—A 4 we have that the optimal test statistic IC is equal

$$\left\{f_1' \circ F(\Delta^*) - f_0' \circ \left[1 - F(\Delta^*)\right]\right\} / 2\tau + \Delta^* ,$$

where $\Delta^* = d_0 \vee \Delta \wedge d_1$ for some uniquely determined constants d_0, d_1 and f' denotes the derivative of f.

Proof. From Bednarski (1983) we have that the optimal IC can be constructed as follows:

The first step is to minimize over z, for each $t \in R$, the expression

$$g(z,t) = f_0 \circ P_0(\Delta > z) + f_1 \circ P_0(\Delta \leq z) - 2\tau \int_{\Delta > z} \Delta\, dP_0 + \begin{cases} tP_0(\Delta > z) & \text{for } t \geq 0 \\ -tP_0(\Delta \leq z) & \text{for } t < 0 . \end{cases}$$

Under Condition A 2 the distribution function F is strictly increasing on the support of F, therefore one easily check that if inf g (z, t) $\leq$ O, then there is a unique z (t) for which the infimum z is attained. Let $t_0 < O$ and $t_1 > O$ be such that g (z (t_0), t_0) = g (z(t_1), t_1) = O. Then the function z (·) is continuous and strictly increasing on [t_0, t_1].
In the second step we define the family $\{A_t\}_t \in R$ of measurable sets by he formula

$$A_t = \begin{cases} \emptyset & \text{for } t \geq t_1 \\ \{\Delta > z(t)\} & t \in (t_0, t_1) \\ \bigcup\limits_{s>t_0} A_s & t = t_0 \\ \Omega & t < t_0 \end{cases}$$

and finally we put IC (ω) = $(1/2\,\tau)$ inf $\{t: \omega \notin A_t\}$. Therefore we obtain

$$IC(\Delta) = \begin{cases} z(t_1) & \text{for } \Delta > z(t_1) \\ z^{-1}(\Delta) & \Delta \in [z(t_0), z(t_1)] \\ z(t_0) & \Delta < z(t_0) \end{cases}$$

and the problem reduces to finding an explicit formula for $z^{-1}(\Delta)$. Under Conditions A 2 and A 4 we know that the derivative of g (z, t) in z is equal O if we put z = z (t). Hence we obtain that for $t \in [t_0, t_1]$

$$t = f_1' \circ F[z(t)] - f_0' \circ [1 - F(z(t))] + 2\tau z(t).$$

if we put now $d_0 = z(t_0)$ and $d_1 = z(t_1)$ the proof will be completed. In the simple examples given below we shall assume that $f_0 = f_1 = f$ and we put for f typical shapes.

1) If $f(x) = cx$ or $f(x) = c$ for some $c > 0$, then $IC = \Delta^*$.
2) If $f(x) = c\sqrt{x}$, $c > 0$, then

$$IC = (c/4\tau)\{1/\sqrt{F(\Delta^*)} - 1/\sqrt{1 - F(\Delta^*)}\} + \Delta^*.$$

3) If $f(x) = -x^2 + x + c$, where $0 \leq c \leq 1$, then

$$IC = (1/\tau)[1 - 2F(\Delta^*)] + \Delta^*.$$

The interesting feature of the solutions ist that, except for the linear case 1), we smoothly diminish the influence of outlying observations in the region where Δ is not yet truncated. It seems possible that various properties of tests obtained can also be studied via the infinitesimal approach developed by Ronchetti (1982) and Rousseeuw and Ronchetti (1979).

4. **References**

BEDNARSKI, T.
On solutions of minimax test problemes for special capacities.
Z. Wahrscheinlichkeitstheorie und Verw. Gebiete. 58 (1981), 397–405.

BEDNARSKI, T.
A robust asymptotic testing model for special capacities.
Preprint No 268 of the Institute of Mathematics of the Polish Academy of Sciences (1983).

BUJA, A.
Sufficiency, least favourable experiments and robust tests.
PhD thesis, ETH, Zürich (1980).

HUBER-CAROL, C.
Etude asymptotique de tests robustes.
PhD thesis, ETH, Zürich (1970).

HUBER, P. J. and STRASSEN, V.
Minimax tests and the Neyman-Pearson lemma for capacities.
Ann. Statist. 1 (1973), 251–263.

RIEDER, H.
A robust asymptoic testing model.
Anm. Statist. 6 (1978), 1080–1094.

RIEDER, H.
Estimates derived from robust tests.
Ann. Statist. 8 (1980), 106–115.

RONCHETTI, E.
Robust testing in linear models: The infinitesimal approach.
PhD thesis, ETH, Zürich (1982).

ROUSSEEUW, P. J. and RONCHETTI, E.
The influence curve for tests.
Research report 21, Fachgruppe fuer Statistik, ETH, Zürich (1979).

Sektion Mathematik der Wilhelm-Pieck-Universität Rostock, Rostock, GDR

Some Remarks on the Comparison of Means in the Case of Correlated Errors

JÜRGEN BOCK

Abstract

After the discussion of an approximative test for the one-way anlysis of variance with correlated errors an exact method is developed, basing on Hotellings T^2. This procedure can be robustified by means of Tiku's MML-estimators.

1. Introduction

Box (1954 a, b), Ljung and Box (1980) and Tiku (1982) studied the one-way analysis of variance with unequal variances and the two-way analysis of variance, assuming that errors within rows constitute a stationary Gaussian process. It has been shown in a simulation study by Andersen, Jensen and Schou (1981) that the approximations given by Box are excellent, and that disregarding correlations may lead to seriously misleading conclusions. The data have been simulated for AR (1) and MA (1) time series models. But the power of the tests is not investigated in the paper of Andersen et al. On the other hand it is not clear whether the approximations are sufficient good in the case of estimated correlation coefficients. Finally we do not know anything about the robustness against deviations from normality.

In this paper we can not answer all this questions. We will investigate the special case of no row effects (i. e. the one way classification) in more detail. Then we propose an alternative method which can easily be robustified.

Let us assume for the observations y_{tj} at equally spaced time points the model

$$\mathbf{y}_{tj} = \mu_t + \mathbf{u}_{tj} \qquad \begin{matrix}(t=1,\dots,T)\\(j=1,\dots,n)\end{matrix} \tag{1}$$

where the $\mathbf{u}_{tj}$ for fixed j constitute a stationary Gaussian process ($E(u_{tj}=0)$, $\operatorname{cov}(u_{tj}, u_{t+hk}) = \delta_{jk}\,\sigma(h)$), and the rows are independent replications of the same process. We want to test the hypothesis $H_0: \mu_1 = \dots = \mu_T$ by an approximative F-test. The sums of squares are $SSR = \mathfrak{y}'A\mathfrak{y}$, $SSB = \mathfrak{y}'B\mathfrak{y}$ with

$\mathfrak{y}' = (y_{11}, y_{12}, \dots, y_{1n}, y_{21}, y_{22}, \dots, y_{2n}, \dots, y_{T1}, y_{T2}, \dots, y_{Tn})$

and the idempotent matrices $A = I_T \circledast \left(I_n - \frac{1}{n} e_n e_n'\right)$,

$B = \left(I_T - \frac{1}{T} e_T e_T'\right) \circledast \left(\frac{1}{n} e_n e_n'\right)$. The star denotes the Kroneckerproduct, I the identity matrix and e a vector with all components equal to one.

With the positive definite covariance matrix $V = \Sigma_T \circledast I_n$

$$\Sigma_T = \begin{pmatrix} \sigma(0) & \sigma(1) & \dots & \sigma(T-1) \\ \sigma(1) & \sigma(0) & & \sigma(T-2) \\ \vdots & & & \vdots \\ \sigma(T-1) & \dots & \sigma(1) & \sigma(0) \end{pmatrix}$$

one gets $AVB = 0$, therefore SSR and SSB are uncorrelated (see e. g. Rasch (1976)).

Box has approximated the distributions of the sums of squares under the null hypothesis $H_0: \mu_1 = \dots = \mu_T$ by Gamma distributions with the same expectations and variances.

For further discussions we give the expectations and variances for the nonnulldistribution too:

$E(\mathbf{SSR}) = \operatorname{tr}(AV) = T(n-1)\,\sigma(0)$.

$E(\mathbf{SSB}) = \operatorname{tr}(BV) + \mu' B \mu =$

$$= \sigma(0)\left[T - \frac{1}{T}\sum_{s,t=1}^{T} \varrho(t-s)\right] + n \sum_{t=1}^{T} (\mu_t - \bar{\mu})^2 \tag{2}$$

with $\mu' = (\mu_1, \dots, \mu_1, \mu_2, \dots, \mu_2, \dots, \mu_T, \dots, \mu_T)$

$$\bar{\mu} = \frac{1}{T}\sum_{t=1}^{T} \mu_t, \quad \varrho(t-s) = \sigma(t-s)/\sigma(0).$$

Calculating the moments by the derivatives of the characteristic function one can show, that for a normal distributed vector

$\mathfrak{z} \sim N(\mu^*, \Sigma_{\mathfrak{z}})$, the variance of $\mathfrak{z}'\mathfrak{z}$ is equal to

$$V(\mathfrak{z}'\mathfrak{z}) = 2\operatorname{tr}\left(\Sigma_{\mathfrak{z}}^2\right) + 4\mu^{*\prime}\,\Sigma_{\mathfrak{z}}\,\mu^*. \tag{3}$$

Therefore we get from $\mathbf{SSR} = (A\mathfrak{y})'(A\mathfrak{y})$ and $\mathbf{SSB} = (B\mathfrak{y})'(B\mathfrak{y})$

$$V(\mathbf{SSR}) = 2(n-1)\sigma^2(0)\sum_{s,t=1}^{T} \varrho^2(t-s)$$

$$V(\mathbf{SSB}) = 2\sigma^2(0)\left[\sum_{s,t=1}^{T} \varrho^2(t-s) - \frac{2}{T}\sum_{s=1}^{T}\left(\sum_{t=1}^{T}\varrho(t-s)\right)^2 + \frac{1}{T^2}\left(\sum_{s,t=1}^{T}\varrho(t-s)\right)^2\right] + 4n\sigma^2(0)\sum_{s,t=1}^{T}(\mu_s - \bar{\mu})(\mu_t - \bar{\mu})\varrho(t-s). \tag{4}$$

Following Andersen et al (1981) the null-distribution of

$$\tilde{\mathbf{F}} = c\mathbf{F}, \quad \mathbf{F} = \frac{\mathbf{SSB}}{\mathbf{SSR}}\,\frac{T(n-1)}{T-1} \tag{5}$$

is approximated by an F-distribution with

$$f_B = \frac{2E^2(\mathbf{SSB})}{V(\mathbf{SSB})} \text{ and } f_R = \frac{2E^2(\mathbf{SSR})}{V(\mathbf{SSR})}$$

degrees of freedom, where

$$c = \frac{(T-1)}{(n-1)T}\,\frac{E(\mathbf{SSR})}{E(\mathbf{SSB})}.$$

Therefore

$$f_B = \frac{\left[T - \frac{1}{T}\sum_{s,t=1}^{T}\varrho(t-s)\right]^2}{\left[\sum_{s,t=1}^{T}\varrho^2(t-s) - \frac{2}{T}\sum_{s=1}^{T}\left(\sum_{t=1}^{T}\varrho(t-s)\right)^2 + \frac{1}{T^2}\left(\sum_{s,t=1}^{T}\varrho(t-s)\right)^2\right]} \tag{6}$$

$$f_R = \frac{(n-1)T^2}{\sum_{s,t=1}^{T} \varrho^2(t-s)} \tag{7}$$

$$c = \frac{T-1}{T - \frac{1}{T}\sum_{s,t=1}^{T} \varrho(t-s)}. \tag{8}$$

In practical applikations the autocorrelations in (6), (7), (8) have to be replaced by estimates, so we get the test statistic $\mathbf{F^*} = \hat{\mathbf{c}}\mathbf{F}$ with an estimated c having perhaps an F-distribution approximately. In our case

$$\hat{\sigma}(h) = \frac{\mathbf{SPR}^{(h)}}{(T-h)(n-1)}, \quad \mathbf{SPR}^{(h)} = \sum_{t=1}^{T-h}\sum_{j=1}^{n} (y_{tj} - \bar{y}_{t\cdot})(y_{t+h\,j} - \bar{y}_{t+h\cdot}) \tag{9}$$

is an unbiased estimator of $\sigma(h)$. (This is not true in the two-way classification case.) We use $\hat{\varrho}(h) = \hat{\sigma}(h)/\hat{\sigma}(0)$.
An approximation to the noncentral distribution of $\mathbf{F}$, as in the central case, would have different degrees of freedom for the numerator (see (2), (4)). But there is the same problem, we do not know the goodness of the approximation in the case of estimated autocorrelations. Therefore we started to investigate the power by a simulation study; these results will be given in a following paper.

2. Paired Observations

It is very interesting to look at the case $T=2$. This is the case of paired observations. To test the hypothesis H_0: $\mu_1 = \mu_2$ one uses generally the t-test with the test statistic

$$t = \frac{\bar{\Delta}.}{s_\Delta}\sqrt{n} \tag{10}$$

$$\text{with } \Delta_j = y_{1j} - y_{2j}, \quad \bar{\Delta}. = \frac{1}{n}\sum_{j=1}^{n}\Delta_j, \quad s_\Delta^2 = \frac{1}{n-1}\sum_{j=1}^{n}(\Delta_j - \bar{\Delta}.)^2$$

and n–1 degrees of freedom.
The Box-approximation yields (with $\varrho = \varrho(1)$)

$$f_B = 1, \quad f_R = \frac{2(n-1)}{1+\varrho^2}, \quad c = \frac{1}{1-\varrho} \tag{11}$$

$$\tilde{\mathbf{F}} = \frac{c\,n(\bar{y}_{1.} - \bar{y}_{2.})^2}{2\hat{\sigma}(0)} \tag{12}$$

or, if we replace c by $\hat{c}$

$$\mathbf{F^*} = \frac{(\bar{y}_{1.} - \bar{y}_{2.})^2 n}{2\hat{\sigma}(0)[1 - \hat{\varrho}(1)]} = t^2. \tag{13}$$

If H_0 holds, $\mathbf{F^*} = t^2$ is exactly F-distributed with 1 and n–1 d. f. and the approximation fails in the d. f.
For known correlations one can compute the exact distribution of $\tilde{\mathbf{F}}$.

3. Multiple Comparisons

The easiest way to get a multiple comparison-procedure, which is independent on correlations, seems to construct a t-procedure by means of

$$t_{ik} = \frac{\bar{y}_{i.} - \bar{y}_{k.}}{s_{ik}} \quad (i,k=1,\ldots,T), \quad s_{ik}^2 = \frac{1}{n-1}\sum_{j=1}^{n}(\Delta_{ijk} - \bar{\Delta}_{ik.})^2$$

$$\Delta_{ikj} = y_{ij} - y_{kj}.$$

The distribution of each t_{ik} is a central or noncentral t-distribution with n–1 d. f. Every pairwise comparison is independent on correlations. But the multivariate distribution of all t_{ik} depends on the correlation, therefore the familywise risk too. One way to get a multiple procedure could be, to construct an approximative Scheffé-procedure by means of the Box-Andersen-approximation.

4. A Robust Test

As we have seen, it is not possible to generalize the classical comparison procedure in such a way, that they are entirely independet on correlations, so we look first for a test with a risk of first kind independent on correlations. The basic idea is, to seek for a teststatistic, which does not change under regular linear transformations. Then we can transform to the uncorrelated case.

The random vectors $\mathfrak{z}'_j = (z_{2j}, z_{3j}, \ldots, z_{Tj})$ with $z_{ij} = y_{ij} - y_{1j}$ $(i = 2, \ldots, T)$ are independent and normal distributed with mean vector $\mu'_{\mathfrak{z}} = (\mu_2 - \mu_1, \ldots, \mu_T - \mu_1)$ and covariance matrix GVG′ where

$$G = \begin{pmatrix} -1 & 1 & 0 \ldots 0 \\ -1 & 0 & 1 \quad 0 \\ \vdots & & \vdots \\ \vdots & & 0 \\ -1 & 0 \ldots\ldots & 01 \end{pmatrix}.$$

We can deal with our comparisonproblem as a multivariate test of the nullhypothesis H^*_0: $\mu_{\mathfrak{z}} = 0$. This leads to Hotellings $\mathbf{T}^2(\mathfrak{z})$

$$T^2(\mathfrak{z}) = n\bar{\mathfrak{z}}'.\mathbf{S}^{-1}\bar{\mathfrak{z}}.,\quad \bar{\mathfrak{z}}. = \frac{1}{n}\sum_{j=1}^{n}\mathfrak{z}_j$$

$$\mathbf{S} = \frac{1}{n-1}\sum_{j=1}^{n}(\mathfrak{z}_j - \bar{\mathfrak{z}}.)(\mathfrak{z}_j - \bar{\mathfrak{z}}.)'.$$

The distribution of $\mathbf{T}^2(\mathfrak{z})\,(n-T+1)/(n-1)\,(T-1)$ is noncentral F with $T-1$ and $n-T+1$ d. f. and noncentrality parameter $n\,\mu'_{\mathfrak{z}}\,(GVG')^{-1}\,\mu_{\mathfrak{z}}$ (Anderson (1958)). The teststatistic does not change under a linear transformation $\mathfrak{z} = C\mathfrak{z}^*$ $(|C| \neq 0)$. Choosing C in such a way that $CC' = GVG'$, we get the identity matrix as covariance-matrix of $\mathfrak{z}^* = C^{-1}\mathfrak{z}$, but $\mathbf{T}^2(\mathfrak{z}^*)$ has the same distribution as $T^2(\mathfrak{z})$. The risk of first kind is independent on correlations. while the power depends on correlations through the noncentrality parameter. Due to Tiku and Singh (1982) one can robustify the test in the following manner:
Define $(j = 1, \ldots, n)$

$$w_{2j} = z_{2j}$$
$$w_{3j} = z_{3j} - \hat{b}_{32} z_{2j}$$
$$w_{4j} = z_{4j} - \hat{b}_{43.2} z_{3j} - \hat{b}_{42.3} z_{2j}$$

etc.

where the $\hat{b}$'s are the partial regression coefficients (see Kendall and Stuart, 1973, Chapter 27), and

$$\mathbf{T}_R^{*2} = m\hat{\nu}'\hat{\mathbf{W}}^{-1}\hat{\nu} \tag{16}$$

with $\hat{\nu}' = (\hat{\nu}_2, \ldots, \hat{\nu}_T)$ and the diagonal matrix $\hat{\mathbf{W}} = \mathrm{diag}(\hat{\sigma}^2_{\nu_2}, \ldots, \hat{\sigma}^2_{\nu_T})$ The $\hat{\nu}$'s and $\hat{\sigma}$'s are the MML estimators of mean and standard deviation calculated from typ II censored samples

$$w_{(i,r+1)}, \ldots, w_{(i,n-r)} \quad (i = 2, \ldots, T). \tag{17}$$

The r smallest and the r largest observations are censored, and

$$\hat{\nu}_i = \left\{ \sum_{j=r+1}^{n-r} w_{ij} + r\beta(w_{i\ r+1} + w_{i\ n-r}) \right\} / m$$

$$\hat{\sigma}_{\nu_i} = \left\{ B_i + \sqrt{(B_i^2 + 4A_iC_i)} \right\} / 2\sqrt{\{A_i(A_i - 1)\}}$$

where

$$m = n - 2r + 2r\beta, \ A_i = n - 2r, \ B_i = r\alpha(w_{i\ n-r} - w_{i\ r+1})$$

$$C_i = \sum_{i=r+1}^{n-r} w_{ij}^2 + r\beta(w_{i\ r+1}^2 + w_{i\ n-r}^2) - m\hat{\nu}_i^2.$$

For $n \geqq 10$ the coefficients are obtained from the following equations

$$\beta = -f(t)\{t - f(t)/q\}/q \text{ and } \alpha = \{f(t)/q\} - \beta t,$$

$$q = r/n, \ F(t) = \int_{-\infty}^{t} f(z)dz = 1 - q \text{ and } f(z) = 1/\sqrt{(2\pi)} \cdot \exp(-z^2/2), \ -\infty < z < +\infty.$$

Take r = [1/2 + 0.1n] if no information is available about the tails of the distributions. For details see Tiku and Singh (1982). The test will be robust to most non-normal ditsributions prevalent in practice, where we can assume that the marginal distributions are continous and are of type $(1/\sigma_i)\, f((x - \mu_i)/\sigma_i)$ with existing means and variances.

Only the risk of first kind is independent of correlations, but we can use robust estimators of the variance-covariance matrix (from a foregoing experiment) in the planning stage to ensure good power properties. Such estimators are given by Tiku (1980).

5. References

ANDERSON, T. W.
An Introduction to Multivariate Statistical Analysis.
J. Wiley and Sons, New York, London, Sidney 1958.

ANDERSEN, A. H., JENSEN, E. B. and SCHOU, G.
Two-way Analysis of Variance with Correlated Errors.
Intern. Statistical Review 49 (1981) 153–167.

BOX, G. E. P.
Some Theorems on Quadratic Forms Applied in the Study of Analysis of Variance Problems, I. Effect of Inequality of Variance in the One-way Classification.
Ann. Math. Statist. 25 (1954a) 290–302.

BOX, G. E. P.
Some Theorems on Quadratic Forms Applied in the Study of Analysis of Variance Problems, II. Effects of Inequality of Variance and of Correlation Between Errors in the Two-way Classification.
Ann. Math. Statist. 25 (1954b) 484–498.

KENDALL, M. G. and STUART, A.
The Advanced Theory of Statistics – II.
Hafner Publishing Company, New York, 1973.

LJUNG, G. M. and BOX, G. E. P.
Analysis of variance with auto-correlated observations.
Scandinavian J. Statists. I, (1980) 172–180.

RASCH, D.
Einführung in die mathematische Statistik – I.
VEB Deutscher Verlag der Wissenschaften 1976.

TIKU, M. L.
Robust Estimation of the Variance-covariance Matrix of Symmetric Multivariate Distributions.
Multivariate Statistical Analysis, R. P. Gupta (ed.).
North-Holland Publishing Company, 1980.

TIKU, M. L. and SINGH, M.:
Robust Statistics for Testing Mean Vectors of Multivariate Distributions.
Commun. Statist. – Theor. Meth., 11 (9), (1982) 985–1001.

TIKU, M. L.
A robust statistic for testing that two autocorrelated samples come from identical population.
Time Series Methods in Hydrosciences.
(Ed. A. H. El-Shaarawi) Elsevier Scientific publishing Company, New York (1982)

Computing Centre, Eindhoven University of Technology

Robustness of Multiple Comparisons Against Variance Heterogeneity

JAN B. DIJKSTRA

Abstract

If $H_0: \mu_1 = \ldots = \mu_k$ is rejected for normal populations with classical one way analysis of variance, it is usually of interest to know where the differences may be. If the population variances are equal there are several approaches one might consider:

1. Least Significant Difference test (Fisher, 1935)
2. Multiple Range test for equal sample sizes (Newman, 1939)
3. An adaptation for unequal sample sizes (Kramer, 1956)
4. Multiple F-test (Duncan, 1951)
5. Multiple Comparisons test (Duncan, 1952).

For all these methods (including the one way analysis of variance) alternatives exist that are robust against variance heterogeneity. A modification of (3) has some unattractive properties if the variances and the sample size differ greatly. The adaptations for unequal variances of (4) and (5) seem better than (1) for cases with many samples. Test (2) is rather robust in itself if the variances are not too much different. Modifications exist that allow slight unequalities in the sample sizes.

1. **Introduction**

In 1981 Werter and the author published a study on tests for the equality of several means when the population variances are unequal. The problem can be stated as follows:

$$H_0: \mu_1 = \ldots = \mu_k$$

$$x_{ij} \sim N(\mu_i, \sigma_i^2) \text{ for } i = 1, \ldots, k$$
$$j = 1, \ldots, n_i.$$

The conclusion of this study was that the second order method of James (1951) gives the user better control over the size than some other tests [Welch (1951), Brown and Forsythe (1974)], so it is to be preferred since none of the tests in the study was uniformly most powerful.

The test statistic t is defined as:

$$t = \sum_{i=1}^{k} w_i (x_i - \bar{x})^2, \text{ where } w_i = \frac{x_i}{s_i^2}, \; x_i = \frac{1}{n_i} \sum_{j=1}^{n_i} x_{ij}, \; \bar{x} = \frac{1}{w} \sum_{i=1}^{k} w_i x_i \text{ and } w = \sum_{i=1}^{k} w_i.$$

For some chosen size α this test statistic is to be compared with a critical level $h_2(\alpha)$, given by:

$$h_2(\alpha) = \chi^2 + \frac{1}{2}(3\chi_4 + \chi_2) \sum_{i=1}^{k} \frac{1}{\upsilon_i}\left(1 - \frac{w_i}{w}\right)^2 + \left\{\frac{1}{16}(3\chi_4 + \chi_2)^2 \left(1 - \frac{k-3}{\chi^2}\right)\left(\sum_{i=1}^{k} \frac{1}{\upsilon_i}\left(1 - \frac{w_i}{w}\right)^2\right)^2\right.$$

$$+ \frac{1}{2}(3\chi_4 + \chi_2)\Big[(8R_{23} - 10R_{22} + 4R_{21} - 6R_{12}^2 + 8R_{12}R_{11} - 4R_{11}^2) + (2R_{23} - 4R_{22} + 2R_{21} - 2R_{12}^2$$

$$+ 4R_{12}R_{11} - 2R_{11}^2)(\chi_2 - 1) + \frac{1}{4}(-R_{12}^2 + 4R_{12}R_{11} - 2R_{12}R_{10} - 4R_{11}^2 + 4R_{11}R_{10} - R_{10}^2)(3\chi_4 - 2\chi_2 - 1)\Big]$$

$$+ (R_{23} - 3R_{22} + 3R_{21} - R_{20})(5\chi_6 + 2\chi_4 + \chi_2) + \frac{3}{16}(R_{12}^2 - 4R_{23} + 6R_{22} - 4R_{21} + R_{20})(35\chi_8 + 15\chi_6 + 9\chi_4 + 5\chi_2)$$

$$+ \frac{1}{16}(-2R_{22} + 4R_{21} + R_{20} + 2R_{12}R_{10} - 4R_{11}R_{10} + R_{10}^2)(9\chi_8 - 3\chi_6 - 5\chi_4 - \chi_2)$$

$$\left. + \frac{1}{4}(-R_{22} + R_{11}^2)(27\chi_8 + 3\chi_6 + \chi_4 + \chi_2) + \frac{1}{4}(R_{23} - R_{12}R_{11})(45\chi_8 + 9\chi_6 + 7\chi_4 + 3\chi_2)\right\}$$

Here $\chi^2 = \chi^2(\alpha)$ is the percentage point of a χ^2-distributed variate with $r = k - 1$ degrees of freedom, having a tail probability α. The other basic items in the formula are given by:

$$\chi_{2s} = [\chi^2(\alpha)]^s (k-1)(k+1) \ldots (k + 2s - 3)$$

$$\text{and } R_{st} = \sum_{i=1}^{k} \frac{1}{\nu_i^s}\left(\frac{w_i}{w}\right)^t, \text{ where } \nu_i = n_i - 1$$

This method is an approximation of order -2 in the ν_i to an "ideal" method. Brown and Forsythe (1974) considered the first order method of James (order -1 in the ν_i). Their conclusion was that for unequal variances the difference between the nominal size and the actual probability of rejecting the null hypothesis when it is true can be quite impressive. Werter and the author found that this difference almost vanishes if one takes into account the second order terms.

The test as stated gives only the binary result that H_0 is accepted or rejected. If one prefers the tail probability of the test the equation $t = h_2(\alpha)$ has to be solved. Because $h_2(\alpha)$ is monotonous in α this can be done in about ten function evaluations with an acceptable precision of 0.001 in α. In the formula for $h_2(\alpha)$ the terms R_{st} are independent of α, so it is only necessary to recompute the $\chi_2{}^s$ for every iteration. This version of the test was used on a Burroughs B 7700 computer. The average amount of processing time for common cases was about 0.026 sec, so the very complicated formula does not yield an expensive algorithm.

If H_0 is accepted this usually means the end of the analysis. Otherwise it may be of interest to know where the differences lie. For this one has to perform a simultaneous test and it would be nice if this could be done in such a way that α means "The accepted probability of declaring any pair μ_i, μ_j different when in fact they are equal". In the following sections some strategies are worked out for this kind of simultaneous statistical inference.

2. Least Significant Difference Test

The method consists of two stages. First $H_0: \mu_1 = \ldots = \mu_k$ is to be tested with classical one way analysis of variance. If H_0 is rejected a t-test is to be performed for every pair. This idea originates from Fisher (1935) and it presupposes the variances to be equal.

Fisher suggested using the same α for the t-tests as for the overall analysis of variance. Of course this is not safe in the sense mentioned in the introduction. An alternative to be considered is the Bonferroni idea $\beta = \alpha/\binom{k}{2}$ that is mentioned in Miller (1966). For this the probability that no error is made under H_0 is limited as followes:

$$P = \left(1 - \frac{\alpha}{\binom{k}{2}}\right)^{\binom{k}{2}} \leq 1 - \binom{k}{2}\frac{\alpha}{\binom{k}{2}} = 1 - \alpha$$

For unequal variances the one way analysis of variance can be replaced by the James second order test. For comparing the pairs there are several possibilities. The situation iscalled the Behrens-Fisher (1929) problem, and one of the best approximate solutions is Welch's modified t-test (1949). This test has been evaluated by Wang (1971) and he concluded that it gives the user excellent control over the size, whatever the value of the nuisance parameter $\Theta = \sigma_i^2 \, \sigma_j^2$ may be. The test statistic is

$$t = \frac{x_i - x_j}{\left[\left(\frac{s_i^2}{n_i}\right) + \left(\frac{s_j^2}{n_j}\right)\right]^{1/2}}$$

and the critical level for some chosen size β is given by Students t-distribution with a parameter ν that takes the pattern of the variances into account:

$$\nu_{ij} = \frac{\left(\frac{s_i^2}{n_i} + \frac{s_j^2}{n_j}\right)^2}{\frac{s_i^4}{n_i^2(n_i - 1)} + \frac{s_j^4}{n_j^2(n_j - 1)}}$$

In most cases ν_{ij} is not an integer, so it has to be replaced by the nearest one. Ury and Wiggins (1971) suggested using this test with the Bonferroni β. The simultaneous confidence intervals for this approach are given by:

$$\mu_i - \mu_j \in \left[x_i - x_j \mp t_{\nu_{ij}}^{1/2\beta} \left(\frac{s_i^2}{n_i} + \frac{s_j^2}{n_j}\right)^{1/2}\right]$$

There are some alternatives mentioned in the literature. Hochberg (1976) suggested using:

$$\mu_i - \mu_j \in \left[x_i - x_j \mp \gamma_\alpha \left(\frac{s_i^2}{n_i} + \frac{s_j^2}{n_j}\right)^{1/2}\right]$$

where γ_α is the solution of $\sum_{i=1}^{k} \sum_{j=i+1}^{k} P\{|t_{\nu_{ij}}| > \gamma\} = \alpha$,
in which ν_{ij} comes from Welch's modified t-test.

Tamhane (1977) suggested using Banerjee's (1961) approximate solution of the Behrens-Fisher problem with $\gamma = 1 - (1-\alpha)^{\frac{1}{k-1}}$. This γ has some history and will also be mentioned in the following sections. The confidence intervals become:

$$\mu_i - \mu_j \in \left[x_i - x_j \mp \left\{\left(t_{\nu_i}^{1/2\gamma}\right)^2 \frac{s_i^2}{n_i} + \left(t_{\nu_j}^{1/2\gamma}\right)^2 \frac{s_j^2}{n_j}\right\}^{1/2}\right]$$

Tamhane also suggested using Welch's test with this γ. In the literature the author has found nine different approximate solutions of the Behrens-Fisher problem and five ideas concerning the size of the separate tests. Every combination can be made, so there is quite a lot of methods one can consider for pairwise comparisons. But to be really safe, in the sense that the probability of declaring any pair different when in fact they are equal should be limited by α, the pairwise size β will become very small. For $k = 15$ and $\alpha = 0.05$ the Bonferroni approach willi yield $\beta = 0.00048$, so it becomes almost impossible to reject any pairwise comparison.

Another disadvantage of this approach is the fact that the results have to be represented by a matrix containing symbols for acceptance and rejection. Working at a terminal, as is usually done in applied statistics nowadays, one has to swallow an enormous lot of information in one glance if k exceeds the region of very small values. The next sections will suggest approaches that are better in this respect.

3. Multiple Range Tests

In this section a strategy will be pointed out that was originated by Newman (1939), Duncan (1951) and Keuls (1952). At first it will be necessary for the sample sizes to be equal ($n_i = n$ for $i = 1, \ldots, k$). Also variance heterogeneity will not be allowed. Later on these limitations will be dropped.

Let $x_{(1)}, \ldots, x_{(k)}$ be the sample means, sorted in non-decreasing order. The first hypothesis of interest is $H_0: \mu_1 = \ldots = \mu_k$, where the μ_i's are renumbered so that their ordering becomes the same as the sample means which are their estimates.

Then H_0 can be tested with:

$$\mu_1 - \mu_k \in \left[x_1 - x_k \mp q_{k,\nu}^{\alpha} \frac{s}{n^{1/2}}\right]$$

where q is the studentized range distribution, $\nu = k(n-1)$ and the residual variance is estimated by:

$$s^2 = \frac{1}{\nu} \sum_{i=1}^{k} \sum_{j=1}^{n} (x_{ij} - x_i)^2.$$

If H_0 is rejected, the next stage is to test $\mu_1 = \ldots = \mu_{k-1}$ and $\mu_2 = \ldots = \mu_k$. Proceeding like this until every hypothesis is accepted will yield a result that can be represented as followes:

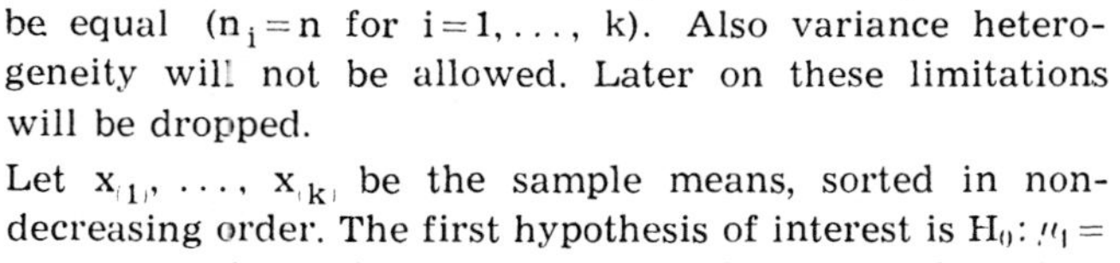

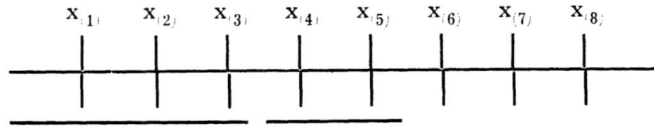

The interpretation of this figure is that $\mu_i = \mu_j$ has to be rejected if there is no unbroken line that underscores $x_{(i)}$ and $x_{(j)}$. For instance:

$\mu_4 = \mu_5$: accepted
$\mu_5 = \mu_6$: accepted
$\mu_4 = \mu_6$: rejected.

If a candidate for the splitting process contains p means then $q_{p,\nu}^{\alpha p}$ is to be used instead of $q_{k,\nu}^{\alpha}$. Newman and Keuls suggested $\alpha_p = \alpha$ and Duncan preferred $\alpha_p = 1 - (1-\alpha)^{p-1}$. Now the equality of the sample sizes will be dropped, but for the moment the variances will still have to be equal. Miller (1966) suggested using the median of $n_1, \ldots, n_R$. Winer (1962) considered the harmonic mean H

$$\left(\frac{1}{H} = \frac{1}{k}\sum_{i=1}^{k}\frac{1}{n_i}\right).$$

Kramer (1956) modified the formula of the test to this situation:

$$\mu_i - \mu_j \in \left[x_i - x_j \mp q_{p,\nu}^{\alpha_p} s \left\{ 1/2 \left(\frac{1}{n_i} + \frac{1}{n_j} \right) \right\}^{1/2} \right],$$

where $\nu = N - k$ and $N = \sum_{i=1}^{k} n_i$

Only in Kramer's case does the studentized range distribution hold. For Miller and Winer the approximation will be reasonable if the sample sizes are not too different. Kramer's test contains a trap that can be shown in the following figure:

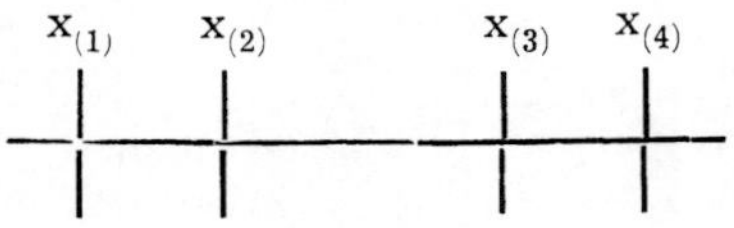

Suppose n_1 and n_4 are much smaller than n_2 and n_3. Then $\mu_1 = \ldots = \mu_4$ can be accepted while μ_2 and μ_3 are significantly different. But the strategy will make sure that this difference will never be found.

From here on the variances will be allowed to be unequal. For equal sample sizes Ramseyer and Tcheng (1973) found that the studentized range statistic is remarkably robust against variance heterogeneity. So for almost equal sample sizes it seems reasonable to use the Winer or Miller approach and ignore the differences in the variances.

Unfortunately, the robustness of Kramer's test is rather poor [Games and Howell (1976)], so if the sample sizes differ greatly one might be tempted to consider:

$$\mu_i - \mu_j \in \left[x_i - x_j \mp q_{p,\nu_{ij}}^{\alpha_p} \left\{ 1/2 \left(\frac{s_i^2}{n_i} + \frac{s_j^2}{n_j} \right) \right\}^{1/2} \right]$$

where only the variances of the extreme samples are taken into account. This idea was mentioned by Games and Howell (1976) with Welch's ν_{ij}. The studentized range distribution does not hold for these separately estimated variances, but the approximation seems reasonable though a bit conservative.

The context in which Games and Howell suggested using this method was one of pairwise comparisons with other parameters for q. But it looks like a good start for the construction of a "Generalized Multiple Range test".

This test, however attractive it may seem, still contains the trap that was already mentioned for Kramer's method. But there is more:

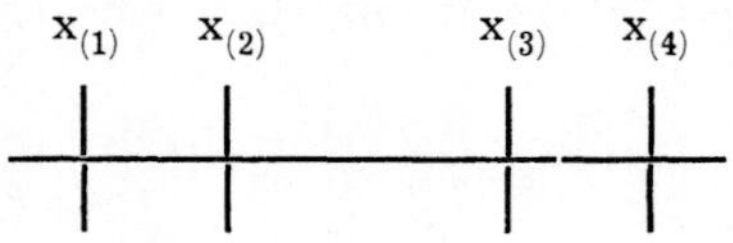

Suppose s^2_2 and s^2_3 are (much) smaller than s^2_1 and s^2_4, Then a significant difference between μ_2 and μ_3 can easily be ignored.

The author has not found in the literature other approaches to variance heterogeneity within the strategy of multiple range tests. Some other α's have been suggested, but since the choice of α_p has almost nothing to do with robustness against variance heterogeneity, their merits will not be discussed in this paper.

The representation of the results with underscoring lines seems very attractive since this simple figure contains a lot of information, and also the artificial consistency that comes from the ordered means has some appeal. However the whole idea of a Generalized Multiple Range test seems wrong. One simply cannot afford to take only the extreme means into account if the sample sizes and the variances differ greatly.

4. **Multiple F-Test**

This test was proposed by Duncan (1951). In the original version the population variances must be equal. The procedure is the same as for the Multiple Range test, only the q-statistic is replaced by an F, so that the first stage becomes classical one way analysis of variance. At first Duncan proposed using $\alpha_p = 1 - (1-\alpha)^{p-1}$, but later he found $\alpha_p = 1 - (1-\alpha)^{(p-1)/(k-1)}$ more suitable [Duncan (1955)]. The nature of the F-test allows unequal sample sizes. This seems to make this approach more attractive than the Multiple Range test, but there is a problem:

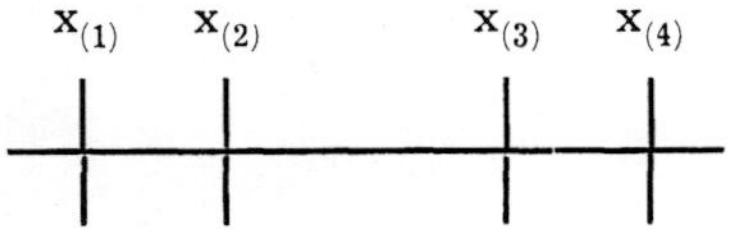

Suppose $\mu_1 = \ldots = \mu_4$ is rejected. The next two hypotheses to be tested are $\mu_1 = \ldots = \mu_3$ and $\mu_2 = \ldots = \mu_4$. So μ_1 and μ_4 will always be called different. But if n_1 and n_4 are much smaller than n_2 and n_3 it is possible that a pairwise test for μ^1 and μ_4 would not yield any significance. Duncan (1952) saw this problem and suggested using a t-test for the pairs that seemed significant as a result of the Multiple F-test. This approach he called the Multiple Comparisons test. Nowadays this term has a more general meaning and it seems to cover every classifying procedure one might consider after rejecting $\mu_1 = \ldots \mu_k$. Now the equality of the variances will be dropped. It is well known that the F-test is not robust against variance heterogeneity [Brown and Forsythe (1974), Ekbohm (1976)]. So it seems reasonable to use the non-iterative version of the second order method of James, thus making a "Multiple James test". One could use Duncan's α_p, but the author prefers $\alpha_p = 1 - (1-\alpha)^{p/k}$ [Ryan (1960)] as a consequence of some arguments pointed out by Einot and Gabriel (1975). This α_p was mentioned in another context, but the arguments are not much shaken by the unequality of the variances.

This new test contains the same problem as the Multiple F-test, but that is not all:

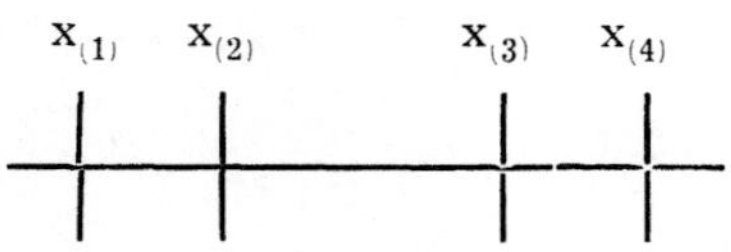

μ_1 and μ_4 will always be called different if $\mu_1 = \ldots = \mu_4$ is rejected. Now suppose that s^2_2 and s^2_3 are much smaller than s^2_1 and s^2_4. Then the difference between μ_1 and μ_4 may not be significant in a pairwise comparison. Here the structural difference between this test and the approach mentioned in the previous section comes into the picture:

If extreme means coincide with big variances and small samples, then the Generalized Multiple Range test can ignore important differences, while the Multiple James test can wrongly declare means to be different.
One can of course apply Welch's test for the Behrens-Fisher problem to the pairs that seem significant as a consequence of the Multiple James test. This combination should be called the "Generalized Multiple Comparisons test". A lot of extra work may be asked for, so it is of interest to know if this extension can have any serious influence on the conclusions.
Werter and the author have examined this by adding another member to the family: the "Leaving One Out test". This is a Multiple James test in which after rejection of $\mu_1 = \ldots = \mu_k$ not only $\mu_1 = \ldots = \mu_{k-1}$ and $\mu_2 = \ldots = \mu_k$ are considered but all the subsets of $\mu_1, \ldots, \mu_k$ where one μ_i is left out. The same α_p is used and the acceptance of a hypothesis means that the splitting process for this subset stops. The Leaving One Out strategy is not limited to $\mu_1 = \ldots = \mu_k$ but is applied to every subset that becomes a candidate. This approach will avoid the classical trap of the Multiple F-test and also the specific problem that comes from variance heterogeneity.
The Multiple James test and the Leaving One Out test were applied to 7 case studies, containing 277 pairs. Only 2 different pairwise conclusions were reached, where the Leaving One Out test did not confirm the significance found by the Multiple James test. But since the Multiple Comparisons test is considered a useful extension of the Multiple F-test, this may not be representative.
The Leaving One Out test can be very expensive. In the worst case situation where all the means are isolated the number of tests will be $2 \;\; -(k+1)$ instead of only $\frac{1}{2}k(k-1)$ for the Multiple James test and any member of the Least Significant Difference family. For k = 15 this means 32 752 tests instead of only 105.
For values of k that make the Least Significance Difference approach unattractive, the Multiple James test is recommended with Ryan's α_p. A terminal oriented computer program such as BMDP should not only give the final result but also the mean, variance and number of observations for every sample. An interesting pairwise significance can be verified by Welch's test for the Behrens-Fisher problem. This should be considered if the sample variances involved are relatively big or if the samples contain only a few observations.

5. Final Remark

This small study on robustness of multiple comparisons against variance heterogeneity only just touches some of the major problems. They are dealt with separately in a simplified example of four samples. In reality one has to deal with them simultaneously which makes the problems much more difficult. Also there are some well known disturbing effects that are not mentioned in this paper.

6. Acknowledgements

Prof. dr. R. Doornbos has been willing to discuss this study in two stages of its development. For his comments I am very grateful.
I wish to thank Paul Werter for his simulation study on the robustness of the q-statistic as used in the Generalized Multiple Range test. And also for comparing the behaviour of the robust tests of the last section in seven case studies.

Further I wish to express my gratitude for the helpfulness of Marjan van Rooij who typed this manuscript in a hurry.

7. References

1. DIJKSTRA, Jan B. and WERTER, Paul S. P. J.
Testing the equality of several means when the population variances are unequal.
Communications in Statistics (1981) B 10, 6.
2. BROWN, M. B. and FORSYTHE, A. B.
The small sample behavior of some statistics which test the equality of several means.
Technometrics 16 (1974), 129—132.
3. WELCH, B. L.:
The comparison of several mean values: an alternative approach.
Biometrika 38 (1951), 330—336.
4. JAMES, G. S.
The comparison of several groups of observations when the ratios of the population variances are unknown.
Biometrika 38 (1951), 324—329.
5. FISHER, R. A.
The Design of Experiments.
Oliver & Boyd (1935) Edinburgh and London.
6. MILLER, R. G.
Simultaneous Statistical Inference.
McGraw Hill Book Company, New York (1966).
7. BEHRENS, W. V.:
Ein Beitrag zur Fehlerberechnung bei wenigen Beobachtungen.
Landwirtschaftliche Jahrbücher 68 (1929), 807—837.
8. WELCH, B. L.
Further note on Mrs. Aspin's tables and on certain approximations to the tabled function.
Biometrika 36 (1949), 293—296.
9. WANG, Y. Y.
Probabilities of type I errors of the Welch tests for the Behrens-Fisher problem.
Journal of the American Statistical Association 66 (1971).
10. URY, H. K. and WIGGINS, A. D.
Large sample and other multiple comparisons among means.
British Journal of Mathematical and Statistical Psychology 24 (1971), 174—194.
11. HOCHBERG, Y.
A modification of the T-method of multiple comparisons for a one-way lay-out with unequal variances.
Journal of the American Association 71 (1976), 200—203.
12. TAMHANE, A. C.
Multiple Comparisons in model-1 one way ANOVA with unequal variances.
Comunications in Statistics A 6 (1) (1977), 15—32.
13. BANERJEE, S. K.
On confidence intervals for two-means problem based on separate estimates of variances and tabulated values of t-variable.
Sankhya, A 23 (1961).
14. NEWMAN, D.
The distribution of the range in samples from a normal population, expressed in terms of an independent estimate of standard deviation.
Biometrika 31 (1939), 20—30.
15. KEULS, M.
The use of the "studentized range" in connection with an analysis of variance.
Euphytica 1 (1952), 112—122.
16. WINER, B. J.
Statistical principles in experimental design.
New York, McGraw-Hill (1962).
17. KRAMER, C. Y.
Extension of multiple range tests to group means with unequal numbers of replications.
Biometrics 12 (1956), 307—310.

18. RAMSEYER, G. C. and TCHENG, T.
The robustness of the studentized range statistic to violations of the normality and homogeneity of variance assumptions.
American Educational Research Journal 10 (1973).
19. GAMES, P. A. and HOWELL, J. F.
Pairwise multiple comparison procedures with unequal N's and/or Variances: a Monte Carlo Study.
Journal of Educational Statistics 1 (1976), 113–125.
20. DUNCAN, D. B.
A significance test for differences between ranked treatments in an analysis of variances.
Virginia Journal of Science 2 (1951).
21. DUNCAN, D. B.
Multiple range and Multiple F-tests.
Biometrics 11 (1955), 1–42.
22. DUNCAN, D. B.
On the properties of the multiple comparisons test.
Virginia Journal of Science 3 (1952).
23. EKBOHM, G.
On testing the equality of several means with small samples.
The Agricultural College of Sweden, Uppsala (1976).
24. RYAN, T. A.
Significance Tests for multiple comparison of proportions, variances and other statistics.
Psychological Bulletin 57 (1960), 318–328.
25. EINOT, I. and GABRIEL, K. R.
A study of the powers of several methods of multiple comparisons.
Journal of the American Statistical Association 70 (1975), 574–583.

Institute of Econometrics and Statistics, University of Łódź

The Power of Run Test Verifying the Hypothesis of Model Linearity

CZESŁAW DOMANSKI

Abstract

The paper concerns the studies of empirical power of some tests based on the number and length of runs, which verify the linearity hypothesis of a model. The obtained results of the power of run tests are compared with the power of the F test.

1. Introduction

The paper presents the results of studies on the power of tests based on the length of runs and on the number of runs, verifying the linearity hypothesis of a model with two explanatory variables in the form

$$Y = \alpha_0 + \alpha_1 X_1 + \alpha_2 X_2 + \varepsilon . \qquad (1)$$

Model (1) is considered at usually postulated assumptions concerning ε ($[\varepsilon_1, \ldots, \varepsilon_n]^T \sim N(0, \sigma^2 I)$) – cf. e.g. Goldberger (1966). Let a sample consisting of n independent observations (x_{1i}, x_{2i}, y_i) be given for $i = 1,2,\ldots,n$. On the basis of this sample a hypothesis $H_0: E(Y|X_1, X_2) = \alpha_0 + \alpha_1 X_1 + \alpha_2 X_2$ should be verified taking into account the runs of residual signs (cf. Domański (1980))

$$e_i = y_i - a_0 - a_1 x_{1i} - a_2 x_{2i} \qquad (2)$$

where a_0, a_1, a_2 are the o. l. s. estimates of parameters α_0, α_1, α_2, respectively. While analyzing the model with two explanatory variables usually many criteria of ordering of e_i can be given. Let h be some function of two variables, and w – some permutation ordering the numbers $h_i = h(x_{1i}, x_{2i})$, i. e. $h_{w(1)} \leq h_{w(2)} \leq \ldots \leq h_{w(n)}$.
Let us consider such criteria for ordering the residuals e_i for which function h is determined by one of the following formulae

(i) $h(x_{1i}, x_{2i}) = a_0 + a_1 x_{1i} + a_2 x_{2i}$

(ii) $h(x_{1i}, x_{2i}) = x_{1i}$,

(iii) $h(x_{1i}, x_{2i}) = x_{2i}$,

(iv) $h(x_{1i}, x_{2i}) = x_{1i} + x_{2i}$,

(v) $h(x_{1i}, x_{2i}) = x_{1i}^2 + x_{2i}^2$.

Assume that the explanatory variables are standardized, thus making the above mentioned criteria independent of linear transformations of these variables.
Let us note that the four first criteria are special cases of the function of the form

$$h(x_{1i}, x_{2i}) = \gamma_0 + \gamma_1 x_{1i} + \gamma_2 x_{2i}$$

(the coefficient in criterion (i) being random), while criterion (v) is a special case of the function

$$h(x_{1i}, x_{2i}) = \gamma_1 x_{1i}^2 + \gamma_2 x_{2i}^2 .$$

We gave up, however, this type of generalization in the present study.

2. The Range of Study

The subject of our study is the evaluation of several variants of a run test and F test verifying the hypothesis of linearity for the model with two explanatory variables. Let us note that run tests are based on the tests with discrete distributions, while the F test has a continuous distribution. That is why the randomized run tests based both on the number and length of runs have been investigated.

Critical values and randomizing probabilities for the tests based on numbers of series have been taken from Domański (1979) while those for the tests based on the length of series – from not published Domański, Tomaszewicz (1980). The F statistic used for verification of hypothesis that some of the regression coefficients are equal to zero is described in Goldberger (1966).

To evaluate the power of tests being studied, we used the Monte Carlo experiment with the following procedure. For determined sample sizes n = 10, 20, 30 the values of x_{1i} were generated from the uniform distribution and the values x_{2i} from the normal distribution in such a way that the correlation coefficient between sequences $\{x_{1i}\}$ and $\{x_{2i}\}$ be equal to the fixed number r. In turn, for each sequence of pairs (x_{1i}, x_{2i}) treated further on as already stated, sequences $\{y_i\}$ were generated, where

$$y_i = g(x_{1i}, x_{2i}) + \xi_i, \quad i = 1,2,\ldots,n); \quad \xi_i : N(0, \sigma_\xi). \qquad (3)$$

The variance σ_ξ^2 determines the dispersion of empirical points on the area defined by function g. Function g is also as follows

$$g(x_1, x_2) = c_0 + c_1 x_1 + c_2 x_2 + c_3 x_1^2 + c_4 x_2^2 + c_5 x_1 x_2 . \qquad (4)$$

Without a loss of generality it can be assumed that $c_0 = 0$ and $c_5 = 0$. It is possible to reach it by isometric transformation of (x_1, x_2) plane. Thus

$$g(x_1, x_2) = c_4(-2v^2 u_1 x_1 - 2u_2 x_2 + v^2 x_1^2 + x_2^2) \qquad (5)$$

where $v = \sqrt{c_3/c_4}$. In the experiment some variants of parameters u_1, u_2, v and the following value were considered:

$$\psi^2 = \frac{\sigma_\xi^2}{S_\Theta^2 + \sigma_\xi^2} \qquad (6)$$

where

$$S_\Theta^2 = \frac{1}{n} \sum_i \left(g(x_{1i}, x_{2i}) - l(x_{1i}, x_{2i})\right)^2$$

and l is a linear approximation function of g minimizing S^2_Θ.
A significant problem in this experiment is the choice of alternative distributions of random component ξ_i. As the alternatives of normal distribution, the Pareto and double exponential distributions as well as the uniform, log-normal and exponential distributions were considered (cf. Domański and Tomaszewicz (1983)).
Densities of these distributions are defined as follows:

a) uniform (UNIF)

$$f(x)=\frac{1}{\sqrt{12}} \quad \text{for} \quad -\sqrt{3}\le x\le\sqrt{3}$$

b) normal (NORM)

$$f(x)=\exp\left(-x^2/2\right)/\sqrt{2\pi}$$

c) log-normal (LNOR)

$$f(x)=\exp\left(-\ln\left(x+\sqrt{e}\right)^2/2\right)/\sqrt{2\pi} \quad \text{for} \quad x\ge-\sqrt{e}$$

d) exponential (EXP)

$$f(x)=\exp(-x-1) \quad \text{for} \quad x\ge-1$$

e) double exponential (2EXP)

$$f(x)=\exp\left(-|x|\sqrt{2}\right)/\sqrt{2}$$

f) Pareto (PAR)

$$f(x)=3(1+|x|)^{-4}/2$$

The experiment covered over 70 combinations of parameters r, u_1, u_2, v, ψ^2. The results of some of them are presented in Tables 1—4.

3. Conclusions

The following main conclusions can be formulated on the basis of the experimental results:

1. With an increase of the correlation coefficient r the power of all tests increases, except the run test based on ordering criterion (v) for which it decreases.
2. The shift of (u_1, u_2) usually does not affect significantly the power of tests being considered.
3. The F test proved to be the strongest in almost all cases.
4. The F test appeared to be the most robust to non-normal distributions.
5. The powers of all tests being considered, are similar for symmetrical distributions NORM, UNIF and 2EXP. In the case of symmetrical distribution PAR and asymmetrical distributions LNOR and EXP significant differences can be observed in the powers of tests as compared with the normal distribution.
6. The power of tests based on the length of run is usually higher than that of tests based on the number of runs.
7. The power of tests based both on the number and length of runs is usually the highest for variants (ii) or (v).

Table 1
Empirical power of tests (in %) for $r = 0.9$, $v = 3$, $\psi^2 = 0.5$, $u_1 = u_2 = 0$, $\alpha = 0.1$

Test based on	Distribution																	
	UNIF			NORM			LNOR			EXP			2EXP			PAR		
	n= 10	n= 20	n= 30	n= 10	n= 20	n= 30	n= 10	n= 20	n= 30	n= 10	n= 20	n= 30	n= 10	n= 20	n= 30	n= 10	n= 20	n= 30
number of runs (i)	6	8	6	9	11	5	6	25	9	14	16	15	4	8	8	4	12	6
number of runs (ii)	33	41	49	45	46	73	71	64	96	19	26	28	51	60	76	73	73	95
number of runs (iii)	19	30	20	28	36	57	25	68	41	17	23	20	25	42	25	34	57	43
number of runs (iv)	39	35	40	47	42	61	70	72	85	19	20	25	53	48	57	70	64	84
number of runs (v)	29	37	51	38	45	65	59	90	90	19	23	27	42	63	70	57	85	86
length of runs (i)	15	10	13	16	7	12	9	2	14	28	28	35	13	8	15	11	3	13
length of runs (ii)	29	42	58	43	57	75	64	83	91	34	46	50	43	61	81	57	81	96
length of runs (iii)	28	42	32	38	50	51	54	80	67	36	42	43	36	58	47	49	80	67
length of runs (iv)	32	47	49	47	55	74	65	82	88	35	42	48	43	65	69	57	83	86
length of runs (v)	46	57	67	49	67	88	67	94	97	37	38	49	59	78	84	71	90	94
F statistic	51	96	99	54	97	100	83	94	97	18	31	44	56	94	99	73	92	99

Table 2
Empirical power of tests (in %) for $r = 0.9$, $v = 3$, $\psi^2 = 0.9$, $u_1 = u_2 = 0$, $\alpha = 0.1$

Test based on	Distribution																	
	UNIF			NORM			LNOR			EXP			2EXP			PAR		
	n= 10	n= 20	n= 30	n= 10	n= 20	n= 30	n= 10	n= 20	n= 30	n= 10	n= 20	n= 30	n= 10	n= 20	n= 30	n= 10	n= 20	n= 30
number of runs (i)	9	11	9	10	12	8	15	14	17	14	11	9	10	10	9	4	10	8
number of runs (ii)	15	15	17	14	18	21	31	48	56	49	67	83	22	21	24	31	37	47
number of runs (iii)	12	15	12	14	12	13	26	32	31	29	54	29	16	19	14	20	26	19
number of runs (iv)	13	16	11	14	17	19	32	36	45	48	52	67	21	20	22	33	30	38
number of runs (v)	15	13	15	17	12	17	25	37	46	40	67	78	18	18	23	27	33	45
length of runs (i)	20	19	19	17	22	19	27	16	51	27	3	13	21	17	18	16	11	15
length of runs (ii)	26	25	23	21	26	33	51	64	69	48	63	83	28	30	39	32	44	63
length of runs (iii)	22	23	20	21	27	26	48	54	57	41	61	57	24	26	22	28	40	34
length of runs (iv)	23	24	19	22	26	36	56	60	67	48	63	79	27	31	29	33	46	47
length of runs (v)	29	27	25	25	28	36	47	57	68	53	77	86	33	36	38	46	52	64
F statistic	13	21	30	12	30	42	31	50	61	56	93	99	16	24	30	28	41	53

Table 3
Empirical power of tests (in %) for $r = 0.5$, $v = 1$, $\psi^2 = 0.5$, $u_1 = u_2 = 0$, $\alpha = 0.1$

Test based on	Distribution														
	UNIF			NORM			LNOR			EXP			2EXP		
	n= 10	n= 20	n= 30	n= 10	n= 20	n= 30	n= 10	n= 20	n= 30	n= 10	n= 20	n= 30	n= 10	n= 20	n= 30
number of runs (i)	13	12	22	15	15	21	8	13	28	7	15	30	10	.	.
number of runs (ii)	21	10	32	18	10	40	17	4	65	14	7	48	32	.	.
number of runs (iii)	8	20	28	5	17	29	1	44	34	5	40	35	4	24	28
number of runs (iv)	22	30	33	26	30	42	49	40	67	35	33	44	31	40	49
number of runs (v)	9	13	38	10	14	47	6	16	75	6	15	51	8	17	60
length of runs (i)	22	22	37	29	23	46	44	27	60	34	23	56	22	.	.
length of runs (ii)	25	29	52	28	29	59	52	47	82	41	38	70	37	.	.
length of runs (iii)	13	27	34	17	23	41	13	43	49	22	35	46	11	.	.
length of runs (iv)	26	41	33	31	44	39	59	65	55	45	50	31	35	.	.
length of runs (v)	27	30	53	29	35	62	37	44	83	32	35	66	27	.	.
F statistic	49	97	100	52	94	100	80	91	96	62	92	99	58	94	98

Table 4
Empirical power of tests (in %) for $r = 0$, $v = 1$, $\psi^2 = 0.5$, $u_1 = u_2 = 0$, $\alpha = 0.1$

Test based on	Distribution														
	UNIF			NORM			LNOR			EXP			2EXP		
	n= 10	n= 20	n= 30	n= 10	n= 20	n= 30	n= 10	n= 20	n= 30	n= 10	n= 20	n= 30	n= 10	n= 20	n= 30
Number of runs (i)	21	42	26	25	32	32	29	80	46	1	54	37	29	.	.
number of runs (ii)	3	5	30	2	6	35	1	1	55	1	2	43	3	.	.
number of runs (iii)	31	35	21	35	35	23	66	65	38	42	46	29	43	47	23
number of runs (iv)	3	20	28	2	23	41	1	40	58	1	24	40	4	19	44
number of runs (v)	17	22	37	10	24	48	7	32	51	13	21	47	13	27	58
length of runs (i)	19	48	42	30	46	49	55	74	71	3	56	63	35	.	.
length of runs (ii)	4	18	49	7	19	58	1	2	78	3	13	71	4	.	.
length of runs (iii)	26	48	38	37	47	45	62	74	63	46	57	52	41	.	.
length of runs (iv)	4	14	20	4	18	28	1	12	28	2	17	26	6	.	.
length of runs (v)	37	32	56	27	30	68	39	41	77	31	94	72	20	.	.
F statistic	46	94	100	55	95	100	79	92	96	60	90	99	56	93	98

4. References

DOMAŃSKI, C., MARKOWSKI, K., TOMASZEWICZ, A.
Run Test for Linearity Hypothesis of Economitric Model with Two Explanatory Variables.
Przegląd Statystyczny 25 (1978), 87–93.

DOMAŃSKI. C.
Quantities of Uncondition of Numbers of Runs.
Prace Institutu Ekonometrii i Statystyki UŁ 30 (1979).

DOMAŃSKI, C.
Notes on the Theil Test for the Hypothesis of Linearity for the Model with two Explanatory Variables.
Colloquia Mathematica Societatis Janos Bolyai, 32 Nonparametric Statistical Inference. Budapest (1980), 213–220.

DOMAŃSKI, C., TOMASZEWICZ, A.
Variants of Tests Based on the Length of Runs.
VII Conference Problems of Building and Estimation of Large Econometric Models, Polanica Zdrój (1980).

DOMAŃSKI, C., TOMASZEWICZ. A.
An Empirical Power of Some Tests for Linearity.
Transaction of the Ninth Prague Conference, Prague (1983), 191–197.

GOLDBERGER, A. S.
Econometric Theory.
John Wiley and Sons. Inc. New York (1966).

Sektion Mathematik der Wilhelm-Pieck-Universität, Rostock, GDR

The Robustness of Selection Procedures Investigated by a Combinatorial Method

HARTMUT DOMRÖSE

Abstract

Bechhofer's indifference-zone-approach in selection with selection rules based on the sample mean, the α-trimmed mean an the median is investigated. The author investigates the robustness of these selection rules against deviation from an assumed distribution by assuming the distributions belonging to the populations are threepoint distributions. This method yields exact results. The selection rule based on the sample mean is investigated for robustness against deviation from the normal distribution.

1. Introduction

We consider the problem of selecting the t "best" from a given populations $\Pi_1, \dots, \Pi_a$ with the means $\mu_1, \dots, \mu_a$. The "best" ones are defined as populations with the t greatest means. Let $\mu_{(1)} \leq \dots \leq \mu_{(a)}$ and $\Pi_{(i)}$ be the population with mean $\mu_{(i)}$. In our paper we use the Bechhofer's indifference-zone-approach, that is

$$\left| \min_{j > a-t} \mu_{(j)} - \max_{i \leq a-t} \mu_{(i)} \right| \geqq d > 0,$$

and assume the least favourable case for continuous distributions

$$\mu_{(1)} = \dots = \mu_{(a-t)} = \mu_{(a-t+1)} - d = \dots = \mu_{(a)} - d. \qquad (1)$$

For every population Π_i an estimation of μ_i is calculated from a sample $\mathbf{y}_i$ of size n by an estimator $h(\mathbf{y}_i, n)$, and the populations yielding the t greatest estimations are selected. The distributions belonging to $\Pi_1, \dots, \Pi_a$ are assumed to be threepoint distributions with the same variance $\sigma^2 = 1$, the same skewness γ_1 and the same kurtosis γ_2. Then the cardinality of the set of possible values of $h(\mathbf{y}_i, n)$ is the same limited number $N(h, n)$ for all populations. The probabilities of these values are easily computed. If $\{h_i, i = 1, \dots, N(h, n)\}$ is the set of possible values of $h(\mathbf{y}_{(a)}, n)$, the probability of a correct selection (CS) can be calculated because of (1) as

$$P(CS) = \sum_{i=1}^{N(h,n)} P\left(\min_{j>a-t} h(\mathbf{y}_{(j)}, n) = h_i \right) \cdot P\left(h(\mathbf{y}_{(1)}, n) < h_i \right)^{a-t} \qquad (2)$$

A correct selection is defined as the selection of the variates with the t greatest means. We will now denote the selection rule described above by SR (h, n, a, t, d).

2. Description of the Method Used

In our investigations we consider only equidistant three-point distributions $\begin{pmatrix} x & x+\Delta & x+2\Delta \\ p_1 & p_2 & p_3 \end{pmatrix}$. Therefore the difference between two possible values $d_n = h_{i+1} - h_i$ is constant for $1 \leq i \leq N(h, n)$ and $N(h, n)$ is less than or the same as in the case of nonequidistance. Let $D_n = [d/d_n]$ the maximal number of possible values of $h(\mathbf{y}_{(a)}, n)$ in an interval of length d. Then we have

$$P(h(\mathbf{y}_{(1)}, n) < h_i) = P(h(\mathbf{y}_{(a)}, n) \leq h_{i+D_n}) \qquad (3)$$

From the binomial theorem follows

$$P\left(\min_{j>a-t} h(\mathbf{y}_{(j)}, n) = h_i \right) = \sum_{i=1}^{t} \binom{t}{i} P\left(h(\mathbf{y}_{(a)}, n) = h_i\right)^i P\left(h(\mathbf{y}_{(a)}, n) > h_i\right)^{t-i}$$
$$= P\left(h(\mathbf{y}_{(a)}, n) \geq h_i\right)^t - P\left(h(\mathbf{y}_{(a)}, n) > h_i\right)^t. \qquad (4)$$

With (3) and (4) we obtained from (2)

$$P(CS) = \sum_{i=1}^{N(h,n)} \left[P\left(h(\mathbf{y}_{(a)}, n) \geq h_i\right)^t - P\left(h(\mathbf{y}_{(a)}, n) > h_i\right)^t \right] \cdot P\left(h(\mathbf{y}_{(a)}, n) \leq h_{i+D_n}\right)^{a-t}.$$

This formula was used in our computer calculations. We now see that we can calculates P(CS) under the above conditions if we know the distribution of the best population.

3. Special Problems of the Method Used

The assumed discrete distribution yields a discrete distribution of $h(\mathbf{y}_i, n)$ for each sample size n. This is why in our investigations there exists more than one least favourable case, in contrast to the continuous distributions, and why P(CS) is constant fore the same n but different values of d. For example we consider $\begin{pmatrix} -\sqrt{3} & 0 & \sqrt{3} \\ 1/6 & 2/3 & 1/6 \end{pmatrix}$ with $\mu = 0$, $\sigma^2 = 1$, $\gamma_1 = \gamma_2 = 0$.
For $n = 2$ the sample mean $\bar{\mathbf{y}}$ can take the five possible values $\{-\sqrt{3}, -\sqrt{3}/2, 0, \sqrt{3}/2, \sqrt{3}\}$ with the difference $d_n = \sqrt{3}/2$.
For SR($\bar{\mathbf{y}}$, 2, 2, 1, d) we obtain $P(CS) = 0.676$ if $0 < d < \sqrt{3}/2$ and $P(CS) = 0.909$ if $\sqrt{3}/2 < d < \sqrt{3}$.
Moreover P(CS) does not increase continuously while n increases. We choose the above example and $d = 1$ and consider Table 1.

Table 1

SR($\bar{\mathbf{y}}$, n, 2, 1, 1)

a	t	n	d_n	$D_n = d/d_n$	P(CS)
2	1	2	$\sqrt{3}/2$	1	.909
2	1	3	$\sqrt{3}/3$	1	.860
2	1	4	$\sqrt{3}/4$	2	.939
2	1	5	$\sqrt{3}/5$	2	.917
2	1	6	$\sqrt{3}/6$	3	.961

P(CS) decreases when N(h, n) increases and D_n remains constant. Therefore the probability of every possible value h_i decreases and, because of (3), $P(h(\mathbf{y}_{(1)}, n) < h_i)$ decreases also.

4. Robustness Against Deviation From an Assumed Distribution

4.1. *Definition of Robustness*

Definition 1:

SR(h, n, a, t, d) is called $\varepsilon(\beta)$-robust against deviation from an assumed distribution G in a set $\mathfrak{F}$ of distributions, if it follows from $P(CS/G) = 1 - \beta$ that $\min_{F \in \mathfrak{F}} P(CS/F) \geq 1 - \beta - \varepsilon(\beta)$.

Definition 2:

SR($\bar{\mathbf{y}}$, n, a, t, d) is called $\varepsilon(\beta)$-robust against deviation from the normal distribution $N(\mu, \sigma^2)$ in a set $\mathfrak{F}$ of distributions, if it follows from $P(CS/N(\mu,\sigma^2)) = 1 - \beta$ that $\min_{F \in \mathfrak{F}} P(CS/F) \geq 1 - \beta - \varepsilon(\beta)$.

4.2. *The Sample Mean*

In this paper we investigated SR($\bar{\mathbf{y}}$, n, a, t, d) with $a = 2(1)20$, $t = 1(1)\ [a/2]$, $n = 3(1)20(5)50(10)100$, $d = 1$ and 0.5 with the method described in chapter 2 in the (γ_1, γ_2)-points given in Table 2.

Table 2

γ_1	γ_2	p_1	p_2	p_3
0	0	1/6	2/3	1/6
0	−1	1/4	1/2	1/4
0	6	1/18	8/9	1/18
1	0	.6222	.3333	.0444
2	6	.0099	.9008	.0892

Because of the problems described in chapter 3 we restricted our attention to sample sizes n planned under the normal distribution for $P(CS) \geq 0.90$, that means the computer program printed P(CS) for a configuration n, a, t, d only if $P(CS) \geq 0.85$ by using $\varepsilon(\beta)$ as small as possible. Table 3 gives an example.

Table 3

SR($\bar{\mathbf{y}}$, n, 20, 1, 1) in the point $\gamma_1 = \gamma_2 = 0$

n	10	11	12	13	14	15
P(CS)	.8545	.9170	.8959	.9408	.9676	.9581
$P(CS/N(\mu, \sigma^2))$		.90			.95	
n	16	17	18	19	20	21
P(CS)	.9770	9706	.9838	.9794	.9887	.9939
$P(CS/N(\mu, \sigma^2))$						.99

SR($\bar{\mathbf{y}}$, n, 20, 1, 1) is 0.2β-robust against deviation from $N(\mu, \sigma^2)$ in the equidistant threepoint distribution with $\gamma_1 = \gamma_2 = 0$. Table 4 compares our results with the exact values under $N(\mu, \sigma^2)$ given for instance in Rasch (1984). Rasch et al. (1978) and Rasch et al. (1981). We denote by n_β the minimum sample size with $P(CS/N(\mu, \sigma^2)) \geq 1 - \beta$. A "−" means that the selection rule is not $\varepsilon(\beta)$-robust in the sense of definition 2 for $n_{0 \cdot 01}$, and "+" means that the procedure is still robust with an $\varepsilon(\beta)$ less than that considered. Robustness is given in $\gamma_1 = \gamma_2 = 0$. We therefore conclude that our investigation method is practicable for an exact robustness study.

It seems that the skewness influences the robustness more than kurtosis does. With most configurations the least favourable distribution was the threepoint distribution with $\gamma_1 = 2$ and $\gamma_2 = 6$. But nevertheless in most cases we obtained $\varepsilon(\beta)$-robustness with $\varepsilon(\beta) \leq \beta$ in all five (γ_1, γ_2)-points for all $n \geq n_{0 \cdot 10}$.

Table 4

Sample sizes n where SR(**y**, n', a, t, d) is $\varepsilon(\beta)$-robust against deviation from $N(\mu, \sigma^2)$ for all $n' \geq n$. n denotes the minimum sample size with $P(CS/N(\mu, \sigma^2)) \geq 1 - \beta$.

d	$\varepsilon(\beta)$	γ_1	γ_2	a=2 t=1	a=3 t=1	a=4 t=1	a=4 t=2	a=5 t=1	a=5 t=2	a=6 t=1	a=6 t=2	a=6 t=3	a=7 t=1	a=7 t=2	
1				4	5	6	7	7	9	8	9	10	8	10	$n_{0\cdot 1}$
				11	14	15	16	16	17	17	18	19	17	19	$n_{0\cdot 01}$
	0.4β	0	0	4	6	6	7	7	9	8	9	10	8	10	
		0	−1	4	5	8	8	8	9	8	9	10	8	10	
		0	6	4	5	8	7	7	9	8	9	10	8	10	
		1	0	4	6	6	7	7	9	8	9	10	8	10	
		2	6	7	14	14	—	—	—	—	—	—	—	—	
	β	2	6	4	5	6	7	7	9	8	9	10	8	10	
0.5				14	20	25	28	28	33	30	36	38	32	39	$n_{0\cdot 1}$
				44	53	58	62	62	68	65	72	74	67	75	$n_{0\cdot 01}$
	0.2β	0	0	14	20	25	25	30	30	30	35	40	35	40	
		0	−1	14	20	25	25	30	30	30	35	40	35	40	
		0	6	14	20	25	25	30	30	30	35	40	35	40	
		1	0	14	20	25	25	30	30	50	50	40	35	40	
		2	6	20	30	30	30	50	50	60	50	40	60	50	
	0.4β	1	0	+	+	+	+	+	+	30	35	+	+	+	
		2	6	14	20	25	30	30	30	30	35	+	35	40	

d	$\varepsilon(\beta)$	γ_1	γ_2	a=7 t=3	a=8 t=1	a=8 t=2	a=8 t=3	a=8 t=4	a=9 t=1	a=9 t=2	a=9 t=4	a=10 t=1	a=10 t=2	a=10 t=5	
1				11	9	11	11	12	9	11	12	9	11	13	$n_{\cdot 1}$
				20	18	20	21	21	18	20	22	18	21	22	$n_{\cdot 01}$
	0.4β	0	0	11	8	10	10	11	9	11	12	9	11	13	
		0	−1	11	8	10	10	11	9	11	12	9	11	13	
		0	6	11	8	10	10	11	9	11	12	9	11	13	
		1	0	11	13	10	10	11	13	11	12	13	11	13	
		2	6	—	—	—	—	—	—	—	—	—	—	17	
	β	1	0	+	8	+	+	+	9	+	+	9	+	+	
		2	6	11	8	10	10	11	9	11	12	9	11	13	

0.5				42	33	41	44	45	35	43	48	36	44	51	$n_{.1}$
				78	69	77	81	82	71	79	85	73	81	88	$n_{.01}$
	0.2β	0	0	40	35	40	45	45	35	40	50	35	45	50	
		0	—1	40	35	40	45	45	35	40	50	35	45	50	
		0	6	40	35	40	45	45	35	40	50	35	45	50	
		1	0	40	50	40	45	45	50	40	50	35	50	50	
		2	6	40	—	—	50	50	—	50	50	—	60	50	
	0.4β	1	0	+	35	+	+	+	35	+	+	+	45	+	
		2	6	+	35	40	45	45	35	40	+	35	45	+	

d	ε(β)	γ_1	γ_2	a=11 t=1	a=11 t=3	a=11 t=5	a=12 t=1	a=12 t=5	a=14 t=1	a=14 t=5	a=16 t=1	a=20 t=1	
1				10	13	14	10	14	10	15	11	11	$n_{.1}$
				19	22	24	19	24	20	24	20	21	$n_{.01}$
	0.4β	0	0	9	12	14	10	14	10	15	11	11	
		0	—1	9	12	14	10	14	10	15	11	11	
		0	6	12	12	14	10	14	10	15	11	11	
		1	0	13	13	14	13	14	14	15	13	13	
		2	6	—	—	17	—	—	—	17	—	—	
	β	1	0	9	12	+	10	+	10	+	11	11	
		2	6	14	13	14	10	14	14	15	14	17	

d	ε(β)	γ_1	γ_2	a=11 t=1	a=11 t=3	a=11 t=5	a=12 t=1	a=12 t=5	a=14 t=1	a=14 t=5	a=16 t=1	a=20 t=1	
0.5				37	50	55	38	55	40	59	41	44	$n_{.1}$
				74	87	93	75	93	77	96	79	82	$n_{.01}$
	0.2β	0	0	35	50	50	35	50	40	60	40	45	
		0	—1	35	50	50	35	50	40	60	40	45	
		0	6	35	50	50	40	50	40	60	40	45	
		1	0	35	50	50	35	50	40	60	60	50	
		2	6	50	50	50	60	50	60	60	60	60	
	0.4β	1	0	+	+	+	+	+	+	+	40	45	
		2	6	35	+	+	50	+	50	+	60	50	
	β	2	6	+	+	+	35	+	40	+	40	45	

4.3. *The α-trimmed Mean*

Let $y_i = (y_{i1}, \ldots, y_{in})$ be a sample from Π_i with $y_{i(1)} \leq \ldots \leq y_{i(n)}$. Than $\bar{y}_\alpha = 1/(n-2\alpha) \sum_{j=\alpha+1}^{n-\alpha} y_{i(j)}$ is called the α-trimmed mean of the sample. We investigated $SR(\bar{y}_{0.1n}, n, a, t, d)$ with $n = 10(1)20(5)50$, $a = 2(1)20$, $t = 1(1)\,[a/2]$, $d = 1$ and 0.5, and $SR(\bar{y}_{0.2n}, n, a, t, d)$ with $n = 6(1)20(5)50$ and the values of a, t, and d given above in the same five distributions as in chapter 4.2. We were looking for robustness in the sense of definition 1 when the assumed distribution G is the threepoint distribution with $\gamma_1 = \gamma_2 = 0$. Table 5 gives the minimum sample sizes required by $SR(\bar{y}_\alpha, n, a, t, d)$ to yield $P(CS) \geq 1 - \beta$ for some chosen configurations. In the point $\gamma_1 = \gamma_2 = 0$ we obtained nearly the same values of n as those required by $\bar{y}$. We found that $\gamma_1 = 0$, $\gamma_2 = -1$ and $\gamma_1 = 1$, $\gamma_2 = 0$ are clearly the two least favourable distributions. While the 0.1n-trimmed mean is still robust against deviation from G with $\varepsilon(\beta) \leq 1.5\beta$ in all five points in most configurations, the 0.2n-trimmed means has not this property. Trimming the mean reduces roubstness in the sense of definition 1 in our investigations. Special results verifying this statement can be found in Domröse (1984).

Table 5

Sample sizes required by $SR(\bar{y}_\alpha, n, a, t, d)$ to yield $P(CS) \geq 1 - \beta$

			α = 0.1 n		α = 0.2 n	
1 — β	γ_1	γ_2	a=10 t=1	a=20 t=1	a=10 t=1	a=20 t=1
0.9	0	0	10	13	11	11
	0	—1	12	14	13	18
	0	6	10	11	6	8
	1	0	11	15	17	20
	2	6	10	12	9	10
0.95	0	0	15	16	13	15
	0	—1	15	18	20	25
	0	6	11	11	8	8
	1	0	15	16	19	25
	2	6	12	16	11	11
0.99	0	0	25	25	20	25
	0	—1	30	30	30	35
	0	6	14	14	11	11
	1	0	20	25	30	35
	2	6	19	20	14	16

4.4. *The Median*

In this chapter we want to give a short comment regarding the selection rule based on the median y_{med}. We investigated this rule for $n = 3(2)39$ and the same values a, t, d in the same five (γ_1, γ_2)-points considered in chapter 4.1. In contrast to the sample mean and the α-trimmed means, the median has only three possible values and therefore the same P(CS) for all d less than the difference Δ between the points of the threepoint distribution. Table 6 gives P(CS) for a chosen configuration and $0 < d < \Delta$. A gap means that $P(CS) < 0.85$ or $P(CS) > 0.99$ respectively. When $\gamma_1 = \gamma_2 = 0$, $SR(y_{med}, n, a, t, d)$ yields nearly the probabilities of $SR(\bar{y}, n, a, t, 1.05)$. But in the point $\gamma_1 = 1$, $\gamma_2 = 0$ we found no meaningful robustness against deviation from the equidistant threepoint distribution in (0, 0).

Table 6

P(CS) of SR($\mathbf{y}_{med}$, n, 2, 1, d) with $0 < d < \Delta$

n	$\gamma_1 =$ 0 $\gamma_2 =$ 0	0 —1	0 6	1 0	2 6
3	.868		.982		.977
5	.932		.996		.993
7	.965				
9	.982	.909			
11	.990	.934			
13		.953		.851	
15		.966		.862	
17		.975		.872	
19		.982		.882	
21		.987		.890	
23		.990		.898	
25				.906	
27				.912	
29				.919	
31				.924	
33				.930	
35				.935	
37				.939	
39				.944	

The least favourable discrete distribution we found is the symmetric twopoint distribution. In this case P(CS) of SR($\mathbf{y}_{med}$, n, a, t, d) would be constant for all odd sample sizes. Summerising the median cannot be proposed for selection rules for arbitrary distributions since it was not robust in the sense of definition 1 in the chosen discrete example.

5. References

DOMRÖSE, H.
Dissertation A, 1984.

HERRENDÖRFER, G. (ed.)
Robustheit I.
Probleme der angewandten Statistik.
FZ für Tierproduktion Dummerstorf-Rostock, Heft 4, (1980).

RASCH, D.
Einführung in die mathematische Statistik, Bd. 2.
VEB Deutscher Verlag der Wissenschaften, Berlin, 1984.

RASCH, D., HERRENDÖRFER, G (ed.)
Robustheit III.
Probleme der angewandten Statistik, Heft 7.
FZ für Tierproduktion Dummerstorf-Rostock (1982).

RASCH, D., HERRENDÖRFER, G., BOCK, J., BUSCH, K.
Verfahrensbibliothek, Bd. 2.
VEB Deutscher Landwirtschaftsverlag, Berlin 1978.

RASCH, D., HERRENDÖRFER, G., BOCK, J., BUSCH, K.
Verfahrensbibliothek, Bd. 3.
VEB Deutscher Landwirtschaftsverlag, Berlin 1981.

Results of Comparisons Between Different Random Number Generators

FEIGE, K.-D.; GUIARD, V.; HERRENDÖRFER, G.; HOFFMANN, J.; NEUMANN, P.; PETERS, H.;
RASCH, D.; VETTERMANN, Th.

Staff from various research establishments in the GDR have investigated generators of equidistributed pseudorandom numbers in regard of their statistical properties. The tests were performed on the following computers:

ES 1040	by Ms. Peters (Rostock)
BESM-6	by Dr. Neumann (Dresden)
CDC-1604 A	by Mr. Vettermann (Böhlitz-Ehrenberg)
KRS 4200	by Dr. Feige (Dummerstorf-Rostock).

Apart from investigating several multiplicative generators of the type

$$x_{n+1} = (a\,x_n) \bmod M$$

and generators of the mixed type

$$x_{n+1} = (a\,x_n + b) \bmod M$$

(fixed point arithmetic in some cases but floating point arithmetic others), the group also studies a few special techniques for generating random numbers.

The parameters of the generators are given in table 1.

In case 4a) the algorithm certainly seems quite complicated, but in fact it represents merely a bit shift generator: Bits 16 and 32 of a 32-place binary number are combined in such a way that they yield a new bit information which is then attached to the preceding random number. The possible overflow at the beginning of the random number is ignored. When programmed in an assembler language, generators of this type are usually very fast.

Thumhart's generator consists of four multiplicative components that are combined in such a manner that, by means of a special transformation, the index for the next generator is calculated from the preceding random number.

Following up an idea expressed by McLaren and Prof. Marsaglia in 1965, Mr. Teuscher found in 1979 new parameter combinations for a random number generator which in practical terms consists of two independent, hierarchically arranged multiplicative generators. The generator works on the principle that 128 random numbers are first calculated with the first generator. The second generator is then used to select random numbers pseudorandomly from the 128 numbers and to replace each selected number by the next one calculated by the first generator.

Most generators were programmed in the assembler language of the computer concerned, although FORTRAN was used in some cases (only FORTRAN was used for the BESM-6).

With each generator 10 000 random numbers (RN) were generated, the same 10 starting points (x_0) being used in each case. The generators were subjected to the following tests by using these random numbers:

G — χ^2-test for equi-distribution in 100 equidistant classes in (0,1),

R — χ^2-test for the two-dimensional distribution of consecutive pairs of numbers (lag 1, 2, 3 and 4) in 10×10 classes,

I — χ^2-interval test in respect of omission of a certain interval $I_i \subset (0,1)$, (I_1: (0, 0.2), I_2: (0.2, 0.4), .. I_5: (0,8, 1)),

M — χ^2-maximum test for evaluating the distribution that might be expected on the assumption of the equidistribution of the generator,

$\bar{x}$, s^2, γ_1, γ_2 — the mean, variance, skewness and kurtosis of the RN sequence was put as an additional, descriptive from of generator evaluation,

— some members of the group ascertained the number of ascending or descending sequences (v_1, v_2) of RN and the number of RN above or below the arithmetic mean (n_0) of a sample of 10 000 RN. n_1 or n_2 are the numbers of sequences below or above the arithmetic mean of the sample.

The different results are presented in tables 2—5. If you find in the tables an A, then initial test shows that the properties of this particular generator are bad.

We propose that the following best generators be used in the subroutine versions we have elaborated for the different computers:

CDC-1604 A

$$x_{n+1} = (5^{15} x_n + 1) \bmod 2^{35} \qquad \widehat{=}\ (2e)$$

ES 1040

$$x_{n+1} = \left(10^{14} \pi\, x_n + 10^{14} \sqrt{3}\right) \bmod 10^{14} e \qquad \widehat{=}\ (2f)$$

$$x_{n+1} = (5^{13} x_n) \bmod 2^{39} \qquad \widehat{=}\ (1f)$$

$$x_{n+1} = (7^{13} x_n) \bmod 2^{31} \qquad \widehat{=}\ (5h)$$

and with good results

$$x_{n+1} = (2^{10} x_n) \bmod 1049399 \qquad \widehat{=}\ (4c)$$

$$y_{n+1} = (8323\, y_n) \bmod 2^{28}$$

KRS 4200

$$x_{n+1} = (2^{10} x_n) \bmod 1049399 \qquad \widehat{=}\ (4c)$$

$$y_{n+1} = (8323\, y_n) \bmod 2^{28}$$

BSM-6

$$x_{n+1} = (3141592221\, x_n + 1) \bmod 2^{35} \qquad \widehat{=}\ (2b)$$

$$x_{n+1} = (5^{17} x_n) \bmod 2^{42} \qquad \widehat{=}\ (1e)$$

While testing the uniformly distributed random number generators, the group at Dummerstorf also analysed the properties of transformers which, at least approximately, transform the random numbers of a (0,1)-equidistribution into a normal distribution.

We tested the following transformations:

$$\text{a)} \quad u_n = \sqrt{\frac{\pi}{8}}\, \ln \frac{1+x_n}{1-x_n}$$

$$\text{b)} \quad |u_n| = \frac{2.30753 + 0.27061\, z_n}{1 + 0.99229\, z_n^2} - z_n$$

with

$$z_n = \sqrt{-2 \ln x_n} \qquad x_n < 0.5$$

$$z_n = \sqrt{-2 \ln (1 - x_n)} \qquad x_n \geq 0.5$$

c) $$u_1 = x_1 \sqrt{\frac{-2 \ln z}{z}}$$

$$u_2 = x_2 \sqrt{\frac{-2 \ln z}{z}}$$

with

$$z = x_1^2 + x_2^2 \text{ and } 0 < z < 1$$

d) $$u_1 = \sqrt{-2 \ln x_1} \quad \cos 2\pi x_2$$

$$u_2 = \sqrt{-2 \ln x_1} \quad \sin 2\pi x_2$$

e) $$|u_n| = z_n + \frac{(((z_n \cdot p_4 + p_3) z_n + p_2) z_n + p_1) z_n + p_0}{(((z_n \cdot q_4 + q_3) z_n + q_2) z_n + q_1) z_n + q_0}$$

with

$$z_n = \sqrt{-2 \ln x_n} \qquad \text{for } x_n < 0.5$$

and

$$z_n = \sqrt{-2 \ln (1 - x_n)} \qquad \text{for } x_n \geq 0.5$$

$p_0 = -0.322232431088$	$q_0 = 0.99348462606 \cdot 10^{-1}$
$p_1 = -1.$	$q_1 = 0.588581570495$
$p_2 = -0{,}342242088547$	$q_2 = 0.531103462366$
$p_3 = -0.0204231210245$	$q_3 = 0.10353775285$
$p_4 = -0.453642210148 \cdot 10^{-4}$	$q_4 = 0.385607006341 \cdot 10^{-2}$

The class limits for the χ^2-test were selected so that the same expected frequencies E_i can be assumed for the 25 classes on the condition of an N(0,1)-normal distribution. In the table 6 we have included a couple of selected observed frequences. The χ^2-test of fit, finally, also permits only the second and fifth transformation to be performed on the KRS 4200 and on the ES 1040 again.

Table 1

Generators

1. Multiplicative Generators

$x_{n+1} = (a\, x_n) \bmod M$

	a	M
a)	82027	2^{32}
b)	$2^{13} + 6$	$2^{23} + 1$
c)	$5 + 8^6$	2^{35}
d)	5^{15}	2^{35}
e)	5^{17}	2^{42}
f)	5^{13}	2^{39}
g)	65539	2^{31}

2. Generators of mixed typ

$x_{n+1} = ((a\, x_n) + b) \bmod M$

	a	b	M
a)	$2^{10} + 1$	101	2^{36}
b)	3141592221	1	2^{35}
c)	2718281821	1	2^{35}
d)	5^{13}	1	2^{35}
e)	5^{15}	1	2^{35}
f)	$10^{14}\,\pi$	$10^{14}\sqrt{3}$	$10^{14}\,e$

3. Mixed generators (floating point arithmetic)

	a	b	M
a)	c	$\sqrt{3}$	10.
b)	e	0.	1.
c)	e	$\sqrt{3}$	1.

4. Special techniques

a) $$x_{n+1} = 2x_n \bmod 2^{32} + \left(\left(\frac{x_n}{2^{31}} + \frac{x_n}{2^{16}} \bmod 2 \right) + 1 \right) \bmod 2$$

b) Thumhart's generator (Roe (1970))

4 multiplicative generators with
$a_1 = 252\,114\,903\,917$
$a_2 = 8\,064\,131\,757$
$a_3 = 282\,629$
$a_4 = 4\,357$
$M = 2^{35}$

the index I_n for the next generator is calculated with

$$I_n = \frac{(a_{I_n} x_n) \bmod 2^{38} - (a_{I_n} x_n) \bmod 2^{36}}{2^{36}}$$

c) McLaren — Marsaglia (1965)
(special parameters Teuscher (1979))

$$x_{n+1} = 2^{10} x_n \bmod 1049339$$

$$y_n + 1 = 8323\, y_n \bmod 2^{38}$$

$$I_{n+1} = y_{n+1} / 2^{21} \Rightarrow 7 \text{ bit}$$

5. Additional generators
(only tested on ES 1040 computer)

a) $$x_{n+1} = 10^7 x_n \bmod 10006499$$

(but other starting points)

b) $$x_{n+1} = (10^{14} \pi x_n + 10^{14} \sqrt{3}) \bmod 10^{14} e$$

(and then like 4c) — gen.)

c) $$x_{n+1} = 5 x_n$$

if $x_{n+1} > A \Rightarrow x_{n+1} = x_n - A$

$x_{n+1} > B \Rightarrow x_{n+1} = x_n - B$

$x_{n+1} > C \Rightarrow x_{n+1} = x_n - C$

else $x_{n+1} = x_n$

with $A = 137\,438\,953\,472$
$B = 68\,719\,476\,736$
$C = 34\,359\,738\,368$
and $x_0 = 11\,919\,641\,733$

	a	b	M
d)	97	788 675	100 001
e)	3^{26}	—	10^{13}
f)	7^5	—	$2^{31} - 1$
g)	7^9	—	$2^{31} - 1$
h)	7^{13}	—	$2^{31} - 1$
i)	3^{17}	—	10^{10}
j)	3^{19}	—	10^{10}
k)	10 006 499	—	$2^{31} - 1$

Table 2
Results for CDC-1604 A

Gen.	G	R	I	M
1a)	95.5	87.7— 105.8	5.5— 7.9	79.8—102.1 (*)
1b)	99.7	85.8— 98.4	7.5— 9.9	29.5— 38.6 (*)
1c)	96.1	85.3— 106.7	6.5—11.8	83.2— 90.8 (*)
1d)	96.5	96.3— 110.8 (*)	6.1— 7.5	83.0— 96.9
1e)	104.6	97.0— 104.0 (*)	7.9— 9.8	90.6—112.0 (*)
1f)	98.1	85.3— 102.5	5.9—11.0	64.4— 95.3 (*)
2a)	89.9	92.0— 109.8	7.3— 9.1	86.2— 99.9 (*)
2b)	89.9	94.0— 104.1	5.7—10.3	66.3— 88.4 (*)
2c)	100.8	94.5— 103.2	7.3—10.7 (*)	60.8— 85.1
2d)	95.8	92.8— 102.4	5.3— 9.9	84.3— 87.5 (*)
2e)	88.7	86.1— 97.6	5.6— 9.2	66.5— 82.3
3a)	105.5	94.5— 119.9	7.8—11.2 (*)	85.0— 90.0 (*)
4a)	101.9 (*)	968.9—40045. (*)	1353.0 (*)	500000 (*)
4b)	102.8 (*)	92.6— 108.0 (*)	6.2— 8.3	70.8— 85.5
4c)	101.4	97.8— 108.8 (*)	6.6—10.6	48.9— 94.4
theor. values	124.3	124.3	16.9	124.4

Table 2
(continuation)

Gen.	u_1	u_2	$\bar{x}$	s^2	γ_1	γ_2
1a)	—	—	—	—	—	—
1b)	—	—	—	—	—	—
1c)	—0.07	—0.25	0.5014	0.0835	—0.0039	—1.2070
1d)	—0.76	—0.22	0.4988	0.0835	0.0003	—1.1994
1e)	0.76	0.15	0.4992	0.0826	—0.0004	—1.1905
1f)	0.17	0.27	0.5003	0.0831	0.0003	—1.1962
2a)	—	—	—	—	—	—
2b)	0.24	1.05	0.4999	0.0829	0.0044	—1.1981
2c)	0.18	0.60	0.4990	0.0834	0.0044	—1.1981
2d)	—0.21	—0.12	0.4982	0.0827	0.0099	—1.1923
2e)	0.31	—0.27	0.5003	0.0829	0.0002	—1.1975
3a)	—0.08	0.63	0.4970	0.0838	0.0120	—1.2065
4a)	—	—	—	—	—	—
4b)	—	—	—	—	—	—
4c)	—	—	—	—	—	—
theor. values	±1.96	±1.96	0.5000	0.0833	0.0	—1.2

Table 3
Results for ES 1040

Gen.	G	R	I	M	n_0	$v_1 \approx v_2$
1a)	118.1	131.1	14.3	159.1	4975—5034	3291—3369
1b)	112.3	144.1	17.6	383.9	4974—5050	3282—3351
1c)	123.0	116.0	24.9	218.4	4983—5023	3301—3355
1d)	150.8	127.6	18.3	140.1	4949—5045	3305—3375
1e)	116.9	136.5	22.6	2244.1	4975—5031	3300—3350
1f)	115.7	124.6	17.8	108.1	4959—5046	3303—3364
2a)	128.5	124.2	19.3	129.2	4918—5005	3264—3340
2b)	155.3	127.8	20.4	149.0	4922—5021	3281—3364
2c)	144.5	138.0	19.0	170.0	4904—5071	3286—3388
2d)	108.4	134.4	19.8	246.4	4945—5028	3289—3350
2e)	114.3	121.4	28.5	642.6	4978—5032	3306—3385
2f)	133.4	112.3	16.0	119.0	4971—5073	3290—3370
4c)	102.4	135.0	17.2	130.0	4966—5044	3289—3373
5a)	115.26	110.9	23.0	126.0	4947—5050	3294—3353
5b)	112.6	116.8	17.4	257.0	4949—5043	3303—3364
5c)	122.3	10121.0	19.7	24540.0	4953—5063	2982—3047
5d)	115.2	115.7	22.4	183.9	4960—5063	3309—3362
5e)	112.2	126.0	19.6	117.2	4952—5012	3302—3355
5f)	114.0	117.6	15.4	171.8	4961—5067	3293—3361
5g)	123.9	127.5	13.5	415.3	4951—5028	3293—3369
5h)	128.5	150.5	19.2	113.1	4966—5028	3313—3377
5i)	127.5	115.6	15.7	127.0	4915—5049	3285—3359
5k)	130.3	115.8	16.4	179.5	4963—5045	3302—3381
5l)	120.8	144.1	19.8	103.1	4949—5049	3284—3372
theor. values	123.2	123.2	15.5	123.2	5000	3333

Table 3
(continuation)

Gen.	$n_1 \approx n_2$	x	s^2	γ_1	γ_2
1a)	2454—2535	0.5004	0.0828	—0.0248	—1.211
1b)	2427—2540	0.4985	0.0833	0.0277	—1.172
1c)	2465—2528	0.4975	0.0835	0.0331	—1.232
1d)	2450—2551	0.5006	0.0840	—0.0245	—1.217
1e	2447—2539	0.4998	0.0831	0.0248	—1.218
1f	2465—2530	0.5005	0.0833	—0.0298	—1.170
2a)	2447—2528	0.4970	0.0828	0.0443	—1.178
2b)	2474—2548	0.4987	0.0838	0.0421	—1.215
2c)	2454—2539	0.5009	0.0824	0.0409	—1.160
2d)	2455—2508	0.5003	0.0832	0.0198	—1.171
2e)	2464—2528	0.5001	0.0837	—0.0230	—1.225
2f)	2461—2514	0.4992	0.0838	—0.0267	—1.224
4c)	2469—2531	0.5004	0.0834	—0.0329	—1.181
5a)	2438—2533	0.4999	0.0832	—0.0278	—1.230
5b)	2469—2528	0.4997	0.0833	0.0833	—1.170
5c)	1973—2080	0.5009	0.0836	—0.0306	—1.232
5d)	2428—2531	0.5004	0.0829	—0.0325	—1.182
5e)	2473—2543	0.5000	0.0834	—0.0834	—1.214
5f)	2444—2523	0.5005	0.0831	—0.0344	—1.176
5g)	2473—2551	0.4995	0.0828	—0.0189	—1.166
5h)	2455—2520	0.5008	0.0835	—0.0179	—1.213
5i)	2475—2527	0.4983	0.0834	0.0372	—1.213
5k)	2472—2528	0.4995	0.0834	0.0351	—1.229
5l)	2443—2532	0.4999	0.0832	0.0363	—1.174
theor. values	2500	0.5	0.0833	0.0	—1.2

Table 4
Results for BESM-6

Gen.	G	R	I	M	$\bar{x}$	s^2	γ_1	γ_2
1c)	122.4	125.6	22.2	299.1	0.4970—0.5049	0.0824—0.0846	—0.0268	—1.23
1d)	120.6	143.0	15.0	262.1	0.4952—0.5028	0.0821—0.0842	0.0300	—1.17
1e	111.2	132.3	14.2	517.5	0.4980—0.5036	0.0826—0.0847	0.0144	—1.18
1f)	122.1	116.1	22.5	227.7	0.4958—0.5049	0.0830—0.0850	—0.0310	—1.23
2b)	107.1	119.0	16.9	174.8	0.4964—0.5046	0.0818—0.0838	—0.0229	—1.18
2c)	134.6	130.2	19.2	118.9	0.4943—0.5033	0.0823—0.0846	—0.0235	—1.18
2d)	116.1	144.4	16.8	235.9	0.4943—0.5018	0.0818—0.0834	0.0318	—1.18
3a)	131.9	142.3	31.8	115.9	0.4933—0.5048	0.0816—0.0833	—0.0274	—1.17
theor. values	123.2	123.2	15.5	123.2	0.5000	0.0833	0.0	—1.2

Table 5
Results for KRS 4200

Gen.	G	R	I	M	$\bar{x}$	s^2
1a)	2064—2346	209.0—294.1	1.96—26.1	160.3—401.6	0.483—0.489	0.083—0.0852
1b)	106.3—174.3	101.6—259.4	7.6—16.2	35.7—114.0	0.498—0.503	0.083—0.0835
1c)	98.1—127.2	79.3—2611.5	9.5—117.1	100.9—546.6	0.498—0.507	0.0822—0.0835
1d)—1f)	A	A	A	A	A	A
1g)	493.1—512.3	1082—35839	A	A	0.544—0.557	0.0786—0.082
2a)	53147	25879—50107	2687	388.4—547.3	0.340—0.3411	0.1051—0.1061
2b)—2e)	A	A	A	A	A	A
3a)	210.8—269.7	23560	225.3—828.1	255—17342.5	A	A
3b)	246.4—344.3	527—82473	315.1—1014.9	500—193020	0.460—0.474	0.0769—0.0777
3c)	553.7—728.1	1078—24707	23.9—866.3	259.5—1245.0	0.498—0.503	0.0926—0.0951
4a)	A	A	A	A	A	A
4b)	34689	27245—31652	756.4—1761.9	3039.5—3356.2	0.308—0.310	0.0986—0.098
4c)	72.6—109.1	78.8—124.0	4.1—17.1	45.5—94.1	0.496—0.503	0.0823—0.0842
theor. values	123.2	123.2	15.5	123.2	0.5000	0.08333

Table 6
Results of the comparison of N(0,1)-transformations

transf.	intervals							χ^2-value	time
	(—2.054)	(2.054,1.751)	(1.751,1.555) min.	...	(0.1509,0.107) max.	(0.107—0.0502)	(0.0502—0.)		
a	7200	4262	3680	,	3991	3997	4000	2900.4	1.00
b	4006	3992	3980	,	4032	4036	4045	2.25	1.95
c	4066	4015	3945	,	4014	3974	4247	18.2	2.20
d	3848	3841	3942	,	3981	4250	4507	80.6	2.85
e	3918	4034	3998	,	4009	4000	3999	2.1	3.15
E_i	4000	4000	4000		4000	4000	4000		

in all cases N = 100 000

References

Mac LAREN, M. D., MARSAGLIA, G., BRAY, I. A.
A fast procedure generating exponential random variables.
Communications of the ACM 7 (1964) 5, 298—300.

TEUSCHER, F.
Ein hierarchischer Pseudo-Zufallszahlengenerator.
Unveröffentlichte Praktikumsarbeit, Dummerstorf (1979).

Academy of Agricultural Sciences of the GDR
Research Centre of Animal Production Dummerstorf-Rostock

Robustness of the Two-sample Sequential t-test

DIETMAR FRICK

Abstract

Hajnal's pooled variance sequential two-sample and a Welch type sequential two-sample t-test proposed by Reed and Frantz are compared for different forms of violation of the underlying normal assumption. For Monte-Carlo studies eight alternative distributions of the Fleishman System with given values of skewness γ_1 and kurtosis γ_2 were used. The results show that both tests are robust for α and β when the violation of the normal assumption in the X-population and Y-population is of the same nature. The influence of unequal variances is also considered. The results show that for small deviations of σ_x/σ_y from 1 both tests can be used if we take one x- and one y-observation at each stage of the sequential test, or if we take x- and y-observations with the same probability $\pi_x = \pi_y = 1/2$. For $\pi_x \neq \pi_y$ only the Welch type test can be used because Hajnal's test tends to have uncontrolled type I or type II error in this case.

1. Introduction

The model assumptions for Hajnal's (1961) pooled variance sequential two-sample t-test are that observations are taken from two normal populations with unknown means μ_x and μ_y and with common unknown variance σ^2. Each observation comes from the X- or Y-population according to the constant probabilities π_x and $\pi_y = 1 - \pi_x$, respectively. The hypothesis tested is $H_0 : \mu_x = \mu_y$ and the alternative is $H_A : (\mu_x - \mu_y)^2/\sigma^2 = d^2$, where d is a fixed constant. The sequential probability ratio criterion is

$$Q = \exp\left(-\frac{d^2}{2a}\right) H\left(\frac{f+1}{2}, \frac{1}{2}, \frac{d^2 t^2}{2a(f+t^2)}\right)$$

where

$$t = \frac{\bar{x} - \bar{y}}{s\,a},\quad a^2 = \frac{1}{n_x} + \frac{1}{n_y},\quad f = n_x + n_y - 2,$$

$$s^2 = \frac{(n_x - 1)\, s_x^2 + (n_y - 1)\, s_y^2}{f}$$

and H is the confluent hypergeometric function. Sampling and computation of Q proceeds as long as $A = \beta/(1-\alpha) < Q < (1-\beta)/\alpha = B$. If $Q < A$ we accept H_0 and if $Q > B$ we reject H_0. Reed and Frantz (1979) proposed a Welch modification of this test for the case $\sigma_x \neq \sigma_y$.

Now we have to compute

$$Q_w = \exp\left(-\frac{d_w}{2a}\right) H\left(\frac{f_w+1}{2}, \frac{1}{2}, \frac{d_w^2\, t_w^2}{2a(f_w + t_w^2)}\right),$$

where

$$t_w = \frac{\bar{x} - \bar{y}}{\sqrt{\frac{s_x^2}{n_x} + \frac{s_y^2}{n_y}}},\quad f_w = \frac{(s_x^2/n_x + s_y^2/n_y)^2}{\frac{(s_x^2/n_x)^2}{n_x - 1} + \frac{(s_y^2/n_y)^2}{n_y - 1}}$$

and $d_w = \dfrac{(\mu_x - \mu_y)^2}{\sigma_w^2}$.

In the latter equation we find a weighted average of the population variances $\sigma_w^2 = \pi_y \sigma_x^2 + \pi_x \sigma_y^2$.

2. The Simulation Experiment

Each simulation experiment consisted of 10,000 samples in order to satisfy precision requirements for the estimations of the probabilities α and β. The cases $d = d_w = 1$ and $d = d_w = 1.5$ were considered for $\alpha = 0.05$ and $\beta = 0.10$ and various forms of sampling. Only the results of the case $d = d_w = 1.5$ are presented here. Smaller values of d and d_w or of α and β have in general influence on the ASN-values. Without loss of generality we put $\mu_x = 0$ and $\sigma^2 = \sigma_w^2 = 1$ for computer simulations. The confluent hypergeometric function was computed by a series approximation.

3. The results

3.1. *Influence of nonnormality*

Both tests are robust if the violation of the normal assumption in the X- and Y-population is the same. The tests are conservative in α when γ_2 increases. That means

$\hat{\alpha}(\gamma_2 > 0) < \hat{\alpha}(\gamma_2 = 0) < \alpha$ (nominal)

and for $\gamma_2 < 0$:

$\hat{\alpha}(\gamma_2 = 0) < \hat{\alpha}(\gamma_2 < 0) < \alpha$

in general ($\hat{\alpha}$ denotes the estimation of the type I error α). The values of β increase when γ_2 increases. An influence of γ_1 does not exist when at each stage of the tests $n_x \sim n_y$ holds.

Some extreme cases of differences between the types of violation are presented too. An influence of kurtosis exists as described above. (The α- and β-values depend on the sum of the γ_2-values of the populations.) It seems that the values of the skewness

$$\gamma_{1v} = \frac{\sigma_x^3 \gamma_{1x}/n_x^2 - \sigma_y^3 \gamma_{1y}/n_y^2}{(\sigma_x^2/n_x + \sigma_y^2/n_y)^{3/2}}$$

of $v = \bar{x} - \bar{y}$ have an important influence on the α- and β-values. That means that the differences in skewness of the populations and/or differences in sample size at each stage (for instance in the case $\pi_x \neq \pi_y$) are serious. If $|\gamma_{1v}(n_x, n_y)| > 0$ on each stage then we can find greater $\hat{\alpha}$-values. For $\gamma_{1v} > 0$ we can find greater $\hat{\beta}$-values in the case $\mu_y = \mu_x + d$ $(d > 0)$ and the opposite for $\gamma_{1v} < 0$.

Table 1

Empirical values of α and β for $n_x = n_y$ (pairwise sampling) (upper values) and for $\pi_x = 0.75$, $\pi_y = 0.25$ $d = d_w = 1.5$, $\alpha = 0.05$, $\beta = 0.10$

γ_{1x}	γ_{2x}	γ_{1y}	γ_{2y}	$\hat{\alpha}_H$	$\hat{\beta}_H$	$\hat{\alpha}_W$	$\hat{\beta}_W$
0	0	0	0	0.0339	0.0614	0.0308	0.0552
				0.0386	0.0670	0.0686	0.0615
0	1.5	0	1.5	0.0365	0.0650	0.0271	0.0635
				0.0387	0.0704	0.0614	0.0661
0	3.75	0	3.75	0.0296	0.0671	0.0232	0.0670
				0.0368	0.0686	0.0499	0.0650
0	7	0	7	0.0225	0.0696	0.0199	0.0681
				0.0372	0.0700	0.0436	0.0677
0	−1	0	−1	0.0376	0.0544	0.0347	0.0546
				0.0414	0.0646	0.0905	0.0496
0.5	0	0.5	0	0.0352	0.0583	0.0319	0.0586
				0.0405	0.0619	0.0754	0.0433
1	1.5	1	1.5	0.0360	0.0633	0.0264	0.0656
				0.0401	0.0576	0.0704	0.0361
1.5	3.75	1.5	3.75	0.0321	0.0623	0.0236	0.0708
				0.0405	0.0585	0.0804	0.0354
2	7	2	7	0.0247	0.0702	0.0188	0.0696
				0.0391	0.0599	0.0771	0.0337
2	7	0	0	0.0464	0.0911	0.0388	0.0929
				0.0685	0.0913	0.0712	0.0766
0	0	2	7	0.0429	0.0312	0.0376	0.0313
				0.0329	0.0254	0.0955	0.0165
2	7	−2	7	0.0950	0.1147	0.0850	0.1187
				0.0856	0.1078	0.1223	0.1097
−2	7	2	7	0.0910	0.0041	0.0824	0.0043
				0.0873	0.0058	0.1274	0.0033

3.2. *Influence of unequal variances*

For $\pi_x = \pi_y = 0.5$ (pairwise sampling or alternative sampling is possible too) both tests are robust for $1/4 < \sigma_y^2/\sigma_x^2 < 4$. For extrem deviations in variances Hajnal's test has increasing values of α but the values of β are stabil (see Lee and Fung (1980)). For $\pi_x \neq \pi_y$ and $\sigma_x^2 \neq \sigma_y^2$ ($\pi_x = 0.75$ was considered only) Hajnal's test has uncontrolled type I and type II errors. Note that Welch type test has for $\pi_x = 0.75$ and for $\sigma_x^2 = \sigma_y^2$ the value $\hat{\alpha} \approx 0.07$ but for unequal variances the test has relativ constant error rates ($\hat{\alpha} + \hat{\beta} \sim$ const.). If we have informations about $c = \sigma_y^2/\sigma_x^2$ we can also use Hajnal's test in the case $\pi_x \neq \pi_y$ with the following modification:

$$t' = t\left(\frac{n\,(f_x c + f_y)}{f\,(n_x + n_y c)}\right)^{1/2}$$

where $n = n_x + n_y$, $f_x = n_x - 1$ and $f_y = n_y - 1$. The estimations of α and β of the proposed modification are denoted by $\hat{\alpha}_{H'}$, and $\hat{\beta}_{H'}$.

Table 2

Empirical values of α and β for $d = d_w = 1.5$, $\alpha = 0.05$, $\beta = 0.10$, $\pi_x = 0.5$

σ_y^2/σ_x^2	$\hat{\alpha}_H$	$\hat{\beta}_H$	$\hat{\alpha}_W$	$\hat{\beta}_W$
1	0.0398	0.0697	0.0437	0.0696
2	0.0450	0.0682	0.0437	0.0658
4	0.0583	0.0629	0.0465	0.0672
9	0.0854	0.0608	0.0481	0.0698

$\pi_x = 0{,}75$ σ_y^2/σ_x^2	$\hat{\alpha}_H$	$\hat{\beta}_H$	$\hat{\alpha}_W$	$\hat{\beta}_W$	$\hat{\alpha}_{H'}$	$\hat{\beta}_{H'}$
1/9	0.0107	0.2251	0.0430	0.0898	0.0497	0.0917
1/4	0.0096	0.1522	0.0457	0.0796	0.0462	0.0818
1/2	0.0167	0.0981	0.0575	0.0796	0.0439	0.0732
1	0.0385	0.0670	0.0686	0.0615	0.0385	0.0670
2	0.0970	0.0463	0.0788	0.0576	0.0408	0.0655
4	0.1861	0.0315	0.0837	0.0544	0.0478	0.0625
9	0.3038	0.0249	0.0809	0.0537	0.0638	0.0589

References

FLEISHMAN, A. I.
A method for simulating non-normal distributions.
Psychometrika **43** (1978), 521–532.

HAJNAL, J.
A two-sample sequential t-test.
Biometrika **48** (1961), 65–75.

LEE, H. und FUNG, K. Y.
A Monte Carlo Study on the Robustness of the Two-Sample Sequential t Test.
J. Statist. Comput. Simul. **10** (1980), 297–307.

RASCH, D.
Einige Bemerkungen zum sequentiellen Zweistichproben-t-Test (unpublished).

REED, A. H. und FRANTZ, M. E.
A Sequential Two Sample t Test using Welch Type Modification for unequal variance.
Comm. Statist.-Theor. Meth., A8 **(14)** (1979) 1459–1471

WELCH, B. L.
The generalisation of "Student's" Problem when several different population variances are involved.
Biometrika **34** (1947) 28–35.

Institut für Statistik und Wirtschaftsmathematik, RWTH Aachen, BRD

Optimal Designs for Contaminated Linear Regression

NORBERT GAFFKE

Abstract

One notion of robustness of linear regression designs refers to moderate deviations of the regression function from the ideal linear regression setup, which may be modelled by including additive contamination functions. These may be caused for example in polynomial regression by higher order terms, which have not been included in the ideal model. The paper shows how the concepts of optimal design theory for linear regression can be extended to contaminated linear regression and points out the main problems arising in the contaminated case.

1. Introduction

Consider a regression problem with a real valued regression function y (x) which is more or less unknown. The controlled variable x can be chosen by the experimenter within the experimental region $\mathfrak{X}$ without random error, whereas an observation of the regression function y at x is affected by random error. More precisely, let d = $(x_1, \ldots, x_n)$ be an exact design of size n with points $x_i \in \mathfrak{X}$ which are not necessarily distinct. The observations under d are represented by real valued random variables $Y_{d1} \ldots, Y_{dn}$ with

$$E Y_{di} = y(x_i),\ \operatorname{Var} Y_{di} = \sigma^2,\ 1 \leq i \leq n,$$
$$\operatorname{Cov}(Y_{di}, Y_{dj}) = 0,\ 1 \leq i \neq j \leq n. \tag{1.1}$$

The variance $\sigma^2 > 0$ may be known or unknown and is independent of d and i. An important special case is given, if $Y_{d1}, \ldots, Y_{dn}$ are assumed to be independent and normally distributed with expectations $y(x_1), \ldots, y(x_n)$ and variance σ^2. But we will only be concerned with the theory of linear estimation, so the normality assumption will not be imposed. A linear regression setup specifies the regression function y to be a member of a given finite dimensional space of real functions,

$$\text{(LR)} \qquad y \in \left\{a'f : a = (a_1, \ldots, a_k)' \in \mathbb{R}^k\right\},$$

where $f = (f_1, \ldots, f_k)' : \mathfrak{X} \to \mathbb{R}^k$ is given. A′ denotes the transpose of the vector or the matrix A. In the linear regression setup (LR) least squares estimation under a given design d provides best linear unbiased estimators for the parameters $a_1, \ldots, a_k$ and for the whole regression function y. The theory of optimal linear regression designs, mainly initiated by Kiefer, aims at minimizing (w. r. t. the design d) the dispersions of these estimators. Of course, the resulting optimal designs are heavily based on the setup (LR), which is somewhat contrary to the common occurrence in practice, that there is often uncertainty about the specification of the functions $f_1, \ldots, f_k$. Box and Draper (1959) were the first, who tried to take into account the possibility of deviations from the ideal model (LR). A situation which shows well the relevance of such considerations arises, when (LR) is a polynomial regression of some specified degree m, which is thought of as a reasonable Taylor approximation of the unknown regression function. Here it might occur, that m was chosen too small and higher order terms not included in the model are present. In fact, Box and Draper and others considered these polynomial models, and they gave some attention to the simple, but relevant, case of fitting a straight line, when the actual regression function is quadratic. Although the conclusions of Box and Draper (1959) turned out to be generally not acceptable, (cf. Stigler (1971), Galil and Kiefer (1977)), there seems to be agreement, that it may sometimes be favourable, to take into account the possibility of deviations from the ideal linear regression model in the design and analysis of the experiment. This is indicated by the numerous articles on this subject after the Box-Draper paper.

Generally deviations from the ideal setup (LR) may be modelled assuming the presence of a contamination function γ (x) from some specified set Γ, and the "contaminated linear regression setup" is given by

$$\text{(CLR)} \qquad y \in \left\{a'f + \gamma : a \in \mathbb{R}^k,\ \gamma \in \Gamma\right\}.$$

The set Γ may be thought of as a neighbourhood of the constant zero. Γ may be a parametric family as in the Box-Draper approach,

$$\Gamma \subset \left\{b'g : b \in \mathbb{R}^p\right\},\ \text{e.g.}$$
$$\Gamma = \left\{b'g : b'b \leq \varepsilon^2\right\},$$

where $g = (g_1, \ldots, g_p)' : \mathfrak{X} \to \mathbb{R}^p$ and $\varepsilon > 0$ are given, or Γ may be nonparametric, e. g. $\Gamma = \{\gamma : \mathfrak{X} \to \mathbb{R} : |\gamma(x)| \leqq \varphi(x),\ x \in \mathfrak{X}\}$, with a given function $\varphi \geqq 0$ on $\mathfrak{X}$, (cf. Marcus and Sacks (1977), Sacks and Ylvisaker (1978), Pesotchinsky (1982), Li and Notz (1982)).

In Section 2 we will briefly outline the concept of optimal linear regression design, which will be extended in Section 3 to the contaminated case. There are two major differences to the linear regression setup (LR): Firstly, the use of least squares estimators as obtained from (LR) can no longer be justified, and other linear estimators, which are unbiased under (LR), should be taken into consideration. Secondly, optimal linear estimators in (CLR) and also optimal designs for (CLR) will generally depend on the variance σ^2, which will mostly be unknown. Although in some cases the optimal design and optimal estimator are the same for any σ^2, by the latter fact the theory remains unsatisfactory. For parametric families of contaminations there are some alternative approaches, which will not be considered here. The all-bias designs of Box and Draper (1959) minimizing the integrated bias of the regression function estimator from the ideal model, which could be improved by Karson, Manson, and Hader (1969) by using more suitable estimators. The efficiencies of a design under two or more rival linear regression models may be combined to yield a global criterion, so that an "optimal" design performs well under any of the possible models. This was done by Stigler (1971), Atkinson (1972),

Läuter (1976), Cook and Nachtsheim (1982). Recently Studden (1982) obtained further results for polynomial regression using Stigler's approach.

2. Linear Regression Designs

In the linear regression setup (LR) a least squares estimator $\hat{a}_d$ under a fixed design d provides best linear unbiased estimators for the parameters $a_1, \ldots, a_k$ (or for linear functions of these parameters), and hence for the regression function y (x). If K is a given (s×k)-matrix, and if Ka is estimable (identifyable) under d, then $K\hat{a}_d$ is the best linear unbiased estimator for Ka, i.e. it minimizes in the Löwner semiordering the matrix risk (relative to variance)

$$\sigma^{-2} E\left\{\left(\hat{b}_d - Ka\right)\left(\hat{b}_d - Ka\right)'\right\} \qquad (2.1)$$

over the set of all linear unbiased estimators $\hat{b}_d$ for Ka. The Löwner semiordering on the set of all symmetric (s×s)-matrices is defined by $A \leqq B$ iff $B - A$ is nonnegative definite.

Of course, the restriction to linear estimators is critical, if there are further assumptions on the underlying distributions than those in (1.1.). The requirement of a linear estimator $\hat{b}_d$ to be unbiased is supported by the fact, that $\hat{b}_d$ is unbiased if its matrix risk (2.1) is bounded, which is a consequence from the unbounded parameter set in (LR). In fact, if the ratio a/σ can be restricted to a bounded subset of IR^k, then least squares estimation may be improved using biased linear estimation, cf Hoffmann (1977). But we will consider here the simpler case (LR) and least squares estimation. The dispersion of the best linear unbiased estimator $K\hat{a}_d$ is given by

$$\sigma^{-2} E\left\{\left(K\hat{a}_d - Ka\right)\left(K\hat{a}_d - Ka\right)'\right\} = \frac{1}{n} K M^-(d) K'. \qquad (2.2)$$

where

$$M(d) = \frac{1}{n} \sum_{i=1}^{n} f(x_i) f'(x_i)$$

is the information matrix (per observation) of $d = (x_1, \ldots, x_n)$, and $M^-(d)$ denotes a generalized inverse of M (d).
If $x \in \mathfrak{X}$ and $K = f'(x)$, then (2.2) reduces to the variance (relative to σ^2) of $\hat{y}_d(x) = \hat{a}_d' f(x)$, the best linear unbiased estimator for y (x),

$$\sigma^{-2} E(\hat{y}_d(x) - y(x))^2 = \frac{1}{n} f'(x) M^-(d) f(x). \qquad (2.3)$$

The dispersions (2.2) and (2.3) provide a basis for the selection of an "optimal" design in the linear regression setup (LR):

Suppose that the experiment aims at the estimation of the linear transformation Ka of the parameter vector, then d should be chosen to make the dispersion matrix in (2.2) "small". Or suppose that the global performance of the regression function is to be explored, then d should be chosen to make the variance function in (2.3) $1/n \cdot f'M^-(d) f$ "small". Too measure the "size" of a dispersion matrix or a variance function one has to introduce an optimality criterion. In the case of parameter estimation this is a real function Ψ on the set of all positive definite (s×s)-matrices, (here the rows of K are assumed to be linearly independent), which is increasing in the Löwner semiordering, i.e. $A \leq B$ implies $\Psi(A) \leq \Psi(B)$ for any positive definite matrices A and B. A design $d^* \in \Delta(K)$ is called Ψ-optimal for estimating Ka, iff

$$\Psi\left(n_{d^*}^{-1} K M^-(d^*) K'\right) = \min_{d \in \Delta(K)} \Psi\left(n_d^{-1} K M^-(d) K'\right), \qquad (2.4)$$

where $\Delta(K)$ is the set of all designs $d \in \Delta$ under which Ka is estimable, Δ being the set of all designs under consideration, and n_d is the size of d. Algebraically $\Delta(K)$ is characterized by

$$\Delta(K) = \left\{d \in \Delta : \mathfrak{R}(K') \subset \mathfrak{R}(M(d))\right\}, \qquad (2.5)$$

where $\mathfrak{R}(A)$ denotes the range of a matrix A. All the commonly used optimality criteria are homogeneous of some degree with respect to positive scalar factors, so that the factor $1/n_d$ in (2.4) can be omitted, when considering designs of fixed size n. Well-known examples of optimality criteria are the D-, A-, and E-criteria,
$\Psi_D(V) = \det V$, $\Psi_A(V) = \operatorname{tr} V$, $\Psi_E(V) = \lambda_{\max}(V)$ (largest eigenvalue of V). In the case of estimating the whole regression function y an optimality criterion is a function $\eta : \mathfrak{H}_+ \to \mathrm{IR} \cup \{\infty\}$ on the set $\mathfrak{H}_+$ of all nonnegative real functions h on $\mathfrak{X}$, which is increasing in the pointwise semiordering of real functions. i. e. $0 \leq h_1(x) \leq h_2(x)$ for all $x \in \mathfrak{X}$ implies $\eta(h_1) \leq \eta(h_2)$.

Prominent examples are given by

$$\eta_\infty(h) = \sup_{x \in \mathfrak{X}} h(x),$$

$$\eta_w(h) = \int_{\mathfrak{X}} h(x)\, dw(x),\ h \geq 0$$

where w is a given probability measure on $\mathfrak{X}$. A design $d^* \in \Delta(f')$ is called η-optimal for estimating y, iff

$$\eta\left(n_{d^*}^{-1} f' M^-(d^*) f\right) = \min_{d \in \Delta(f')} \eta\left(n_d^{-1} f' M^-(d) f\right). \qquad (2.6)$$

where $\Delta(f')$ denotes the set of all $d \in \Delta$, under which $f'(x)a$ is estimable for all $x \in \mathfrak{X}$. Clearly,

$$\Delta(f') = \left\{d \in \Delta : f(x) \in \mathfrak{R}(M(d)) \text{ for all } x \in \mathfrak{X}\right\}, \qquad (2.7)$$

hence, if the components $f_1, \ldots, f_k$ of f are linearly independent on $\mathfrak{X}$, then $\Delta(f')$ is the set of all designs with regular information matrices. Again, the factor $1/n_d$ in (2.6) can usually be removed, since η is homogeneous of some degree and $n_d = n$ is fixed.

Of course, problems (2.4) and (2.6) can formally be comprised under the more general problem of minimizing $\Phi(n_d M(d))$ or $\Phi(M(d))$ (for fixed sample size), where Φ is an $(\mathrm{IR} \cup \{\infty\})$-valued function on the set of all nonnegative definite (k×k)-matrices which is decreasing in the Löwner semiordering.

For tackling these complex optimization problems the notion of an approximate design has proved to be of high importance. An approximate design ξ is a probability measure on $\mathfrak{X}$ with finite support, or, if a suitable σ-field $\mathfrak{B}$ over $\mathfrak{X}$ has been specified, then ξ may be an arbitrary probability measure on $(\mathfrak{X}, \mathfrak{B})$. The exact designs d of size n are imbedded in the set $\bar{\Delta}$ of all approximate designs as those elements $\xi_n \in \bar{\Delta}$ which have finite support and whose weights are integer multiples of 1/n. The definition of the information matrix (per observation) of a design is extended by

$$M(\xi) = \int_{\mathfrak{X}} f(x) f'(x)\, d\xi(x), \quad \xi \in \bar{\Delta}.$$

For a fixed sample size n problems (2.4) and (2.6) are considered on the larger set $\bar{\Delta}$,

$$\Psi\left(K M^-(\xi^*) K'\right) = \min_{\xi \in \bar{\Delta}(K)} \Psi\left(K M^-(\xi) K'\right), \qquad (2.4a)$$

$$\eta\left(f' M^-(\xi^*) f\right) = \min_{\xi \in \bar{\Delta}(f')} \eta\left(f' M^-(\xi) f\right), \qquad (2.6a)$$

where Δ (K) and $\bar{\Delta}$ (f') are defined analogously to (2.5) and (2.7). The main advantage of the "approximate theory" lies in the tractability of the minimization problems (2.4 a) and (2.6 a), which is gained by the convexity of the sets Δ, $\bar{\Delta}$ (K), Δ (f') and the corresponding sets of information matrices. All the commonly used optimality criteria can be written as convex functions of the information matrices, (i.e. the functions Φ as introduced above are convex), so that convex programming methods are applicable, such as directional derivatives and subgradients, duality and, for iterative procedures, steepest descent algorithms. Two other appealing features of the approximate theory should be mentioned: Firstly, an optimal approximate design ξ^* does not depend on the sample size n. Secondly ξ^* allows an easy interpretation in the exact theory: By some rounding-off procedure of the weights of ξ^* to integer multiples of 1/n an exact design ξ^*_n is obtained, which may be expected to come close to ξ^* with respect to the criterion under consideration, (which behaves continuously). And under the standard assumption of compactness of the effective experimental region $\{f(x) : x \in \mathfrak{X}\}$ this is not affected when admitting general probability measures on $(\mathfrak{X}, \mathfrak{B})$, since by Carathéodory's Theorem for any $\xi \in \Delta$ there exists a $\xi' \in \Delta$ with finite support and $M(\xi') = M(\xi)$, (actually such a ξ' can be found with at most $k(k+1)/2+1$ support points). Of course, a rounding-off procedure as above will yield a good exact design ξ^*_n only if the number of support points of ξ^* is small compared with n.

3. Contaminated Linear Regression

As in Section 2 we distinguish under (CLR) between linear parameter estimation, (estimation of a linear transformation Ka of the parameter vector a), and linear estimation of the regression function y. For parameter estimation the following identifyability condition should be imposed:

If $a'f + \gamma_1 = b'f + \gamma_2$ with $a, b \in \mathrm{IR}^k$, $\gamma_1, \gamma_2 \in \Gamma$, then $Ka = Kb$.

If Γ is convex and such that $-\gamma \in \Gamma$ whenever $\gamma \in \Gamma$, then this condition is equivalent to the following one:

$a'f \in \Gamma$ with $a \in \mathrm{IR}^k$ implies $Ka = 0$.

More specially, if $\Gamma = \{\gamma : \mathfrak{X} \to \mathrm{IR} : |\gamma(x)| \leq \varphi(x), x \in \mathfrak{X}\}$, with $\varphi \geq 0$ given, and $K = I_k$, then the parameter vector a is identifyable in (CLR) iff $|a'f(x)| \leq \varphi(x)$ for all $x \in \mathfrak{X}$ forces $a = 0$, (cf. Sacks and Ylvisaker (1978)). For estimation of the regression function y an identifyability condition is not needed.

We will first consider the case, that under an exact design $d = (x_1, \dots, x_n)$ the ordinary least squares analysis from the ideal model (LR) is used. Under (CLR) the estimators $K\hat{a}_d$ and $\hat{y}_d(x) = \hat{a}_d' f(x)$, $\hat{a}_d$ being a least squares estimator for a (under LR), are biased, and their risks split up into a variance term and a bias term:

For $d \in \Delta$ (K) (as defined in Section 2)

$$\sigma^{-2} E\{(K\hat{a}_d - Ka)(K\hat{a}_d - Ka)'\} = \frac{1}{n} K M^-(d) K' + \sigma^{-2}\big(E(K\hat{a}_d) - Ka\big)\big(E(K\hat{a}_d) - Ka\big)'.$$

Since $EY_d = E(Y_{d1}, \dots, Y_{dn})' = X(d)a + \gamma(d)$, where $X(d) = \big(f_j(x_i)_{1\leq i\leq n, 1\leq j\leq k}\big)$ is the design matrix of d and

$$\gamma(d) = \big(\gamma(x_1), \dots, \gamma(x_n)\big)',$$

we have

$$E K\hat{a}_d = \frac{1}{n} K M^-(d) X'(d)\big[X(d)a + \gamma(d)\big] = Ka + \frac{1}{n} K M^-(d) X'(d)\gamma(d),$$

hence

$$\sigma^{-2} E\{(K\hat{a}_d - Ka)(K\hat{a}_d - Ka)'\} = \frac{1}{n} K M^-(d) K' + \sigma^{-2} K M^-(d)\bar{\gamma}(d)\bar{\gamma}'(d) M^-(d) K', \tag{3.1}$$

where

$$\bar{\gamma}(d) = \frac{1}{n} X'(d) \cdot \gamma(d) = \frac{1}{n}\sum_{i=1}^{n} \gamma(x_i) f(x_i).$$

Similarly, for $d \in \Delta$ (f') (as defined in Section 2)

$$\sigma^{-2} E(\hat{y}_d(x) - y(x))^2 = \frac{1}{n} f'(x) M^-(d) f(x) + \sigma^{-2}\big(E(\hat{y}_d) - y(x)\big)^2,$$

and

$$E\hat{y}_d = \frac{1}{n} f'(x) M^-(d) X'(d) E Y_d = f'(x)a + \frac{1}{n} f'(x) M^-(d) X'(d)\gamma(d)$$

hence

$$\sigma^{-2} E(\hat{y}_d(x) - y(x))^2 = \frac{1}{n} f'(x) M^-(d) f(x) + \sigma^{-2}\big[f'(x) M^-(d)\bar{\gamma}(d) - \gamma(x)\big]^2. \tag{3.2}$$

Let $R_K(d, \sigma^{-1}\gamma)$ denote the matrix risk in (3.1), and $r(d, \sigma^{-1}\gamma)(x)$ the mean squared error function in (3.2), and let Ψ and η be optimality criteria as introduced in Section 2. A design $d^* \in \Delta$ (K) might be called Ψ-optimal or Ψ-minimax for estimating Ka, iff

$$\sup_{\gamma \in \Gamma} \Psi\big(R_K(d^*, \sigma^{-1}\gamma)\big) = \min_{d \in \Delta(K)} \sup_{\gamma \in \Gamma} \Psi\big(R_K(d, \sigma^{-1}\gamma)\big). \tag{3.3}$$

and $d^* \in \Delta$ (f') might be called η-optimal or η-minimax for estimating y, iff

$$\sup_{\gamma \in \Gamma} \eta\big(r(d^*, \sigma^{-1}\gamma)\big) = \min_{d \in \Delta(f')} \sup_{\gamma \in \Gamma} \eta\big(r(d, \sigma^{-1}\gamma)\big), \tag{3.4}$$

Of course, these criteria depend on σ^2, so that in general they will lead to the selection of a single optimal design, only if the variance σ^2 can be specified in advance. It might be tempting to consider contaminations relative to standard deviation σ and to assume in (CLR) that $\sigma^{-1}\gamma \in \Gamma$ (instead of $\gamma \in \Gamma$), so that in (3.3) and (3.4) the supremum can be taken over $\sigma^{-1}\gamma \in \Gamma$ and σ is ruled out. But then the problem of an unknown variance σ^2 is merely transfered to the problem of specifying the set Γ. In the case of a parametric family of contaminations

$$\Gamma = \{b'g : b \in B\}$$

where $g = (g_1, \dots, g_p)' : \mathfrak{X} \to \mathrm{IR}^p$ and $B \subset \mathrm{IR}^p$ are given, and if $g_1, \dots, g_p$ are linearly independent on $\mathfrak{X}$, then we may write

$$R_K(d, \sigma^{-1}bg) = R_K(d, \sigma^{-1}b) \text{ and}$$

$$r(d, \sigma^{-1}bg) = r(d, \sigma^{-1}b).$$

Instead of taking the supremum over $b \in B$ as in (3.3) and (3.4) one may also consider an average with respect to some probability measure β on B:

$$\int_B \Psi\big(R_K(d^*, \sigma^{-1}b)\big)\, d\beta(b) = \min_{d \in \Delta(K)} \int_B \Psi\big(R_K(d, \sigma^{-1}b)\big)\, d\beta(b), \tag{3.5}$$

$$\int_B \eta\big(r(d^*, \sigma^{-1}b)\big)\, d\beta(b) = \min_{d \in \Delta(f')} \int_B \eta\big(r(d, \sigma^{-1}b)\big)\, d\beta(b). \tag{3.6}$$

Criteria (3.3) with $K = I_k$ were considered by Pesotchinsky (1982), criterion (3.4) with $\eta = \eta_w$, w the uniform measure on $\mathfrak{X} = \left[-\frac{1}{2}, \frac{1}{2}\right]$ was considered by Huber (1975), Section 4, and with w the uniform measure on a finite region $\mathfrak{X}$ by Welch (1983), who also worked with an average criterion (3.6). Actually these authors were concerned rather with approximate designs, and the extension of (3.3) — (3.6) to the approximate theory is straightforward:

For $\xi \in \overline{\Delta}$ let

$$\bar{\gamma}(\xi) = \int_{\mathfrak{X}} \gamma(x) f(x) d\xi(x),$$

and extending (3.1) and (3.2)

$$R_K(n, \xi, \sigma^{-1}\gamma) = \frac{1}{n} K M^-(\xi) K' + \sigma^{-2} K M^-(\xi) \bar{\gamma}(\xi) \bar{\gamma}'(\xi) M^-(\xi) K', \quad (3.1a)$$

if $\xi \in \overline{\Delta}$ (K), and

$$r(n, \xi, \sigma^{-1}\gamma)(x) = \frac{1}{n} f'(x) M^-(\xi) f(x) + \sigma^{-2}\left[f'(x) M^-(\xi)\bar{\gamma}(\xi) - \gamma(x)\right]^2, \quad (3.2a)$$

if $\xi \in \overline{\Delta}$ (f').

Thus optimal approximate designs as defined analogously to (3.3) — (3.6) will generally depend on the sample size n, (and of course on σ^2), contrary to the uncontaminated case (LR). Actually they will depend on n and σ^2 through σ^2/n, since Ψ and η are usually homogeneous of some degree. But there are other good points of the approximate theory in (LR) which do not carry over to the contaminated case: The objective functions in (3.3) — (3.6) are generally not convex functions of ξ.

There may be difficulties in interpreting general probability measures ξ on $(\mathfrak{X}, \mathfrak{B})$. For nonparametric contaminations Carathéodory's Theorem is not applicable, which in the setup (LR) ensured the existence of a ξ' with finite support which is equivalent to ξ. It may even be possible that the objective functions in (3.3) — (3.6) behave discontinuously when approximating an optimal design ξ^* with infinite support by designs with finite support. More exactly: If $\mathfrak{X}$ is a compact metric space, (usually a compact subset of $\mathbb{R}^q$), and ξ^* is a Borel probability measure on $\mathfrak{X}$, then the objective functions in (3.3) — (3.6) may fail to be continuous at ξ^* with respect to the vague topology on the set of Borel probability measures. This in fact occurs in Huber (1975), Sec. 4, Sec. 5, who considered the example

$$f(x) = (1, x)', \quad x \in \left[-\frac{1}{2}, \frac{1}{2}\right],$$

$$\Gamma = \left\{\gamma : \min_{\alpha, \beta \in \mathbb{R}} \int_{-1/2}^{1/2} \left(\gamma(x) - \alpha - \beta x\right)^2 dx \le \varepsilon^2\right\}$$

with $\varepsilon^2 > 0$ given. The η_w -minimax design $\xi^* = \xi^*(\sigma^2/(n\varepsilon^2))$, w being the uniform measure on [—1/2, 1/2], is absolutely continuous with respect to Lebesgue measure. But for any design ξ with finite support

$$\sup_{\gamma \in \Gamma} \eta_w\left(r(n, \xi, \sigma^{-1}\gamma)\right) = \infty,$$

so that ξ^* does not admit a reasonable interpretation, which was observed by Li and Notz (1982), p. 136. The same objection pertains to the minimax design for estimating the slope of the regression line given by Huber (1975), pp. 295—296.

So for nonparametric contaminations one should generally restrict to approximate designs with finite support. It can easily be shown that, if the set Γ is uniformly bounded and $\{f(x) : x \in \mathfrak{X}\}$ is compact, then the objective functions in (3.3) and (3.4) are continuous functions of the weights of ξ when the (finite) support of ξ is kept fixed. So when rounding off the weights of an optimal approximate design ξ^* one may still expect to obtain good exact designs ξ_n^*, (under the above assumptions). For a parametric family of contaminations

$$\Gamma = \left\{b'g : b \in B\right\},$$

general probability measures do usually not cause any problems, since $R_K(n, \xi, \sigma^{-1}b)$ and $r(n, \xi, \sigma^{-1}b)$ depend on ξ only through the matrices M (ξ) and $\int f(x) g'(x) d\xi(x)$, and hence Carathéodory's Theorem is applicable, if $\{f(x) : x \in \mathfrak{X}\}$ and $\{g(x) : x \in \mathfrak{X}\}$ are compact.

Criteria (3.3) — (3.6) are based on the assumption that under any design d the least squares analysis from the model (LR) is used. As mentioned in the introduction this is not conclusive, and there may be other linear estimators whose global performances in (CLR) are better than those of the ideal least squares estimators, (cf. Karson, Manson and Hader (1969), Marcus and Sacks (1977), Sacks and Ylvisaker (1978), Agarwal (1981)). These will be considered now.

To ensure that a linear estimator has bounded risk, (matrix risk for parameter estimation, mean squared error for regression function estimation), the unbiasedness condition under the ideal model is still imposed. Let $d = (x_1, \ldots, x_n)$ be a given exact design, and let (1/n) LY_d be a linear estimator for Ka, where K is a given $(s \times k)$-matrix and L an $(s \times n)$-matrix of constants, such that (1/n) LY_d is unbiased for Ka in (LR), i.e.

$$n^{-1} L X(d) = K. \quad (3.7)$$

Then under (CLR)

$$\sigma^{-2} E\left\{(n^{-1}LY_d - Ka)(n^{-1}LY_d - Ka)'\right\}$$
$$= n^{-2}\left\{LL' + \sigma^{-2} L\gamma(d)\gamma'(d)L'\right\} = R(d, L, \sigma^{-1}\gamma), \text{ (say)}. \quad (3.8)$$

So, if Ψ is an optimality criterion measuring the "size" of a positive definite matrix, then the global performance of the estimator in (CLR) may be quantified by

$$\sup_{\gamma \in \Gamma} \Psi\left(R(d, L, \sigma^{-1}\gamma)\right),$$

or in the parametric case $\Gamma = \{bg : b \in B\}$ by an average

$$\int_B \Psi\left(R(d, L, \sigma^{-1}b)\right) d\beta(b),$$

and, clearly, these quantities should be minimized with respect to L subject to (3.7). Let $\mathfrak{L}$ (K, d) denote the set of all matrices L which satisfy (3.7). Then instead of (3.3) and (3.5) one may define a design $d^* \in \Delta$ (K) to be Ψ-optimal for estimating Ka, iff

$$\inf_{L \in \mathfrak{L}(K, d^*)} \sup_{\gamma \in \Gamma} \Psi\left(R(d, L, \sigma^{-1}\gamma)\right)$$
$$= \min_{d \in \Delta(K)} \inf_{L \in \mathfrak{L}(K, d)} \sup_{\gamma \in \Gamma} \Psi\left(R(d, L, \sigma^{-1}\gamma)\right), \quad (3.9)$$

or in the parametric case

$$\inf_{L \in \mathfrak{L}(K, d^*)} \int_B \Psi\left(R(d^*, L, \sigma^{-1}b)\right) d\beta(b)$$
$$= \min_{d \in \Delta(K)} \inf_{L \in \mathfrak{L}(K, d)} \int_B \Psi\left(R(d, L, \sigma^{-1}b)\right) d\beta(b). \quad (3.10)$$

For estimating the regression function one can procede in a similar way: Consider linear estimators for y,

$$\hat{y}_{d.L}(x) = n^{-1}(LY_d)' f(x), \quad x \in \mathfrak{X},$$

where L is a (k × n)-matrix of constants, such that

$$n^{-1} f'(x) L X(d) = f'(x) \quad \text{for all } x \in \mathfrak{X}. \tag{3.11}$$

(3.11) is the unbiasedness condition in (LR), which rewrites as $n^{-1} L X(d) = I_k$, if the components of f are linearly independent on $\mathfrak{X}$. The reason for not admitting any linear unbiased estimator $n^{-1} c'(x) Y_d$, $x \in \mathfrak{X}$, where $c: \mathfrak{X} \to IR^n$, is that the estimated function $x \to n^{-1} c'(x) Y_d$ will generally not be a member of (CLR). The mean squared error function (relative to σ^2) of $\hat{y}_{d.L}(x)$ in (CLR) is given by

$$\sigma^{-2} E(\hat{y}_{d.L}(x) - y(x))^2$$
$$= n^{-2} f'(x) LL' f(x) + \sigma^{-2} [n^{-1} f'(x) L\gamma(d) - \gamma(x)]^2$$
$$= r(d.L.\sigma^{-1}\gamma)(x), \quad \text{(say)}. \tag{3.12}$$

Now, for a given criterion η, an optimal design $d^* \in \Delta(f')$ for estimating the regression function y may be defined analogously to (3.4) and (3.6) to be a minimizer of

$$\inf_{L \in \mathfrak{L}(f',d)} \sup_{\gamma \in \Gamma} \eta(r(d.L.\sigma^{-1}\gamma)), \tag{3.13}$$

or in the parametric case,

$$\inf_{L \in \mathfrak{L}(f',d)} \int_B \eta(r(d.L.\sigma^{-1} b))\, d\beta(b), \tag{3.14}$$

where $d \in \Delta(f')$, and $\mathfrak{L}(f', d)$ denotes the set of all matrices L satisfying (3.11).

Criteria (3.9) were applied by Marcus and Sacks (1977) and Li and Notz (1982), and a criterion of type (3.14) by Agarwal (1981). We will briefly indicate the necessary alterations in (3.7) – (3.14) for the approximate theory.

Firstly, we note, that for a given exact design $d = (x_1, \ldots, x_n)$ one can restrict to estimators $n^{-1} L Y_d$ and $n^{-1} (L Y_d)' f(x)$, respectively, whose matrices L are such that the i-th and j-th columns of L are equal whenever $x_i = x_j$. This can be seen as follows: If $x_1 = x_2$, then replace the first two columns l_1, l_2 of L by $(l_1 + l_2)/2$. The matrix $\bar{L}$, say, still satisfies (3.7) or (3.11), respectively, since

$$L X(d) = \sum_{i=1}^{n} l_i f'(x_i) = \bar{L} X(d),$$

where $L = [l_1, l_2, \ldots, l_n]$. Also

$$L\gamma(d) = \sum_{i=1}^{n} \gamma(x_i) l_i = \bar{L}\gamma(d)$$

for all $\gamma \in \Gamma$. From

$$\left(\frac{1}{2} l_1 + \frac{1}{2} l_2\right)\left(\frac{1}{2} l_1 + \frac{1}{2} l_2\right)' \le \frac{1}{2} l_1 l_1' + \frac{1}{2} l_2 l_2'$$

(in the Löwner semiordering), we get $\bar{L}\bar{L}' \le LL'$, and hence

$$R(d.\bar{L}.\sigma^{-1}\gamma) \le R(d.L.\sigma^{-1}\gamma) \quad \text{for all } \gamma \in \Gamma,$$

or

$$r(d.\bar{L}.\sigma^{-1}\gamma) \le r(d.L.\sigma^{-1}\gamma) \quad \text{for all } \gamma \in \Gamma,$$

respectively. Proceeding in this way the assertion follows. So the matrix L can be viewed as an IR^s- or IR^k-valued function on the support $\{x_1, \ldots, x_n\}$ of d, which may be extended to a function λ on all of $\mathfrak{X}$. If conversely $\lambda: \mathfrak{X} \to IR^s$, ($IR^k$ respektively), and $d = (x_1, \ldots, x_n)$ are given, then the matrix L is reconstructed as $L = [\lambda(x_1), \ldots, \lambda(x_n)]$. For an approximate design ξ (3.7) rewrites as

$$\int_{\mathfrak{X}} \lambda(x) f'(x)\, d\xi(x) = K, \tag{3.7a}$$

and $\mathfrak{L}(K, \xi)$ now is the set of all functions λ satisfying (3.7 a); similarly instead of (3.11) we write

$$f'(x) \int_{\mathfrak{X}} \lambda(z) f'(z)\, d\xi(z) = f'(x) \quad \text{for all } x \in \mathfrak{X}. \tag{3.11a}$$

and $\mathfrak{L}(f', \xi)$ denotes the set of all λ satisfying (3.11 a). Then, modifying (3.8) and (3.12),

$$R(n.\xi.\lambda.\sigma^{-1}\gamma) = \frac{1}{n} \int_{\mathfrak{X}} \lambda(x)\lambda'(x)\, d\xi(x)$$
$$+ \sigma^{-2} \left[\int_{\mathfrak{X}} \gamma(x)\lambda(x)\, d\xi(x)\right] \left[\int_{\mathfrak{X}} \gamma(x)\lambda(x)\, d\xi(x)\right]'. \tag{3.8a}$$

$$r(n.\xi.\lambda.\sigma^{-1}\gamma)(x) = \frac{1}{n} f'(x) \left[\int_{\mathfrak{X}} \lambda(z)\lambda'(z)\, d\xi(z)\right] f(x)$$
$$+ \sigma^{-2} \left[f'(x) \int_{\mathfrak{X}} \gamma(z)\lambda(z)\, d\xi(z) - \gamma(x)\right]^2. \tag{3.12a}$$

so that definitions (3.9), (3.10), (3.13), (3.14) can be extended to approximate designs. Problems arising from nonconvexity and discontinuities of the objective functions are the same as under least squares analysis dicussed above.

4. References

AGARWAL, G. G.
Optimum designs for biased estimation in linear spline regression.
Sankhyā 43 (1981), Ser. B, 198–211.

ATKINSON, A. C.
Planning experiments to detect inadequate regression models.
Biometrika 59 (1972) 275–293.

BOX, G. E. P., DRAPER, N. R.
A basis for the selection of a response surface design.
J. Amer. Statist. Assoc. 54 (1959) 622–654.

COOK, R. D., NACHTSHEIM, C. J.
Model robust, linear-optimal designs.
Technometrics 24 (1982) 49–54.

GALIL, Z., KIEFER, J.
Comparison of Box-Draper and D-optimum designs for experiments with mixtures.
Technometrics 19 (1977) 441–444.

HOFFMANN, K.
Admissibility of linear estimators with respect to restricted parameter sets.
Math. Operationsforsch. Statist., Ser. Statistics 8 (1977) 425–438.

HUBER, P. J.
Robustness and designs.
In: Srivastava, J. N. (ed.): A Survey of Statistical Design and Linear Models. North Holland, 1975 pp. 287–301.

KARSON, M. J., MANSON, A. R., HADER, R. J.
Minimum bias estimation and experimental design for response surfaces.
Technometrics 11 (1969) 461–475.

LÄUTER, E.
Optimal multipurpose designs for regression models.
Math. Operationsforsch. u. Statist. 7 (1976) 51–68.

LI, K. C., NOTZ, W.
Robust designs for nearly linear regression.
J. Statist. Plann. Infer. 6 (1982) 135–151.

MARCUS, M. B., SACKS, J.
Robust designs for regression problems.
Statistical Decision Theory and Related Topics II.
Academic, New York, 1977, 245–268.

PESOTCHINSKY, L.
Optimal robust designs: Linear regression in R^K.
Ann. Statist. 10 (1982) 511–525.

SACKS, J., YLVISAKER, D.
Linear estimation for approximately linear models.
Ann. Statist. 6 (1978) 1122–1137.

STIGLER, S. M.
Optimal experimental design for polynomial regression.
J. Amer. Statist. Assoc. 66 (1971) 311–318.

STUDDEN, W. J.
Some robust-type D-optimal designs in polynomial regression.
J. Amer. Statist. Assoc. 77 (1982) 916–921.

WELCH, W. J.
A mean squared error criterion for the design of experiments.
Biometrika 70 (1983) 205–213.

Academy of Agricultural Sciences of the GDR
Research Centre of Animal Production Dummerstorf-Rostock

Systems of One-Dimensional Continuous Distributions and their Application in Simulation Studies

VOLKER GUIARD

Abstract

One-dimensional continuous distribution systems (Pearson, Johnson, Tadikamalla-Johnson, Burr, Grassia, Gram-Charlier-Edgeworth, generalized Lambda distributions, Schmeiser-Deutsch, Fleishman, truncated normal distribution, double rectangular distribution) are described and their suitability for use in simulation studies for investigating robustness is considered.

1. Introduction

Simulation studies using certain alternative distributions are a common way of investigating the robustness of statistical procedures with respect to violations of the assumed normal distribution.

In such cases the degree of non-normality is usually expressed by the parameters (In the whole paper the μ_k ($k \leq 4$) are assumed to be finite)

skewness: $\gamma_1 = \frac{\mu_3}{\mu_2^{3/2}}$

and

kurtosis: $\gamma_2 = \frac{\mu_4}{\mu_2^2} - 3$

where μ_k is the k-th order central moment.

The parameters γ_1 and γ_2 naturally do not uniquely define the shape of a distribution. (The upper and lower bounds for $P(\mathbf{x} \leq x_0)$ in dependence of x_0 and the first four moments of the distribution of $\mathbf{x}$ are given in Simpson and Welch (1960).) This is why different distributions should be used for each pair γ_1, γ_2 in robustness studies. The γ_1 and γ_2 must satisfy the following inequality:

$$\gamma_2 \geq \gamma_1^2 - 2.$$

The equality sign applies if, and only if, the distribution is a two-point distribution. In some cases only unimodal distributions are of interest as alternative distributions. The inequality to be satisfied by such distributions is

$$\gamma_2 \geq \frac{6}{5}\left[\gamma_1\left(c - \frac{1}{c}\right) - 1\right],$$

where c is given by the equation

$$-8\gamma_1 c = (3 - c^2)^2.$$

The proof for this inequality and the explicit solution of the last equation can be found in Herrendörfer, G. (1980). The same inequality was derived by Johnson and Rogers (1951).

Distribution systems are a useful tool for indentifying distributions with given values for γ_1 and γ_2.

By the term "distribution system" we understand a class of distributions constructed by means of a common rule and covering the whole of the admissible region of the (γ_1, γ_2)-plane or part of this region that is of particular interest.

This paper will present only systems of one-dimensional continuous distributions. A few two-dimensional distribution systems are given, for example, in Mardia (1970), Cook and Johnson (1981) and Johnson and Tenenbein (1981). Johnson, Ramberg and Wang (1982) apply the Johnson system to the multi-dimensional case. Further multi-dimensional systems are described by Johnson and Kotz (1972), for example, and Johnson and Kotz (1982) also discuss discrete distribution systems.

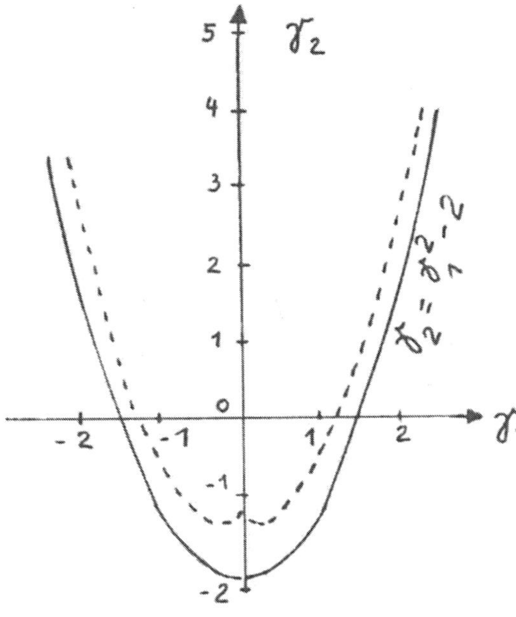

Lower bound of the admissible ranges in the (γ_1, γ_2)-plane for all distributions (——) and for unimodal distributions (- - -)

For the sake of clarity, and to simplify practical manipulation, preference is given to systems which as far as possible contain not more than one distribution for each pair, γ_1, γ_2. The "suprasystem" proposed by Savageau (1982), in which a large variety of different distribution systems are produced by a common construction rule, is unsuitable for use in simulation studies.

We shall, moreover, ignore distribution systems which describe only one line in the (γ_1, γ_2)-plane. Johnson, Tietjen and Beckman (1980), for instance, developed a distribution system containing only symmetric distributions ($\gamma_1 = 0$).

In describing the different distribution systems, we shall discuss the following points:

D:	definition of the system by a construction rule
S:	typical shapes of the distributions
R_γ:	region in the (γ_1, γ_2)-plane covered by the distributions
$C_{p-\gamma}$:	connection between the parameters of the distributions and the values of γ_1 and γ_2
G	Generation of random numbers for these distributions

Random numbers are usually generated by transformation of uniformly or normally distributed variables. It will be assumed that the procedures for generating such random variables are known. This subject is reviewed in Herrendörfer (1980), Guiard (1981) and Rasch and Herrendörfer (1982).

In some cases the construction rule for a distribution system consists of the instructions for transforming appropriate uniformely distributed or normally distributed random variables in order to obtain the random variables required. In this case the point G is contained in point D.

Some distribution systems (for instance the Pearson or the Johnson and Burr system) have not been constructed for simulation purposes but for fitting to empirical distributions. The generation of random numbers for these distribution systems is sometimes quite complicated.

For the sake of simplicity, the location and scale parameters will not be mentioned when discussing the distribution systems. These parameters can, of course, be included, but they do not affect the values γ_1 and γ_2.

2. Systems of Distributions

2.1. *The Pearson System*

D:

The density function, f(x), of a Pearson distribution satisfies the two following conditions:

$$1.\quad \frac{\partial}{\partial x} f(x) = f(x) \frac{a_0 + a_1 x}{b_0 + b_1 x + b_2 x^2}$$

2. The expression

$$\lim_{x \to \text{bound}} x^h (b_0 + b_1 x + b_2 x^2) \cdot f(x) = 0$$

holds for the boundary points of the range of definition of the distribution ($h = 0, 1, 2, 3$).

Pearson's differential equation is usually given in the literature with $a_1 = 1$. In this case, however, the type XII is not a Pearson distribution on account of $a_1 = 0$. There is exactly one Pearson distribution for each pair γ_1, γ_2 ($\gamma_2 > \gamma_1^2 - 2$).

S:

There is no functional form that is common to all Pearson distributions. They are divided into the different types presented in the following table, in which the first and second kind beta distributions are denoted by B1 and B2 respectively.

$$B1:\ f(x) = a x^{n_1} (1-x)^{n_2} \quad (0 < x < 1,\ n_i > -1)$$

$$B2:\ f(x) = a (x+1)^{n_1} x^{n_2} \quad (x > 0,\ n_2 > -1,\ n_1 + n_2 < -5)$$

A normalization constant must be inserted for a.

The table gives only the distributions with $\gamma_1 \geqq 0$. The distributions with $\gamma_1 < 0$ are obtained by the substitution $x \Leftarrow -x$.

Johnson, Nixon and Amos (1963) give a table of the quantiles of all Pearson distributions with $E(x) = 0$ and $V(x) = 1$ for different γ_1 and γ_2.

$$C_{p-\gamma}:\quad \begin{aligned} b_0 &= 4\gamma_2 - 3\gamma_1^2 + 12 \\ b_1 &= \gamma_1(\gamma_2 + 6) \\ b_2 &= 2\gamma_2 - 3\gamma_1^2 \\ a_1 &= -2(5\gamma_2 - 6\gamma_1^2 + 6) \\ a_0 &= -b_1 \end{aligned}$$

The distribution type and the corresponding distribution parameters can be derived from these values by the methods given in Elderton and Johnson (1969), Herrendörfer (1980), Guiard (1981).

Table of the Pearson types

Type No.	Distribution	Shapes
I	B1, $n_1 < n_2$, $n_1 \neq 0$	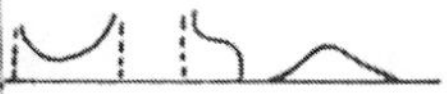
II	B1, $n_1 = n_2$	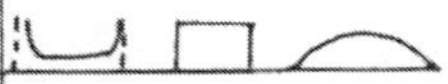
III	Γ-distribution	
IV	$a(x^2+c^2)^d\ e^{f \arctan \frac{x}{c}}$	
V	$a\ x^c\ e^{d/x}$	
VI	B2, $n_2 \neq 0$	
VII	t-distribution	
VIII	B1, $n_1 < n_2 \neq 0$	
IX	B1, $0 = n_1 < n_2$	
X	exponential distribution	
XI	B2, $n_2 = 0$ (Pareto distribution)	
XII	B1, $n_1 < 0 < n_2 < 1$	
N	normal distribution	

G:

Cooper, Davis and Dono (1965) describe a universal generator for all Pearson distributions. Most Pearson distributions can be produced with a generator for beta distributions. Generators for the first kind beta distribution are given in Jöhnk (1964), Ahrens and Dieter (1974) and Cheng (1978). Of these generators, the one presented by Cheng (1978) is the fastest (Schmeiser and Shalaby (1980)). Schmeiser and Babu (1980) constructed a generator which is even faster than that constructed by Cheng (1978). Further very fast generators are given in Atkinson (1979), Atkinson and Pearce (1976) and Atkinson and Whittaker (1979).

R_γ:

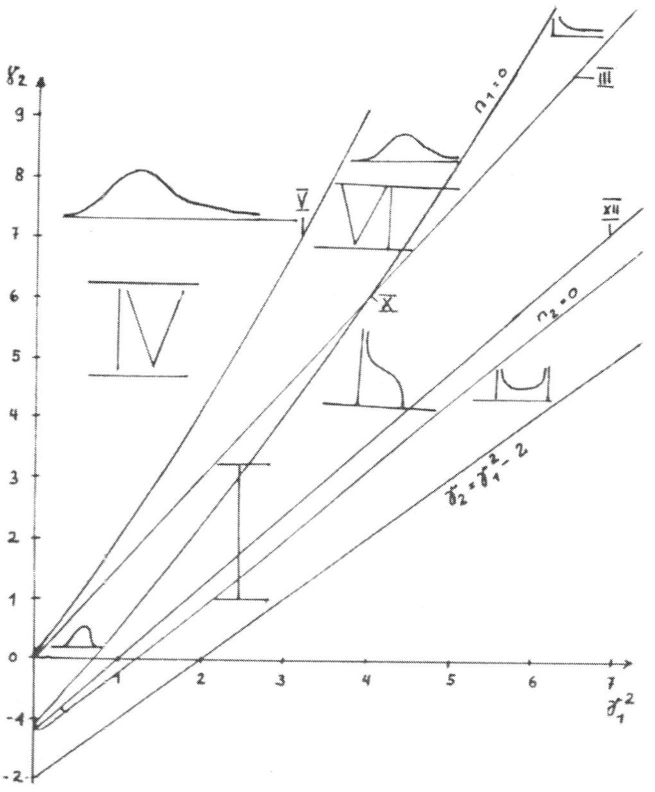

Figure 2
Pearson-types in the (γ_1^2, γ_2)-plane

2.2. *The Johnson System*

D:
If **z** has a standardized normal distribution, it follows that

$$\mathbf{x} = g\left(e^{\frac{z-\mu}{\sigma}}\right)$$

has a Johnson distribution, whereby specific monotonic functions must be inserted for g(u). Depending on g(u), there are three types of Johnson distributions:

S_B: $g(u) = \frac{u}{1+u}$ (bounded, $0 < x < 1$)

S_L: $g(u) = u$ (log-normal distribution, $0 < x < \infty$)

S_U: $g(u) = \frac{1}{2}\left(u - \frac{1}{u}\right)$ (unbounded, $-\infty < x < \infty$)

(Johnson (1949), Elderton and Johnson (1969))

S:

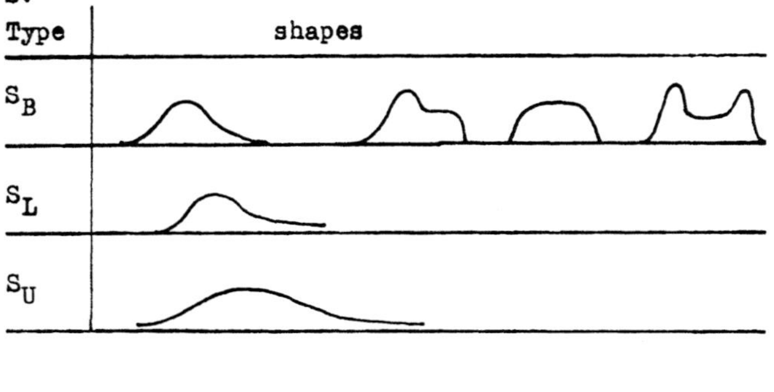

R_γ:
There exists exactly one Johnson distribution for each pair γ_1, γ_2. Type S_L represents the border between types S_B and S_U.

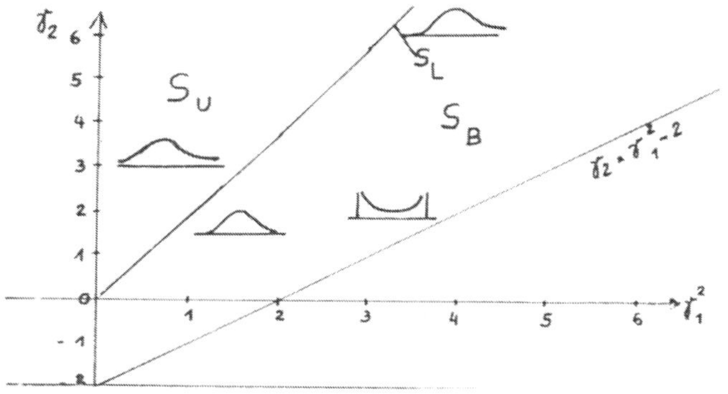

Figure 3
Johnson distributions in the (γ_1^2, γ_2)-plane

$C_{p,\gamma}$:
The connection between the distribution parameters μ and σ^2 and the values of γ_1 and γ_2 is complex. The following possibilities exist for calculating the parameters of the different types:

S_L: For given γ_1, the expressions ($p = e^{\sigma^2}$)

$$\gamma_1^2 = (p-1)(p+2)^2$$

and

$$\gamma_2 = (p-1)(p^3 + 3p^2 + 6p + 6)$$ (Kendall Stuart (Vol. I, 1958))

yield, after solving the cubic equation in p,

$$p = \sqrt[3]{Q + \sqrt{Q^2-1}} + \sqrt[3]{Q - \sqrt{Q^2-1}} - 1 \left(Q = 1 + \frac{\gamma_1^2}{2}\right)$$

$\sigma^2 = \ln p$ and γ_2 (see above).

(In this case μ is a scale parameter.)

S_B: Tables: (Johnson and Kitchen (1971 a, b))
Approximation method: Bowman, Serbin and Shenton (1981)

S_U: Tables: Johnson (1965)
Approximation methods: Leslie (1953), Bowman and Shenton (1980), Shenton and Bowman (1982)

G: See D

2.3. *Systems Analogous to the Johnson System*

Other systems analogous to the Johnson system can be obtained by applying the same functions g( ) if other distributions are assumed for **z**. If, for instance, a Laplace-distributed random variable is used for **z**, the types S'_U, S'_L and S'_B are obtained (Johnson (1954)).
Tadikamalla and Johnson (1982) proposed an additional system:

D: Let **z** have a logistic distribution
$f(z) = e^{-z}(1 + e^{-z})^{-2}$, $F(z) = (1 + e^{-z})^{-1}$.
Then calculate **x** by applying Johnson's transformation functions g( ) to **z**.
This yields the types L_U, L_L and L_B in analogy to S_U, S_L and S_B.

S, R_γ: The forms of the distributions are similar to those of the Johnson distributions. There is exactly one distribution for each pair γ_1, γ_2 ($\gamma_2 > \gamma_1^2 - 2$). The line of the L_L distributions ist situated above the S_S line.

$C_{p-\gamma}$: Tadikamalla and Johnson (1982) mention tables in an unpublished North Carolina report for finding the distribution parameters.

G: Generate a random variable **u** that is (0,1)-uniformly distributed and calculate

$$x = g\left(\left(\frac{u}{1-u}\right)^{\frac{1}{\sigma}} e^{\frac{-\mu}{\sigma}}\right)$$

In other words, simulation is simpler in this case because it is unneccessary to generate a normally distributed random variable.

2.4. *Burr's Distribution Systems*

Intending to fit a theoretical distribution function F(x) to an empirical distribution function, Burr (1942) proposed twelve different types of functions for F(x), which were later designated types I to XII. Since these types cannot all be derived from the same construction rule and, moreover, the corresponding regions in the (γ_1, γ_2)-plane sometimes overlap considerably, we shall not speak of the distribution types belonging to a system but of different systems.
Some of these systems occupy only a line or a point in the (γ_1, γ_2)-plane and are therefore of no interest in our context.
System XII, which has proved to be of considerable practical use, was investigated more thoroughly by Burr (1968, 1973), Burr and Cislak (1968), Hatke (1949), Rodriguez (1977) and Tadikamalla (1980). Tadikamalla stresses the importance of system III and states its relationships to other distributions. System III is identical to the three-parameter Kappa distribution described by Mielke and Johnson (1973).
Only systems III and XII will be described in the following.

D: XII: $F(x) = 1-(1+x^c)^{-k}$

$$f(x) = kc\, x^{c-1}(1+x^c)^{-k-1} \qquad x, c, k > 0 \quad c \cdot k > 4$$

III: $F(x) = (1+x^{-c})^{-k}$

$$f(x) = kc\, x^{-c-1}(1+x^{-c})^{-k-1} \qquad x, k > 0,\ c > 4$$

Relationship between the two systems:
If **x** follows distribution XII with the parameters c and k, then $\mathbf{x}^{-1}$ has the distribution III with the same values of c and k.

S:

XII	III	shape
c < 1	ck < 1	
c = 1	ck = 1	Pareto distribution
c > 1	ck > 1	
k = 1	k = 1	log-logistic distribution

For k = 1 and any c, the corresponding distributions of the two systems are identical.
As shown in Fig. 4, system III occupies a much larger region of the (γ_1, γ_2)-plane than system XII.

R_γ:

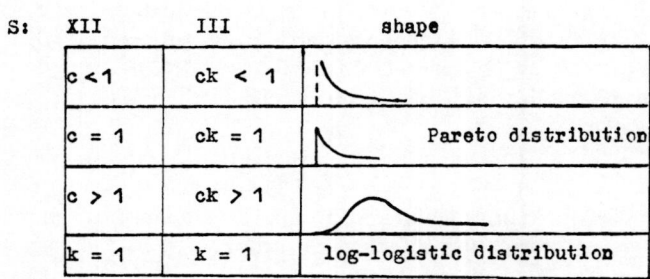

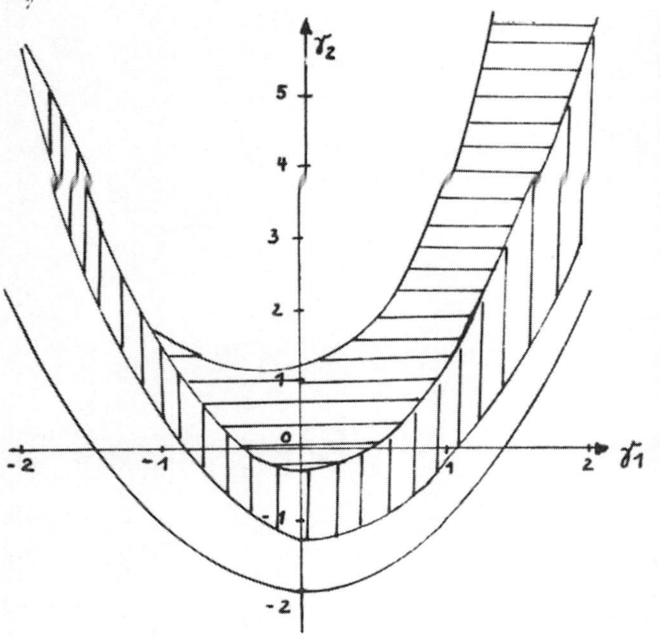

Figure 4
Regions in the (γ_1, γ_2)-plane for
Burr XII (≡)
Burr III (≡ and in addition |||)

$C_{p-\gamma}$:

	XII	III
$E(\mathbf{x}^r) =$	$\dfrac{k\,\Gamma\left(k-\frac{r}{c}\right)\Gamma\left(\frac{r}{c}+1\right)}{\Gamma(k+1)}$	$\dfrac{k\,\Gamma\left(k+\frac{r}{c}\right)\cdot\Gamma\left(1-\frac{r}{c}\right)}{\Gamma(k+1)}$
Table for finding k and c from γ_1 and γ_2	Burr (1973)	

G:

Generate a (0,1)-uniformly distributed random variable **u** and set

$$\mathbf{x} = \left[\mathbf{u}^{-\frac{1}{k}} - 1\right]^{\frac{1}{c}}$$

for system XII and

$$\mathbf{x} = \left[\mathbf{u}^{-\frac{1}{k}} - 1\right]^{-\frac{1}{c}}$$

for system III.

2.5. *Grassia's System*

D:

$\mathbf{x} = e^{-\mathbf{y}}$, let **y** be Γ-distributed with the parameters b and p

$$f(x) = b^p x^{b-1}(-\ln x)^{p-1}/\Gamma(p). \qquad 0 \le x \le 1 \quad p, b > 0$$

(Grassia (1977), Tadikamalla (1981))

S:

Most important types of shape:

p	b	shape
> 1	> 1	
> 1	< 1	
< 1	> 1	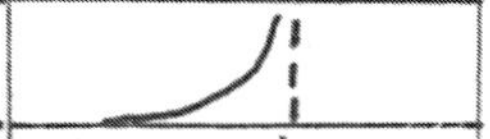
< 1	< 1	

The shapes of these distributions are discussed in greater details by Grassia (1977).

R_γ:
The region occupied by Peason type I distributions is covered in the (γ_1, γ_2)-plane.

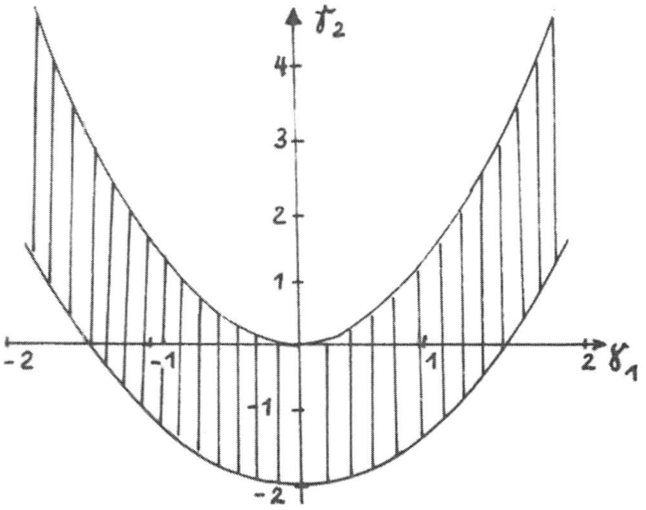

Figure 5
Region of Grassia's distributions in the (γ_1, γ_2)-plane

$$C_{p-\gamma}: \quad E(\mathbf{x}^k) = \left(\frac{b}{b+k}\right)^p$$

Tadikamalla (1981) drew up a program for calculating p and b from γ_1 and γ_2.

G:
See D.
Generators for the Γ-distribution are given by Ahrens and Dieter (1974).

2.6. *Distribution Systems Constructed by Series Expansion*

Gram-Charlier series or Edgeworth series can be used to generate density functions, the coefficients of the series being simple functions of the moments and cumulants, respectively, of the distributions concerned. The Gram-Charlier series has the following form (Kendall, Stuart (1958)):

$$f(x) = \varphi(x) \sum_{k=0}^{\infty} c_k H_k(x)$$

where $\varphi(x)$ is the density function of the standardized normal distribution and $H_k(x)$ is the Hermite polynomial of order k.
We have

$$\frac{\partial^k \varphi(x)}{\partial x^k} = (-1)^k H_k(x) \varphi(x)$$

The orthogonality relationship

$$\int_{-\infty}^{\infty} H_k(x) H_l(x) \varphi(x) \, dx = \begin{cases} 0 & \text{for } l \neq k \\ k! & \text{for } l = k \end{cases}$$

of the Hermite polynomials is used to calculate the coefficient c_k (Kendall, Stuart (1958)). There ist one-to-one correspondence between the first k coefficients of the series and the first k moments (or cumulants) of the distribution.
The Edgeworth series has following form (Kendall, Stuart (1958)):

$$f(x) = \exp\left[\sum_{k=1}^{\infty} d_k D^k\right] \varphi(x)$$

$$\text{where } D = \frac{\partial}{\partial x}$$

$$\text{and } d_k = \begin{cases} \frac{\varkappa_k}{k!}(-1)^k & \text{for } k \neq 2 \\ \frac{\varkappa_2 - 1}{2} & \text{for } k = 2 \end{cases}$$

where $\varkappa_k$ is the cumulant of order k. In other words, the Edgeworth series is calculated by formally applying the exponential function

$$\exp(t) = 1 + t + \frac{t^2}{2!} + \frac{t^3}{3!} + \ldots$$

to a series of differential operators. These differential operators must subsequently be applied to $\varphi(x)$. The result is a Gram-Charlier series with (in general) an infinite number of terms. In other words, an Edgeworth series cannot be described exactly by a finite Gram-Charlier series unless only the first l cumulants are given and the further cumulants are left unknown. In this case the cumulants can be used to calculate the first l moments, which in turn can be used to calculate directly the coefficients of the corresponding Gram-Charlier series. One problem of describing the series in this way is that the generated function f(x) will not in all cases assume only positive values, i. e. it will not always represent a density function.
We shall now give a simple example representing the most important application for both of the series. (The case given here for an Edgeworth distribution was used, for example, by Subrahamaniam (1968 a, b, 1969) to calculate the distribution of quadratic forms and order statistics.)

Gram-Charlier	Edgeworth
D: With $E(\mathbf{x}) = 0$, $V(\mathbf{x}) = 1$ and $c_k = 0$ for $k > 4$ being given, the series obtained is $f(x) = \varphi(x)\,[1 + c_3H_3(x) + c_4H_4(x)]$	D: With $\varkappa_1 = 0$, $\varkappa_2 = 1$, $\varkappa_5 = \varkappa_6 = 0$ $c_k = 0$ for $k > 6$ being given, the series obtained is $f(x) = \varphi(x)\,[1 + c_3H_3(x) + c_4H_4(x) + c_6H_6(x)]$
$C_{p-\gamma}$: $c_3 = \frac{\gamma_1}{3!}$, $c_4 = \frac{\gamma_2}{4!}$	$c_3 = \frac{\gamma_1}{3!}$, $c_4 = \frac{\gamma_4}{4!}$, $c_6 = \frac{1}{2} \cdot c_3^2$

G:
The author knows of no method for generating random numbers.

R_γ:

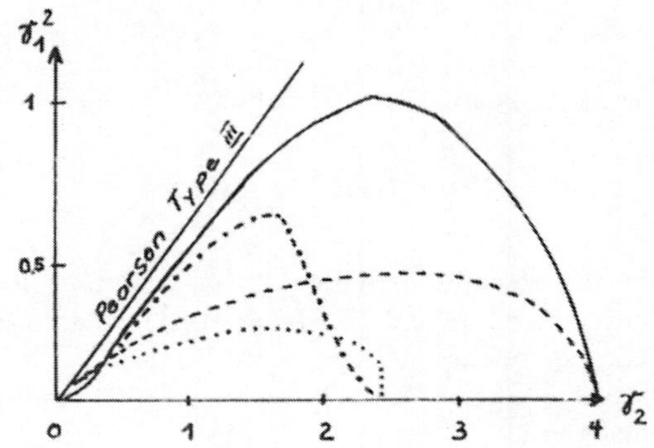

Figure 6
The regions in the (γ_1^2, γ_2)-plane in which all values of f(x) are positive (Gram-Charlier: ——, Edgeworth: - - -) and the regions in which the corresponding distribution is unimodal (Gram-Charlier: - . - . -, Edgeworth: . . . .)
(Barton, Dennis (1952))

These distribution systems cannot be used for simulation purposes until a random generator has been developed. The admissible range in the (γ_1, γ_2)-plane, moreover, is relatively small, although it can be enlarged by choosing suitable coefficients c_k for $k > 4$.

2.7. *The Generalized Lambda Distributions*

D:
Let **u** be uniformly distributed in (0,1). Then construct

$$\mathbf{x} = \lambda_1 + \left[\mathbf{u}^{\lambda_3} - (1-\mathbf{u})^{\lambda_4}\right] / \lambda_2 .$$

The density function is:

$$f(x) = f(x(u)) = \frac{\lambda_2}{\lambda_3 u^{\lambda_3 - 1} + \lambda_4 (1-u)^{\lambda_4 - 1}}$$

S:
Some examples of shapes:

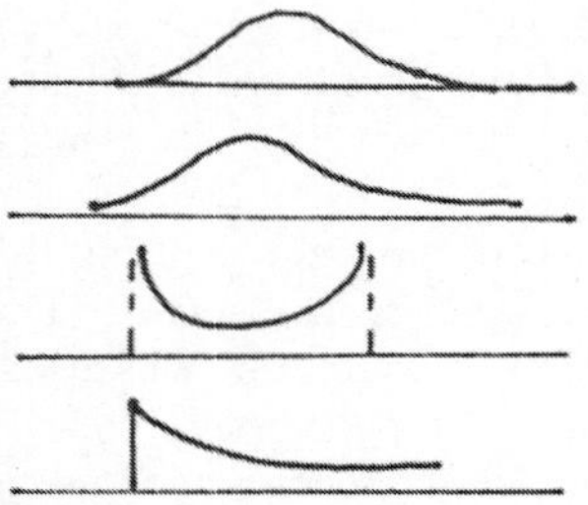

$(1 < \lambda_3, \lambda_4 < 2)$

R_γ:
According to a diagramm of Ramberg et al. (1979) distributions exist for

$$\gamma_2 \gtrsim -1.25 + 1.6625\,\gamma_1^2 .$$

$C_{p-\gamma}$:
Ramberg et al. (1979) give a table showing the values λ_1, λ_2, λ_3 and λ_4 for given γ_1 and γ_2.

G:
See D.

2.8. *The Schmeiser and Deutsch System*

D:
Let **u** be uniformly distributed in (0,1). Then construct

$$x = \begin{cases} -(\lambda_4 - u)^{\lambda_3} & \text{if } u \le \lambda_4 \\ (u - \lambda_4)^{\lambda_3} & \text{if } u > \lambda_4 \end{cases}$$

$$\lambda_3 \ge 0,\ 0 \le \lambda_4 \le 1, \quad -\lambda_4^{\lambda_3} \le x \le (1-\lambda_4)^{\lambda_3}$$

S:

value of λ_3	shape
$\lambda_3 = 0$	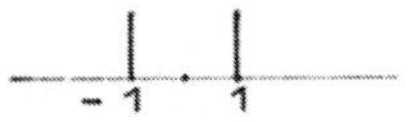
$0 < \lambda_3 < 0.5$	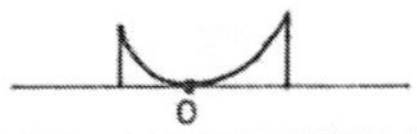
$\lambda_3 = 0.5$	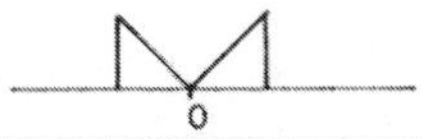
$0.5 < \lambda_3 < 1$	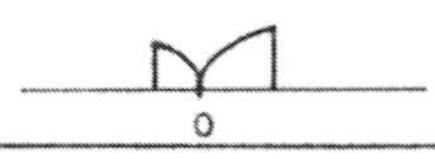
$\lambda_3 = 1$	
$\lambda_3 > 1$	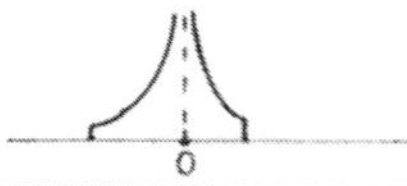

R_γ:

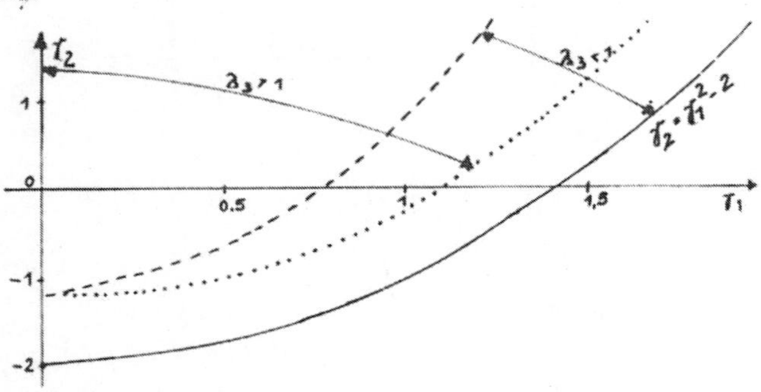

Figure 7
Distributions of Schmeiser and Deutsch in the (γ_1, γ_2)-plane
(—— approximate upper bound for $\lambda_3 < 1$ (bimodal),
(. . . . approximate lower bound for $\lambda_3 > 1$ (unimodal)

The regions for $\lambda_3 > 1$ and $\lambda_3 < 1$ overlap. Two distributions exist for each pair (γ_1, γ_2) in the overlapping region.

$C_{p-\gamma}$:
Schmeiser and Deutsch (1977) constructed nomograms for finding λ_3 and λ_4 from γ_1 and γ_2.

G:
See D.

2.9. *Fleishman's System*

D:
Let **z** have a standardized normal distribution. Then calculate $x = a + bz + cz^2 + dz^3 = P(z)$. (This transformation is the so called power transformation.)
S:

Case 1: $\quad c^2 < 3\,bd \quad \left(\frac{\partial}{\partial z} P(z) \text{ has no point of zero}\right)$

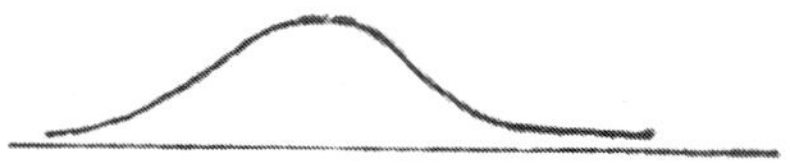

Case 2: $c^2 = 3\,bd$ $\left(\frac{\partial}{\partial z} P(z)\right.$ has exactly one point of zero at $\left.z_0\right)$

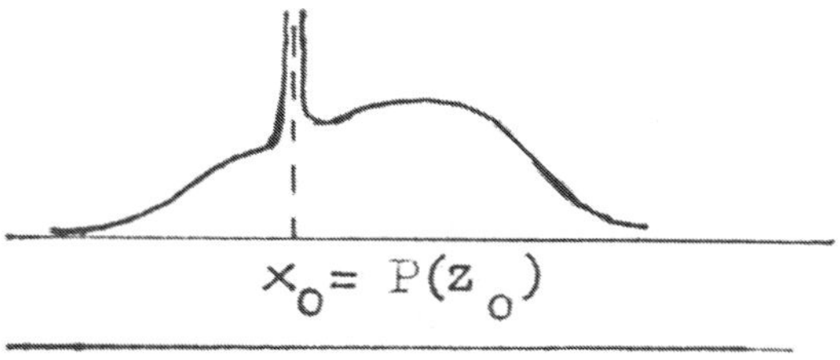

Case 3: $c^2 > 3\,bd$ $\left(\frac{\partial}{\partial z} P(z)\right.$ has two points of zero at z_0 and z_1 respectively$\left.\right)$

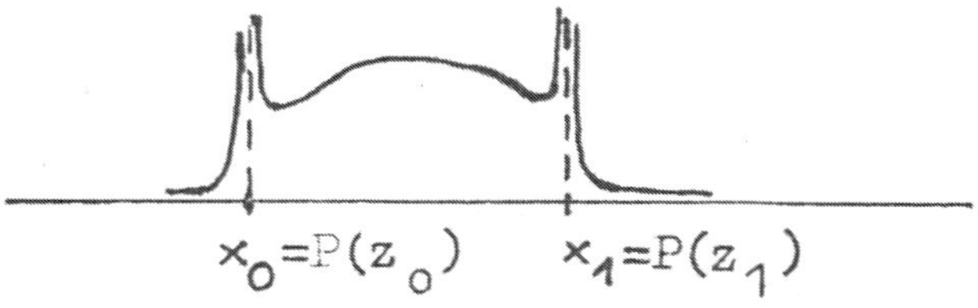

A discussion of the forms with examples is given in Nürnberg (1982).

R_γ:

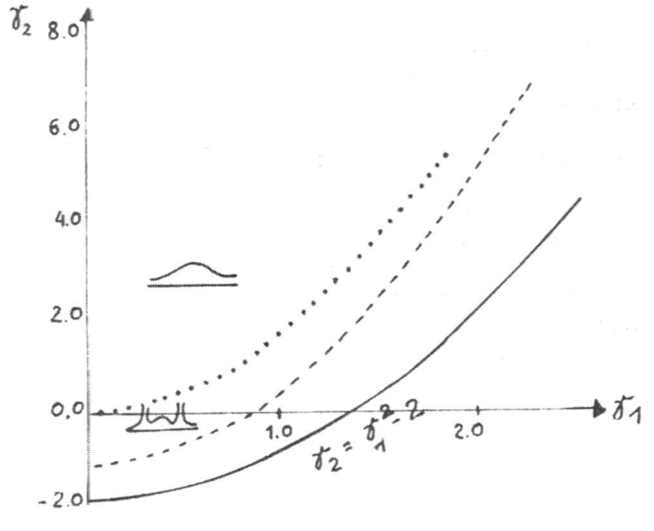

Figure 8
Fleishman distributions in (γ_1, γ_2)-plane
.... approximate border between unimodal and bimodal distributions
- - - approximate lower bound of the possible region in the (γ_1, γ_2)-plane

$C_{p-\gamma}$:
Fleishman (1978) gives a table showing the values of b, c and d (a = —c) for given γ_1 and γ_2.
G:
See D.

2.10. *The Truncated Normal Distribution*

D:
Truncate the standardized normal distribution at the points u and v($u < v$).

R_γ:

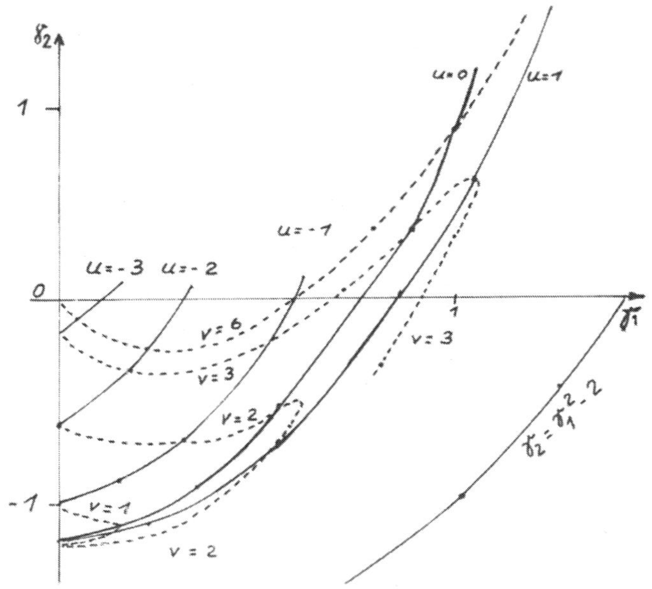

Figure 9
Truncated normal distributions in (γ_1, γ_2)-plane

The limits of the admissible range are still unknown.

$C_{p-\gamma}$:
Rasch and Teuscher (1982) showed how the values of γ_1 and γ_2 depend on u and v and also gave a small table.
With ($\Phi(u)$ = distribution function of the standardized normal distrib.) $\Delta = \Phi(v) - \Phi(u)$
and

$$c_i = \frac{1}{\Delta}\left[v^{i-1}\varphi(v) - u^{i-1}\varphi(u)\right]$$

we have

$$\gamma_1 = \frac{c_1 - 2c_1^3 - 3c_1c_2 - c_3}{(1 - c_2 - c_2^2)^{3/2}}$$

$$\gamma_2 = \frac{4c_1^2 - 6c_1^4 + 3c_2 - 12c_1^2c_2 - c_4 - 4c_1c_3 - 3c_2^2}{(1 - c_2 - c_1^2)^2}$$

G:
If **z** is uniformly distributed in (0,1) construct

$$\mathbf{x} = \Phi^{-1}\left[\Delta \cdot \mathbf{z} + \Phi(u)\right].$$

2.11. *The Double Rectangular Distribution*

D:

$$f(x) = \begin{cases} c_1 & \text{if } x_1 \leq x < x_2 \\ c_2 & \text{if } x_2 \leq x \leq x_3 \\ 0 & \text{otherwise} \end{cases}$$

S:
Example for f(x)

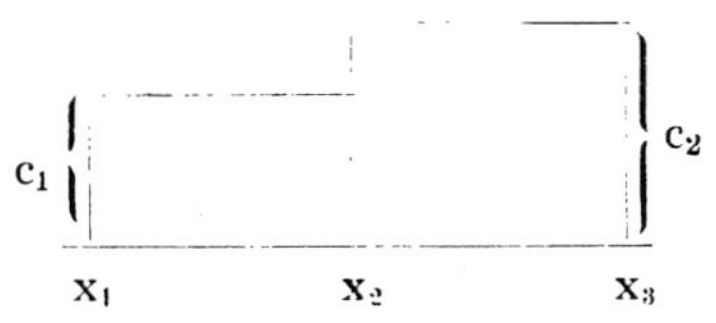

R_γ:
This distribution system covers the whole range of all possible unimodal distributions (cf. chapter 1), without

the distributions with $\gamma_1 = 0$ and $\gamma_2 \neq -1{,}2$. An analogous distribution system which covers the whole admissible range of the (γ_1, γ_2)-plane would, for example, be the system of triple rectangular distributions. The properties of these, however, have yet to be investigated.

$C_{p-\gamma}$:
Guiard (in Herrendörfer (1980)) has shown how the parameters of the distribution depend on γ_1 and γ_2. For given values of γ_1 and γ_2 it is necessary to calculate

$$v = \frac{16\gamma_1^2 - 15\gamma_2 - 18}{4\gamma_1}$$

and

$$w = \frac{5\gamma_2 + 6}{4\gamma_1}$$

The values x_1, x_2 and x_3 are the solutions of the equation

$$x^3 - wx^2 - 3x - v = 0.$$

In order to solve this equation calculate

$$p = -1 - \frac{w^2}{9}, \quad q = -\left(\frac{w}{3}\right)^3 - \frac{w+v}{2}$$

(In the region of the (γ_1, γ_2)-plane being admissible for the system $D = p^3 + q^2 < 0$ must hold.)

Setting $r = \sqrt{|p|}$ sign q and $\alpha = \arccos \frac{q}{r^3}$, we obtain

$$x_1 = -2r \cos \frac{\alpha}{3} + \frac{w}{3}$$

$$x_2 = 2r \cos \left(\frac{\pi + \alpha}{3}\right) + \frac{w}{3}$$

$$x_3 = 2r \cos \left(\frac{\pi - \alpha}{3}\right) + \frac{w}{3}.$$

In order to ensure that $x_1 < x_2 < x_3$, the indices of x_1 and x_2 must be rearranged if $r < 0$. Moreover, we have

$$c_1 = \frac{x_2 + x_3}{(x_3 - x_1)(x_2 - x_1)}$$

$$c_2 = \frac{-(x_1 + x_2)}{(x_3 - x_1)(x_3 - x_2)}$$

G
If **u** is uniformly distributed in (0,1), calculate

$$x = \begin{cases} \dfrac{u}{c_1} + x_1, & \text{if } u < \dfrac{x_2 + x_3}{x_3 - x_1} \\[2ex] \dfrac{u}{c_2} + \dfrac{x_3^2 + x_1 x_2}{x_1 + x_2}, & \text{otherwise} \end{cases}$$

3. Comparison of the Different Distribution Systems

One or more distribution systems can be selected for simulation purposes on the basis, for instance, of the following criteria:

- S — similarity with empirical distributions encountered in practice;
- R_γ: — size of the region covered in the (γ_1, γ_2)-plane
- $C_{p-\gamma}$ — simplicity with which the distribution parameters can be calculated from γ_1 and γ_2 (in the following table the availability of simple procedures for these calculations are denoted by 1 and the availability of tables by 2);
- G — simplicity with which random numbers can be generated;
- GT — computer time required for generating random numbers.

In the following table the author has awardet subjetice "marks" for the different criteria stated above (low marks = good system). The column GT shows the computer times in milliseconds reported by Tadikamalla (1980 a) for calculating 10 000 random numbers on a DEC system 10 computer. In this study the generator $u_{i+1} = 630\,360\,016\, u_i \pmod{2^{31} - 1}$ was used for the uniform distribution and the polar method (Box, Muller (1958), Autorenkollektiv (1980)) for the normal distribution. These generators require a computer time of 18 and 94 microseconds, respectively.

When investigating robustness, it appears advisable to show the robustness of a statistical method also for unfavourable distributions. There seems reason to suspect that "long-tailed" distributions have unfavourable properties in respect of robustness. Pearson, Johnson and Burr (1979) compared the quantiles of systems of Pearson, Johnson, Burr (XII) and of the non-central t-distribution and the non-central χ^2 distribution for several values of γ_1 and γ_2. These comparisons showed that the Burr distributions have extremely long tails.

	S	R_γ	$C_{p-\gamma}$	G	GT
Pearson	1	1	1	4	
Johnson	1	1	2	2	S : 160; S : 165; S : 196
Tadik.-Johnson	1	1	2	1	L : 139; L : 142; L : 162
Burr XII	1	4	2	1	252
Burr III	1	3	4	1	
Grassia	2	4	4	3	
Gram-Charlier		5	1	5	
generalized lambda-distribution	1	2	2	1	245
Schmeiser-Deutsch	3	1	3	1	140
Fleishman	2	2	2	2	110
Truncated normal distribution	2	3	3	2	
Double rectangular distribution	4	2	1	1	

4. References

AHRENS, J. H., DIETER, U.
Computer methods for sampling from gamma, beta, Poisson and binomial distributions.
Computing 12 (1974) 223—246.

AHAJA. J. C., NASH, S. W.
The generalized Gompertz-Verhulst family of distributions.
Sankhya, S. A. 19 (1967) 141—156.

ATKINSON. A. C.
An easily programmed algorithm for generating gamma random variables.
J. R. Statist. Soc. A, 140 (1977) 2. 232—234.

ATKINSON. A. C.
A family of switching algorithms for the computer generation of beta random variables.
Biometrika 66 (1979) 1. 141—145.

ATKINSON. A. C., PEARCE. M. C.
The computer generation of beta, gamma and normal random variables.
J. R. Statist. Soc. A. 139 (1976) 4, 431—467.

ATKINSON, A. C., WITTAKER, J.
The generation of beta random variables with one parameter greater than and one parameter less than 1.
Applied Statistics 28 (1979) 1, 90—93.

BARTON, D. E., DENNIS, K. E.
The condition under which Gram-Charlier and Edgeworth curves are positive definite and unimodal.
Biometrika 39 (1952) 425—427.

BOWMAN, K. O., SHENTON, L. R.
Approximate percentage points for Pearson distributions.
Biometrika 66 (1979) 1, 147—151.

BOWMAN, K. O., SHENTON, L. R.
Further approximate Pearson percentage points and Cornish-Fisher.
Commun. Statist. B — Simul. Comput. 8 (1979) 3, 231—244.

BOWMAN, K. O., SHENTON, L. R.
Evaluation of the parameters of S by rational fractions.
Commun. Statist.-Simula. Computa. B 9 (1980) 2, 127—132.

BOWMAN, K. O., SERBIN, C. A. and SHENTON, L. R.
Explicit approximate solutions for S .
Commun. Statist.-Simula. Computa. B 10 (1981) 1, 1—15.

BOX, G. E. P., MULLER, M. E.
A note on the generation of random normal deviates.
Annals of Mathematical Statistics 29 (1958), 610—611.

BUKAČ, J.
Fitting S curces using symmetrical percentile points.
Biometrica 59 (1972). 688—690.

BURR, I. W.
Cumulative frequency functions.
Annals of Mathematical Statistics 13 (1942), 215.

BURR, I. W.
On a general system of distributions. III. The simple range.
J. Amer. Statist. Assoc. G 3 (1968) 636—643.

BURR, I. W.
Parameters for a general system of distributions to match a grid of α^3 and α^4.
Communications in Statistics 2 (1973) 1—21.

BURR, I. W., CISLAK, P. J.
An a general system of distributions I. Its curve-shape characteristics. II The sample median.
J. Amer. Statist. Assoc. 63 (1968) 627.

CHENG, R. G.
Generating beta variates with non-integral shape parameters.
Communic. of the ACM, 21 (1978) 317—322.

COOK, D., JOHNSON, M. E.
A family of distributions for modelling non-elliptically symmetric multivariate data.
J. R. Statist. Soc. B, 43 (1981) 2, 210—218.

COOPER, J. D., DAVIS, S. A. and DONO, N. R.
Pearson universal distribution generator (PURGE).
Proceedings of the 19th Annual Conference ASQC (1965) 402—411.

ELDERTON, W. P., JOHNSON, N. L.
Systems of frequency curves.
Cambridge, At the University Press (1969).

FLEISHMAN, A. I.
A method for simulating non-normal distributions.
Psychometrika 43 (1978) 521—532.

GRASSIA, A.
On a family of distributions with argument between 0 and 1 obtained by transformation of the gamma and derived compound distributions.
Australian J. Statistcs 19 (1977) 108—114.

GUIARD, V. (ed.)
Robustheit II, Arbeitsmaterial zum Forschungsthema Robustheit.
Probleme der angewandten Statistik, FZ für Tierproduktion Dummerstorf-Rostock Heft 5 (1981).

HARTER, H. L.
A new table of percentage points of the Pearson type II distribution.
Technometrics 11 (1969) 177—187.

HATKE, S. M.
A certain cumulative probability function.
Ann. Math. Statist. 20 (1949) 461—463.

HERRENDÖRFER, G. (ed.)
Robustheit I, Arbeitsmaterial zum Forschungsthema Robustheit.
Probleme der angewandten Statistik, FZ für Tierproduktion Dummerstorf-Rostock Heft 4 (1980).

JÖHNK, M. D.
Erzeugung von betaverteilten und gammaverteilten Zufallszahlen.
Metrika 8 (1964) 1, 5—15.

JOHNSON, M. E., RAMBERG, J. S. and WANG, C.
The Johnson translation system in Monte Carlo Studies.
Comm. in Statistics, Simulation and Compuation 11 (1982) 5, 521—527.

JOHNSON, M. E., TENENBEIN, A.
A bivariate distribution family with specified marginals.
J. Amer. Statist. Assoc. 76 (1981) 373, 198—201.

JOHNSON, M. E., TIETJEN, G. L.
A new family of probability distributions with applications to Monte Carlo studies.
J. Amer. Statist. Assoc. 75 (1980) 370, 276—279.

JOHNSON, N. L.
Systems of frequency curves generated by methods of translation.
Biometrika 36 (1949) 149—176.

JOHNSON, N. L.
Systems of frequency curves derived from the first law of Laplace.
Trab. Estad. 5 (1954) 283—291.

JOHNSON, N. L.
Tables to facilitate fitting S frequency curves.
Biometrika 52 (1965) 547—558.

JOHNSON, N. L., KITCHEN. J. O.
Some notes on tables to facilitate fitting S curves.
Biometrika 58 (1971a) 223—226.

JOHNSON, N. L., KITCHEN, J. O.
Tables to facilitate fitting S curves. II. Both terminals known.
Biometrika 58 (1971b) 657—668.

JOHNSON, N. L., KOTZ. S.
Continuous multivariate distributions.
New York: Wiley & Sons (1972).

JOHNSON. N. L., KOTZ. S.
Developments in discrete distributions (1969—1980).
International Statistical Review 50 (1972) 71—101.

JOHNSON, N. L., NIXON. E., AMOS, D. E. and PEARSON. E. S.
Table of percentage points of Pearson curves.
Biometrika 50 (1963) 459—498.

JOHNSON. N. L., ROGERS. C. A.
The moments problem for unimodal distributions.
Ann. Math. Statist. 22 (1951) 433—439.

KENDALL, M. G., STUART. A.
The advanced theory of statistics.
Griffin, London. Vol. I. 2 end ed. 1958, Vol. II new ed. 1960, Vol. II new ed. 1966.

LESLIE, D. C.
Determination of parameters in the Johnson system of probability distributions.
Biometrika 46 (1959) 229—231.

McGILLIVRAY, H. L.
Skewness properties of asymmetric forms of Tukey lambda distribution.
Comun. Statist.-Theor. Meth. 11 (1982) 20. 2239—2248.

MARDIA, K. V.
Families of bivariate distributions.
Being number Twenty-Seven of Griffin's Statistical Monographs & Courses, Edited by Alan Stuart, D. Sc. (Econ.), 1970.

MIELKE, W. Jr., JOHNSON, E. S.
Three parameter kappa distribution: maximum likelihood estimates and likelihood ration tests.
Monthley Weather Review 101 (1973) 701—707.

NÜRNBERG, G.
Beiträge zur Versuchsplanung für die Schätzung von Varianzkomponenten und Robustheitsuntersuchungen zum Vergleich zweier Varianzen.
Probleme der angewandten Statistik, FZ Dummerstorf-Rostock der AdL der DDR, Heft 6 (1982), siehe Autorenkollektiv.

PEARSON, E. S., JOHNSON, N. L., BURR, I. W.
Comparisons of the percentage points of distributions with the same first four moments, chosen from eight different systems of frequency curves.
Commun. Statist.-Simula. Computa. B 8 (3) (1979) 191—229.

RAMBERG, J. S., SCHMEISER, B. W.
An approximate method for generating asymmetric random variables.
Communic. of the ACM, 107 (1974) 78—82.

RAMBERG, J. S., TADIKAMALLA, P. R., DUDEWICZ, E. J. and MYKYTKA, E. P.
A probability distributions and its uses in fitting data.
Technometrics 21 (1979) 201—214.

RASCH, D. and HERRENDÖRFER, G. (ed.)
Robustheit III, Arbeitsmaterial zum Forschungsthema Robustheit.
Probleme der angewandten Statistik, FZ für Tierproduktion Dummerstorf-Rostock Heft 7 (1982).

RASCH, D., TEUSCHER, F.
Das System der gestutzten Normalverteilungen.
7. Sitzungsbericht der IGMS Mathematische Gesellschaft der DDR (1982) 51—60.

RODRIGUEZ, N.
A guide to the Burr type XII distributions.
Biometrika 64 (1977) 129—134.

SATHE, Y. S., LINGROS, S. R.
Bounds for the Pearson typ IV tail probabilities.
Commun. Statist.-Theor. Meth. A 8 (1979) 6, 533—541.

SAVAGEAU, M. A.
A suprasystem of probability distributions.
Biom. J. 24 (1982) 4, 323—330.

SCHMEISER, B. W., BABU, A. J.
Beta variate generation via exponential majorizing functions.
Operations Research 28 (1980) 4, 917—926.

SCHMEISER, B. W., DEUTSCH, S. J.
A versatile four parameter family of probability distributions suitable for simulation.
AIIE Transactions 9 (1977) 2, 176—182.

SCHMEISER, B. W., SHALABY, M. A.
Acceptance/rejection methods for beta variate generation.
J. Amer. Statist. Assoc. 75 (1980) 371, 673—678.

SHENTON, L. R., BOWMAN, K. O.
A lagrange expansion for the parameters of Johnson's S
J. Statist. Comput. Simul. 15 (1982) 89—95.

SIMPSON, J. A., WELCH, B. L.
Table of the bounds of the probability integral when the first four moments are given.
Biometrika (1960) 399—410.

SUBRAHMANIAM, K.
Some contributions to the theory of non-normality II.
Sankhya S. A, Vol. 30 Part 4 (1968a) 411—432.

SUBRAHMANIAM, K.
Some contributions to the theory of non-normality III.
Sankhya S. B, Vol. 30. Parts 3 & 4 (1968b) 383—408.

SUBRAHMANIAM, K.
Order statistics from a class of non-normal distributions.
Biometrika 56 (1969) 2, 415—428.

TADIKAMALLA, P. R.
On simulating non-normal distributions.
Psychometrika 45 (1980a) 2, 273—279.

TADIKAMALLA, P. R.
A look at the Burr and related distributions.
Internat. Statist. Rev. 48 (1980b) 3, 337—344.

TADIKAMALLA, P. R.
On a family of distributions obtained by the transformation of the gamma distribution.
J. Statist.-Comput. Simul. 13 (1981) 209—214.

TADIKAMALLA, P. R., JOHNSON, N. L.
Systems of frequenzy curves generated by transformations of logistic variables.
Biometrika 69 (1982) 2, 461—465.

WHEELER, R. E.
Quantile estimators of Johnsons curve parameters.
Biometrika 67 (1980) 725—728.

Academy of Agricultural Sciences of the GDR
Research Centre of Animal Production Dummerstorf-Rostock

A Combinatorial Method in Robustness Research and Two Applications

GÜNTER HERRENDÖRFER, KLAUS-DIETER FEIGE

Abstract

After defining robustness for interval estimations and tests, the paper presents an exact method for investigating this property for discrete distributions with finite supports. This method is used to investigate the u- and t-tests in the case of the single sample problem for robustness in respect of two and three point distributions.

1. Introduction

Mathematical statistical procedures can be divided into parametric and nonparametric. All statistical procedures are based on the following assumptions:

— $(y_1, y_2, \ldots, y_n)$ is the realization of a random vector $(\mathbf{y}_1, \mathbf{y}_2, \ldots, \mathbf{y}_n)$
— $\mathbf{y}_i$ is $F_i(y)$-distributed. The moments of $\mathbf{y}_i$ that are of interest are finite and at least partly unknown.

$F_i(y)$ must satisfy one further condition in the case of parametric procedures:

— $F_i(y)$ must be known expect for the parameters.

The known parametric procedures (t-test, u-test, χ^2-test, F-test, etc.) have very desirable properties if all of their assumptions are exactly fulfilled.

The behaviour of these procedures when their assumptions regarding distribution are not fulfilled was investigated as early as the nineteen-twenties (e. g. Rider, 1929). Tukey (1960), Mandelbrot (1962) and Herrendörfer, G., (ed.) (1980) studied a large number of characters and found that the assumptions regarding distribution can be considered justified in only a few cases. Many characters have distributions with "a long tail" or are unilaterally truncated. Moreover, it can generally be assumed that samples include outliers, which are known to have a considerable effect on the properties of parametric procedures. As a consequence of these studies, investigations into robustness proceeded along two lines, which were described by Ray (1978), for instance:

— Within what limits can a given procedure be applied meaningfully, i. e. how "robust" is a procedure against non-fulfillment of the conditions on which it is based?
— How can a "robust" statistical procedure be constructed?

It seems virtually impossible to find a definition of "robustnes" that is simultaneously clear and comprehensive. Bickel (1976) formulated three questions that must be answered whenever robustness is investigated:

— Robustness against what? What is the supermodel?
— What has to be robust? Which procedure is being considered?
— What sort of robustness? What is the aim of the robustness investigations and which criterion of robustness being used?

The robustness investigations presented here deal with known parametric procedures (the u- and t-test statistics) and were performed to find out how they behave if the distribution is not the assumed normal distribution. We assume that all other conditions are satisfied. Let G_1 be the class of distributions for which the statistical procedure d being studied was derived and which give d its "desirable" properties. The supermodel then consists in specifying a larger class, $G_2 \supset G_1$, of distributions. The criterion used to decide whether a statistical procedure is robust or not must be specified separately for each class of decision procedures.

Let d_α be an interval estimation. In the class G_1 (e. g. normal distributions with $|\mu| < \infty$, $0 < \sigma^2 < \infty$), d_α has the real confidence coefficient $1 - \alpha$. If d_α is applied to a sample with a distribution $g \notin G_1$, the real confidence coefficient will become a function of g and the experimental design, V_N, (in the simplest case it will be a function of the sample size, n)

$$\left[1 - \alpha_d(\alpha, V_N, g)\right].$$

The measure used for the deviation from the nominal confidence coefficient, $1 - \alpha$, can be

$$\alpha_d(\alpha, V_N, g) - \alpha \tag{1}$$

or a function of (1).

In the case of intervalestimations it can first only be demanded that this difference is not too great. In other words, the following definition can be used.

Definition 1:

An interval estimation, d_α, for a given nominal confidence coefficient, $1 - \alpha$, which also has the real confidence coefficient $1 - \alpha$ for the class G_1 of distributions is (α, ε)-robust in the class $G_2 \supset G_1$ for the experimental design V_N if

$$\max_{g \in G_2} \left|\alpha_d(\alpha, V_N, g) - \alpha\right| \leq \varepsilon \tag{2}$$

hold for the given values of α and ε.

An analogous definition can also be given for robustness of the first kind risk of a test. In the case of the u-test and t-test, for the one sample problem ($H_0 : \mu = \mu_0$, $H_A : \mu \neq \mu_0$), this would mean finding such a n_0 for a given class, G_2, of distributions that the procedure is (α, ε)-robust for $n \geqq n_0$.

We shall first describe a procedure for calculating discrete distributions with a finite support, r. e. for k-point distributions with $k < \infty$, $\alpha_d(\alpha, n, g)$. We shall then apply this method to investigate the robustness of the u-test and t-test in the sense given in Definition 1.

2. An Exact Method for Investigating Robustness Against Discrete Distributions

The literature dealing with the robustness of the most important statistical procedures was discussed in detail by Posten (1979), Tan (1982), Tiku (1975), Ito (1980) and by Herrendörfer, G., (ed.) (1980), Guiard, V., (ed.) (1981), Rasch, D., and Herrendörfer, G., (ed.) (1982) and will therefore no tbe considered here.

$$\text{Let } g_k = \begin{pmatrix} y_1, y_2, \dots, y_k \\ p_1, p_2, \dots, p_k \end{pmatrix}, \sum_{i=1}^{k} p_i = 1, \ p_i > 0 \qquad (3)$$

be a discrete distribution ($|y_k| < \infty$) with k points in the support, and let

$$\mathfrak{y} = (\mathbf{y}_1, \mathbf{y}_2, \dots, \mathbf{y}_n)'$$

be a random sample whose components $\mathbf{y}_i$ are distributed according to (3). Then the probability for the realization

$$P_{\mathfrak{N}} = \begin{pmatrix} y_1, y_2, \dots, y_k \\ n_1, n_2, \dots, n_k \end{pmatrix}, \sum_{i=1}^{k} n_i = n, n_i \geq 0 \qquad (4)$$

is given by

$$P(\mathfrak{y} = P_{\mathfrak{N}}) = \frac{n!}{n_1! \cdot n_2! \cdot \dots \cdot n_k!} p_1^{n_1} p_2^{n_2} \dots p_k^{n_k}. \qquad (5)$$

The number of possible different samples can be calculated from

$$M = \binom{n+k-1}{k-1}. \qquad (6)$$

The statistic d_α can be calculated for each possible sample. Let, moreover, g_k be known. In this case, $E(\mathbf{y}_i) = \mu$ and $V(\mathbf{y}_i) = \sigma^2$, for instance, are also known and it is possible to test whether a correct or false decision has been made for a particular sample.
The M possible samples are ordered in some (fixed) sequence. We now arrange the probabilities calculated according to (5) in the same order as the samples to obtain the vector P. We shall call the corresponding decision vector H. A component of H is 0 if a correct decision has been made for the corresponding sample; otherwise it is 1. For an interval estimation it is easy, for instance, to calculate

$$\alpha_d(\alpha, n, g_k) = P'H \qquad (7)$$

We shall demonstrate this by estimating a mean as an example. A random sample of size $n = 3$ is given, and

$$g_3 = \begin{pmatrix} -\sqrt{3} & 0 & \sqrt{3} \\ \frac{1}{6} & \frac{4}{6} & \frac{1}{6} \end{pmatrix} \qquad (8)$$

is used as a k-point distribution. For g_3 we have $E(\mathbf{y}) = 0$; $V(\mathbf{y}) = 1$. The interval estimation

$$\left\langle \bar{\mathbf{Y}}. - t(2;0.975)\frac{S}{\sqrt{3}}; \bar{\mathbf{Y}}. + t(2;0.975)\frac{S}{\sqrt{3}} \right\rangle \qquad (9)$$

is calculated for each sample realized. The results of these calculations are shown in table 1.
This yields

$$\alpha_d(0.05; 3; g_3) = \frac{2}{6^3} = \frac{1}{108} \sim 0.01.$$

This interval holds for the class G_1 of normal distributions and $n \geq 2$.

3. Results for the u-test and t-test and the Corresponding Interval Estimations for μ in the Single Sample Problem

3.1. *Results for the Distribution g_3 corresponding to (3)*

We consider distributions whose first four moments are bounded and denote this class by $K(\mu; \sigma^2; \gamma_1; \gamma_2)$ [γ_1 — skewness and γ_2 — kurtosis]. The g_3 given in (8) thus lies within the subclass K(0;1;0;0), which also contains the standardized normal distribution.
Since a k-point distribution is defined by $2k-1$ parameters, K (0;1;0;0) contains an one-parametric family of three-point distributions, of which only g_3 according to (8) will be investigated at first. The results up to $n = 43$ are given in Fig. 1, and show a typical behaviour also for other k-point distributions.
If we set $\varepsilon = 0.2 \cdot \alpha = 0.01$, the real values of α should be between 0.04 and 0.06. These results still give no n_0 for the u-test, so that for $n \geq n_0$ the test can be considered robust for g_3. In respect of the t-test the results are different: according to Def. 1 the test can be considered (0.05;0.01)-robust in respect of g_3 for $n \geq n_0 = 11$.

3.2. *Two-point Distributions*

The inequality

$$\gamma_2 \geq \gamma_1^2 - 2 \qquad (10)$$

hold between γ_1 and γ_2 for a distribution. For two-point distributions,

$$\gamma_2 = \gamma_1^2 - 2 \qquad (11)$$

always holds, so that these distributions are situated at the edge of the permissible region in the (γ_1, γ_2)-plane. It can be shown that exactly one standardized two-point distribution lies at each point of the parabola. The totality of all two-point ditributions with $V(\mathbf{y}) = 1$ can be given by

Table 1
Calculation of α_d (0.05;3;g_3) for the interval estimation (9)

y_i \ Nr.	1	2	3	4	5	6	7	8	9	10
$-\sqrt{3}$	3	2	2	1	1	1	0	0	0	0
0	0	1	0	2	1	0	3	2	1	0
$\sqrt{3}$	0	0	1	0	1	2	0	1	2	3
P'	$\frac{1}{6^3}$	$\frac{12}{6^3}$	$\frac{3}{6^3}$	$\frac{48}{6^3}$	$\frac{24}{6^3}$	$\frac{3}{6^3}$	$\frac{64}{6^3}$	$\frac{48}{6^3}$	$\frac{12}{6^3}$	$\frac{1}{6^3}$
$\bar{y}$	$-\sqrt{3}$	—1.15	—0.58	—0.58	0	0.58	0	0.58	1.15	$\sqrt{3}$
s^2	0	1	4	1	3	4	0	1	1	0
I_L	$\sqrt{3}$	—3.63	—5.54	—3.06	—4.30	—4.38	0	—1.90	—.133	$\sqrt{3}$
I_U	$\sqrt{3}$	1.33	4.38	1.90	4.30	5.54	0	3.06	3.63	$\sqrt{3}$
H'	1	0	0	0	0	0	0	0	0	1

Fig. 1: $\alpha_d(0{,}05;n;g_3)$ for the u- and t-tests $(2 \leq n \leq 43)$

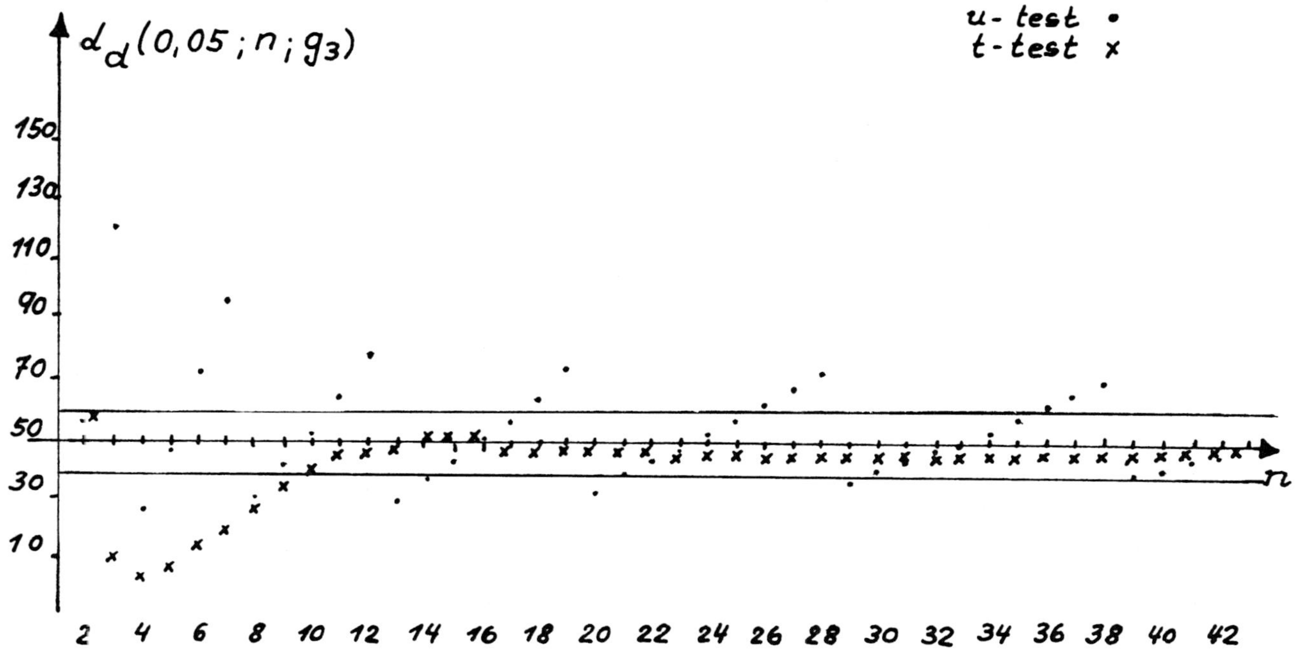

$$g_2(y,\Delta)=\begin{pmatrix} -\frac{1}{y}+\Delta & y+\Delta \\ \frac{y^2}{1+y^2} & \frac{1}{1+y^2} \end{pmatrix} \begin{matrix} y\in R^+ \\ \Delta\in R^1. \end{matrix} \tag{12}$$

The class

$$K\left(\Delta;1;\frac{y^2-1}{y};\frac{1-4y^2+y^4}{y^2}\right). \tag{13}$$

contains $g_2(y,\Delta)$.

Let a random sample, $\mathfrak{y}'=(y_1,\dots,y_n)$, with $y_i \sim g_2(y,\Delta)$ begiven. Its realization have the form

$$\mathfrak{y}_2(y,\Delta,n,n_1)=\begin{pmatrix} -\frac{1}{y}+\Delta & y+\Delta \\ n_1 & n-n_1 \end{pmatrix},\quad 0\leq n_1\leq n \tag{14}$$

with

$$\bar{y}.=y+\Delta-\frac{1+y^2}{y}\cdot\frac{n_1}{n} \tag{15}$$

and

$$s^2=\frac{1}{n-1}\left(\frac{1+y^2}{y}\right)^2\frac{n-n_1}{n}\cdot n_1 \tag{16}$$

The hypotheses

$$\begin{matrix} H_0:\mu=0 & (\Delta=0) \\ H_A:\mu\neq 0 & (\Delta\neq 0) \end{matrix} \tag{17}$$

are to be tested by means of the normal u-test for normal distributions of known variance. This is done with the aid of the statistic

$$\mathbf{u}=\bar{\mathbf{y}}\sqrt{n} \tag{18}$$

and the decision rule

$$\left.\begin{matrix} |\bar{\mathbf{y}}|\cdot\sqrt{n}\leq u_{1-\alpha/2}\rightarrow H_0 \\ |\bar{\mathbf{y}}|\cdot\sqrt{n}< u_{1-\alpha/2}\rightarrow H_A \end{matrix}\right\}\text{ decisions.} \tag{19}$$

The probability of a certain realization of $\mathfrak{y}$ can be calculated from

$$P\left(\mathfrak{y}=\mathfrak{y}_2(y,\Delta,n,n_1)\right)=\binom{n}{n_1}\left(\frac{y_2}{1+y^2}\right)^{n_1}\left(\frac{1}{1+y^2}\right)^{n-n_1} \tag{20}$$

The aim is to find the samples for which H_0 will be rejected:

$$\left|y+\Delta-\frac{1+y^2}{y}\cdot\frac{n_1}{n}\right|>u_{1-\frac{\alpha}{2}}\cdot\frac{1}{\sqrt{n}} \tag{21}$$

After some manipulation, this yields the two inequalities

$$n_1<\frac{\left(y+\Delta-u_{1-\frac{\alpha}{2}}\frac{1}{\sqrt{n}}\right)yn}{1+y^2}=n_u^* \tag{22}$$

and

$$n_1>\frac{\left(y+\Delta+u_{1-\frac{\alpha}{2}}\frac{1}{\sqrt{n}}\right)yn}{1+y^2}=n_0^* \tag{23}$$

Denoting the smallest integer larger than A by [A], we find that (21) is satisfied for all $n_1\leq[n_u^*]=n_u$ and all $n_1\geqq[n_0^*]=n_0$. The power function of the u-test is then given by

$$1-\beta_u[n,g_2(y,\Delta)]=\frac{1}{(1+y^2)^n}\left[\binom{n}{0}+\binom{n}{1}y^2+\dots+\binom{n}{n_u}y^{2n_u}\right.$$

$$\left.+\binom{n}{n_0}y^{2n_0}+\binom{n}{n_0+1}y^{2(n_0+1)}+\dots+\binom{n}{n}y^{2n}\right] \tag{24}$$

The distribution function of the binomial distribution is

$$B(n;p;n_1) = 1 - \sum_{i=n_1}^{n} \binom{n}{i} p^i (1-p)^{n-i} \qquad (25)$$

With

$$p = \frac{y^2}{1+y^2} \qquad (26)$$

we have

$$1 - \beta_u[n, g_2(y, \Delta)] = B(n, p, n_u + 1) + 1 - B(n, p, n_u).$$

It is naturally also possible to describe the distribution function by the F-distribution, and an analogous appraoch can also be used with success for the t-test for testing (17). For $\Delta = 0$ we obtain in each case a real α value for the nominal α and for the given y and n. The values of n_0 are of particular practical interest. These are given in tables 2 and 3. The calculations was performed up to $n = 299$ and $n = 250$.

Table 2
Minimum sample size (n_0) for the u-test for given α and y in the case $\varepsilon = 0.2\,\alpha$

y \ α	0.1	0.05	0.01	γ_1	γ_2
1	95	116	186	0	—2
1.25	64	88	206	0.45	—1.7975
$0.5 + 0.5\sqrt{5}$	104	97	—	1	—1
2	116	131	241	1.5	+0.25
30	> 299	> 299	> 299	29.97	896

Table 3
Minimum sample size (n_0) for the t-test for given α and y in the case $\varepsilon = 0.2\,\alpha$

y \ α	0.1	0.05	0.01	γ_1	γ_2
1	95	151	256	0	—2
1.25	86	87	172	0.45	—1.7975
$0.5 + 0.5\sqrt{5}$	102	186	—	1	—1
2	125	270	> 299	1.5	+0.25
30	> 299	> 299	> 299	29.97	896

As the tables show, the t-test cannot be considered robust in respect of two-point distributions for the selected Σ until n becomes very large.

3.3. *Three-point Distribution*

3.3.1. Results in the Class $K(\mu;\sigma^2;0;0)$

Five parameters are necessary in order to define a three-point distribution. Since we have only four parameters (μ, σ^2, γ_1 and γ_2), $K(\mu/\sigma^2;0;0)$ contains a one-parametric family of three-point distributions. We will consider how these depend on x_3 (the right support point). The values of α have been calculated for $2 \leq n \leq 250$ for the following distributions and are presented in table 4.

$$\begin{pmatrix} -4{,}700230 & -0{,}754315 & 1{,}2 \\ 0{,}004073 & 0{,}601730 & 0{,}394197 \end{pmatrix}$$

$$\begin{pmatrix} -\sqrt{3} & 0 & \sqrt{3} \\ \frac{1}{6} & \frac{4}{6} & \frac{1}{6} \end{pmatrix}$$

$$\begin{pmatrix} -1{,}395867 & 0{,}443486 & 2{,}5 \\ 0{,}294272 & 0{,}658180 & 0{,}047548 \end{pmatrix}$$

$$\begin{pmatrix} -1{,}269762 & 0{,}647540 & 3{,}5 \\ 0{,}357175 & 0{,}629758 & 0{,}013067 \end{pmatrix}$$

$$\begin{pmatrix} -1{,}188165 & 0{,}771498 & 5 \\ 0{,}400561 & 0{,}596254 & 0{,}003185 \end{pmatrix}$$

Table 4
Dependence of n_0 on x_3 for $\varepsilon = 0.2\,\alpha$ for the u-test and t-test in $K(0;1;0;0)$

	u-test			t-test		
x_3 \ α	0.1	0.05	0.01	0.1	0.05	0.01
1.2	80	62	115	87	45	77
3	56	95	164	7	11	24
2.5	21	27	39	16	41	34
3.5	15	41	50	22	29	64
5	43	39	63	41	56	84

If, in the above distributions, x_i is replaced by $-x_i$, we obtain, after rearrangement, for instance $x_3 = 4.700230$ in the first distribution. Due to the special selection of the parameters μ, σ^2, γ_1 and γ_2, this distribution also has the same parameters and thus also belongs to the same family of three-point distributions. It is evident that the real α-values are not affected by this transformation. The results shown in table 4 also hold for distributions resulting from the above transformation. In other words, it would have been sufficient to conduct these investigations only for $x_3 \geq 3$. This can be immediately generalized to the corresponding three-point distributions with $\mu = 0$, $\sigma^2 = 1$ and $\gamma_1 = 0$. Table 4 shows clearly that it is not sufficient to take only one three-point distribution from the one-parametric family as a „representative". The differences in the real α and, consequently, in n_0 are substantial. As shown by Rasch, D., and Herrendörfer, G., (ed.) (1982), the results yielded by the two-point distribution for $y = 1$ can be regarded as results for a three-point distribution with $x_3 = \infty$. If these results are compared with those in table 4, it will be seen that the value of n_0 for the two-point distribution with $y = 1$ is greater than the n_0 values for the three-point distributions. The speculation that the results obtained here for $x_3 = \infty$ must be considered extreme seems to be justified. In other words, for studies on robustness in respect of three-point distributions belonging to the class $K(0;1;0;0)$ it is sufficient to investigate two-point distributions with $y = 1$. The results published in Rasch, D., and Herrendörfer, G., (ed.) (1982) show that the two-point distribution with $y = 1$ can be considered a „borderline distribution" for three-point distributions belonging to the class $K(0;1;0;\gamma_2)$. For the following investigations we choose $\gamma_1 = 7$ and $\gamma_2 = -1.9$.

3.3.2. Results in the Classes $K(0;1;0;7)$ and $K(0;1;0;-1.9)$

In view of the results obtained in 3.3.1., different x_3 values were selected in order to include extreme three-point distributions. The calculations were restricted to $2 \leq n \leq 150$ in order to keep the calculation effort within reasonable limits. In view of this constraint, the investigations into robustness for $\alpha = 0.01$ are no longer sufficient and have therefore been ommitted from the following tables. The distributions have also been omitted to save space. The two following tables do, however, show the two x_3 values.
These results confirm our suspicion.
So far we have considered only distributions with $\gamma_1 = 0$. We shall now take a look at how skewness affects robustness.

3.3.3. Results for $\gamma_1 = 1$

Since there was reason to believe that the values n_0 are increased by skewness, the calculations were performed for $2 \leq n \leq 250$. The two following tables contain the results for three-point distributions from the classes K(0;1;1;0) and K(0;1;1;—0.5).

Comparison of tables 5, 6 and 7 reveals that the effect of $\gamma_1 = 1$ on n_0 is greater than that of $\gamma_2 = 7$ or $\gamma_2 = -1.9$. If, moreover, tables 7 and 8 are compared with the results for two-point distributions with $\gamma_1 = 1$ in tables 2 and 3, it is evident that the two-point distribution can no longer be considered "extreme".

Table 5
Dependence of n_0 on x_3 for $\varepsilon = 0.2\,\alpha$ in the class K(0;1;0;7)

		u-test		t-test	
	α	0.10	0.05	0.10	0.05
x_3	x_3				
1.5	1.021041	12	16	9	23
2	1.017090	31	35	29	38
3	1.012657	61	61	52	66
4	1.010084	72	77	83	84
6	1.007179	35	42	35	41
10	1.004558	48	62	48	65
50	1.000981	62	96	64	94

Table 6
Dependence of n_0 on x_3 for $\varepsilon = 0.2\,\alpha$ in the class K(0;1;0;—1.9)

		u-test		t-test	
	α	0.10	0.05	0.10	0.05
x_3	x_3				
1.5	1.021041	12	16	9	23
2	1.017090	31	33	29	38
3	1.012657	61	61	52	66
4	1.010084	72	77	83	84
6	1.007179	35	42	35	41
10	1.004558	48	62	48	65
50	1.000981	62	96	64	94

Table 7
Dependence of n_0 on x_3 for $\varepsilon = 0.2\,\alpha$ in the class K(0;1;1;0)

	u-test		t-test	
x_3 \ α	0.10	0.05	0.10	0.05
2	19	27	17	32
3	41	42	26	41
6	81	78	60	131
10	77	114	83	211
50	69	142	88	155
100	60	87	101	185

Table 8
Dependence of n_0 on x_3 for $\varepsilon = 0.2\,\alpha$ in the class K(0;1;1;—0.5)

	u-test		t-test	
x_3 \ α	0.10	0.05	0.10	0.05
2	6	12	17	35
3	22	28	32	54
4	40	60	61	111
6	98	103	84	136
10	130	90	76	152
50	60	143	101	185
100	66	93	89	170

4. Concluding Remarks

The calculations presented here for three-point distributions show that the effect of moments of order 5 or more on n_0 is considerable, i. e. that it is not sufficient to restrict investigations in one point of the (γ_1, γ_2)-plane to a single distribution. The results also reveal that the t-test is still not robust in the sense of definition 1 even for large values of n if G_2 includes two-point or three-point distributions.

5. References

BICKEL, P. J.
Another look at Robustness: A review of reviews and some new developments.
Scand. J. Statist., **3** (1976), 145—168.

GUIARD, G. (ed.)
Robustheit II.
Arbeitsmaterial zum Forschungsthema Robustheit.
Probleme der angewandten Statistik (1981) 5.

HERRENDÖRFER, G. (ed.)
Robustheit I.
Arbeitsmaterial zum Forschungsthema Robustheit.
Probleme der angewandten Statistik (1980) 4.

HERRENDÖRFER, G., RASCH, D., FEIGE, K.-D.
Robustness of satistical methods II. Methods for the ons sample problem.
Biom. Journ. **25** (1983) 327—343.

ITO, K.
Robustness of ANOVA and MANOVA test procedures.
Handbook of Statistics, I (P. R. Krishnaiah, Ed.).
North-Holland Publishing Company (1980), 199—236.

MANDELBROT, B.
The role of sufficiency and of estimation in thermodynamics.
Ann. Math. Statist. **33** (1962), 1021—1038.

POSTEN, H. O.
The robustness of the two-sample t-test over the Pearson system.
J. Stat. Comp. Simul., **6** (1978), 295—311.

POSTEN, H. O.
The robustness of the one-sample t-test over the Pearson system.
J. Statist. Comp. Simul. **9** (1979), 133—149.

RASCH, D. und HERRDÖRFER, G. (ed.)
Robustheit III.
Arbeitsmaterial zum Forschungsthema Rebustheit.
Probleme der angewandten Statistik (1982) 7.

REY, W. J. J.
Robust statistical methods.
Springer Verlag Berlin (1978), Berlin — Heidelberg — New York: Springer Verlag 1978.

RIDER, P. R.
On the distribution of the ratio of mean to standard deviation in small samples from non normal universes.
Biometrika **21** (1929), 124—143.

TAN, W. Y.
Sampling distributions and robustness of t, F and variance ratio in two samples and ANOVA models with respect to departure from normality.
Memphis State University, 1982, 28 Seiten.

TIKU, M. L.
Languerre series forms of the distributions of classical test-statistics and their robustness in nonnormal situations.
Applied Statistics.
(R. P. Gupta, Ed.) American Elsevier Pub., Comp., New York, 1975.

TUKEY, J. W.
Where do we go from here?
J. Amer. Statist. Ass. **55** (1960), 80—93.

Central Institut of Microbiology and Experimental Therapy, Jena

Analogy of the Linear Regression and the Two-Sample Problem in Robustness Investigations

MANFRED HORN

Abstract

For the linear model $y_i = a + b x_i + e_i$ with given x-values the robustness of the t-criterion for testing the slope against 0 or of the corresponding confidence interval for the slope may be investigated, among others. When doing this for the simplest case of only two different values of x, the t-test becomes equivalent to the t-test for comparing the two means of y at those two positions of x. And the confidence interval of the slope is simply the interval for the difference of these two means divided by the distance between the two x-values. Thus robustness statements for the 2-sample problem may be transferred to regression.
For the inverse problem of confidence estimation of some x-value corresponding to a given y-value similar considerations are possible.

1. Introduction

The investigations are related to the linear model $y_i = a + b x_i + e_i$ $(i = 1, \ldots, n)$ with given, nonrandom x_i. The e_i are independent random variables with $E e_i = 0$, $\operatorname{Var} e_i = \sigma^2$. If we assume the e_i to be normal distributed, we get by

$$\hat{b} \pm t_{n-2,1-\alpha/2} \frac{\hat{\sigma}}{\sqrt{\sum (x_i - \bar{x})^2}} \tag{1}$$

limits of a $(1-\alpha)$-confidence interval of b,

$$t = \frac{\hat{b}}{\hat{\sigma}} \sqrt{\sum (x_i - \bar{x})^2} \tag{2}$$

a t-distributed variable for testing the hypothesis $H: b = 0$,

$$\left(y_0 - \hat{a} - \hat{b} x_0\right)^2 \leq t^2_{n-2,1-\alpha/2} \hat{\sigma}^2 \left(\frac{1}{n} + \frac{(x_0 - \bar{x})^2}{\sum (x_i - \bar{x})^2}\right) \tag{3}$$

a $(1-\alpha)$-confidence region of an unknown regressor value x_0 for which the expectation $y_0 = a + b x_0$ is given.
The task is to investigate the robustness of (1), (2), (3) against violations of the normality assumption. The most convenient case for practical investigations is that one with only two different values X_1 and X_2 of the regressor $(X_2 > X_1)$, i.e.

$$x_i = \begin{cases} X_1 & i = 1, \ldots, n_1 \\ X_2 & i = n_1 + 1, \ldots, n_1 + n_2 . \end{cases}$$

In this case we write the model in the form

$$y_{ij} = a + b X_j + e_{ij} \left(i = 1, \ldots, n_j;\ j = 1.2;\ n_1 + n_2 = n\right).$$

2. Formulas

$$\text{With } \bar{y}_j = \frac{1}{n_j} \sum_{i=1}^{n_j} y_{ij} \quad (j = 1.2) \text{ we get}$$

$$\hat{b} = \frac{\bar{y}_2 - \bar{y}_1}{X_2 - X_1}, \quad \hat{a} = \frac{X_2 \bar{y}_1 - X_1 \bar{y}_2}{X_2 - X_1}.$$

The estimated line passes trough the points (X_1, y_1), (X_2, y_2). The estimate of σ^2 is

$$s^2 = \frac{\sum_{i=1}^{n} \left(y_i - \hat{a} - \hat{b} x_i\right)^2}{n - 2} = \frac{\sum_{i=1}^{n_1} \left(y_{i1} - \bar{y}_1\right)^2 + \sum_{i=1}^{n_2} \left(y_{i2} - \bar{y}_2\right)^2}{n_1 + n_2 - 2}$$

Because $\bar{x} = \dfrac{n_1 X_1 + n_2 X_2}{n_1 + n_2}$

$$\text{and } \sum_{i=1}^{n} (x_i - \bar{x})^2 = \frac{n_1 n_2}{n_1 + n_2} (X_2 - X_1)^2 .$$

(1) and (2) can be written as

$$\frac{\bar{y}_2 - \bar{y}_1}{X_2 - X_1} \pm \frac{t_{n_1+n_2-2,1-\alpha/2}\, s}{X_2 - X_1} \sqrt{\frac{n_1 + n_2}{n_1 n_2}}$$

and

$$t = \frac{\bar{y}_2 - \bar{y}_1}{s} \sqrt{\frac{n_1 n_2}{n_1 + n_2}} \tag{2'}$$

3. Analogy to the Two-sample Problem

We can consider $y_{11}, \ldots, y_{n_1 1}$ and $y_{12}, \ldots, y_{n_2 2}$ as independent samples. The common variance estimate is

$$\frac{\sum_{i=1}^{n_1} (y_{i1} - \bar{y}_1)^2 + \sum_{i=1}^{n_2} (y_{i2} - \bar{y}_2)^2}{n_1 + n_2 - 2}$$

i.e., it is identical with s^2. Now we can calculate a confidence interval for the difference $E y_{i2} - E y_{i1}$. Its limits are

$$\bar{y}_2 - \bar{y}_1 \pm t_{n_1+n_2-2,1-\alpha/2}\, s \sqrt{\frac{n_1 + n_2}{n_1 n_2}} .$$

This is identical with (1′) apart from the factor $1/(X_2 - X_1)$. Consequently, robustness statements for the t-interval of the difference of expectations of two independent random variables can be transferred to the confidence interval given by (1′).
We can also test the hypothesis $H: E y_{i1} = E y_{i2}$ by the t-criterion using the quantity

$$\frac{\bar{y}_2 - \bar{y}_1}{s} \sqrt{\frac{n_1 n_2}{n_1 + n_2}} .$$

This quantity is identical with (2′). Because $b = (E y_{i2} - E y_{i1}) / (X_2 - X_1)$, both hypotheses $H: b = 0$ and $H: E y_{i1} = E y_{i2}$ are identical. The power function of a t-test depends on H_A (by the noncentrality parameter). For the test of $H: b = 0$ the noncentrality parameter is

$$\frac{b}{\sigma} \sqrt{\sum (x_i - \bar{x})^2} = \frac{E y_{i2} - E y_{i1}}{\sigma} \sqrt{\frac{n_1 n_2}{n_1 + n_2}} .$$

Thus both tests have the same noncentrality parameter and by it identical power functions. Therefore robustness statements concerning the power function of the 2-sample problem can be transferred to the problem examined by (2').

The relation (3) holds under normality with probability $1-\alpha$ for any x_0 or y_0. We suppose that the probability under an alternative distribution will be independent from x_0 or y_0, too. Thus we restrict to the simple case $x_0 = 0$ or $y_0 = a$. Then we get the relation

$$(a-\hat{a})^2 \leq t^2 s^2 \left(\frac{1}{n} + \frac{\bar{x}^2}{\sum (x_i - \bar{x})^2}\right)$$

which reduces to

$$(X_1\bar{e}_2 - X_2\bar{e}_1)^2 \leq t^2 s^2 \left(\frac{X_1^2}{n_2} + \frac{X_2^2}{n_1}\right) \tag{3'}$$

where $\bar{e}_j = \frac{1}{n_j} \sum_{i=1}^{n_j} e_{ij} \quad (j=1.2)$.

Relation (3') can also be derived with the following artificial 2-sample problem. We take $X_1 e_{12}, \ldots, X_1 e_{n_2 2}$ and $X_2 e_{11}, \ldots, X_2 e_{n_1 1}$ as two independent samples and ask for a confidence interval of $E X_1 e_{2i} - E X_2 e_{1i}$. Because

$$\operatorname{Var}(X_1\bar{e}_2 - X_2\bar{e}_1) = X_1^2 \operatorname{Var} \bar{e}_2 + X_2^2 \operatorname{Var} \bar{e}_1$$

$$= \sigma^2 \left(\frac{X_1^2}{n_2} + \frac{X_2^2}{n_1}\right).$$

the limits of a confidence interval of $E X_1 e_{2i} - E X_2 e_{1i}$ are given by

$$X_1\bar{e}_2 - X_2\bar{e}_1 \pm t_{n_1+n_2-2,1-\alpha/2}\, s \sqrt{\frac{X_1^2}{n_2} + \frac{X_2^2}{n_1}}.$$

In this way robustness statements with confidence intervals for a difference of two expectations may be transferred to (3) as well to (1).

Department of Probability and Statistics, Charles University, Prague

Rates of Consistency of Classical One-Side Tests

JANA JUREČKOVÁ

Abstract

One-sided tests of location are considered with the rate of consistency as a measure of performance. It is shown that the classical tests are poor for testing location of a long-tailed distribution and thus are highly non-robust from this point of view. On the other hand, the test based on the sample mean is equally good for every distribution with the exponential tails as for the normal distribution. The situation is similar with the t-test.

1. Introduction

Let $X_1, \ldots, X_n$ be independent random variables, identically distributed according to the distribution function $F(x-\Theta)$, where F belongs to a family $\mathfrak{F}$ of continuous distribution functions (d. f.) such that $0 < F(x) < 1$ for all $x \in R^1$. The problem is that of testing the hypothesis $H: \Theta = \Theta_0$ against $K: \Theta > \Theta_0$.

The power-function of a consistent test tends to 1 as $\Theta - \Theta_0 \to \infty$. The rate of this convergence can be considered as a measure of performance of the test. Jurečková (1980) showed that the rate at which the tails of the distribution F tend to 0 provides a natural upper bound on this convergence. We shall show that this upper bound is attainable by the test based on the sample mean and that the t-test is near to the upper bound, provided F has exponentially decreasing tails. On the other hand, the behaviour of both tests is poor in the case of a long-tailed distribution F. It means that these classical tests are highly non-robust, if we admit long-tailed distributions, with respect to the rate-of-consistency criterion. It was shown in Jurečková (1980, 1982) that the situation is different with the signed-rank tests and with the robust probability ratio tests, which never attain the upper bound but are more robust with respect to the mentioned criterion.

2. Rate of Consistency of $\bar{X}_n$-Test

We shall restrict our attention to the tests of the form

$$\psi_n(\underset{\sim}{X}) = \begin{cases} 1 & \text{if } T_n(X_1-\Theta_0, \ldots, X_n-\Theta_0) > C_n \\ \gamma_n & \text{if } T_n = C_n \\ 0 & \text{if } T_n < C_n \end{cases} \tag{2.1}$$

where $\quad T_n(x_1-t, \ldots, x_n-t)$ is nonincreasing in t $\quad$ (2.2)

and $\quad [X_{(n)} < \Theta_0] \Rightarrow T_n(X_1-\Theta_0, \ldots, X_n-\Theta_0) < C_n$; $\quad$ (2.3)

$X_{(1)} \leqq \ldots \leqq X_{(n)}$ are the order statistics corresponding to $X_1, \ldots, X_n$. We do not impose other conditions on C_n, γ_n in (2.1) but, in special cases of interest, C_n and γ_n are determined so that the test ψ_n is of size α with respect to some fixed distribution $F_0 \in \mathfrak{F}$ $(0 < \alpha < 1/2)$.

Let us denote

$$B(\Theta, \psi_n; F) = \frac{-\log E_\Theta(1-\psi_n(\underset{\sim}{X}))}{-\log F(\Theta_0-\Theta)}, \quad \Theta_0 \leq \Theta < \infty \tag{2.4}$$

where the expectation E_Θ is calculated with respect to $F(x-\Theta)$. The probability of the error of the second kind of the test ψ_n should tend to 0, as $\Theta - \Theta_0 \to \infty$, provided the test ψ_n is consistent for F. The rate of this convergence can be considered as a measure of performance of ψ_n with respect to F. It turns out that the left-hand tail of the distribution F provides a natural upper bound on this rate of convergence. This fact is stated in the following theorem.

THEOREM 2.1. Let $X_1, \ldots, X_n$ be a sample from the distribution $F(x-\Theta)$ $(F \in \mathfrak{F})$. Let ψ_n be the test of $H: \Theta = \Theta_0$ against $K: \Theta > \Theta_0$ satisfying (2.1) – (2.3). Then

$$\overline{\lim_{\Theta \to \infty}}\, B(\Theta, \psi_n; F) \leq n. \tag{2.5}$$

Proof. The theorem was proved in Jurečková (1980) under the assumption of symmetry of F. In fact, the symmetry of F is not necessary, because

$$E_\Theta(1-\psi_n(\underset{\sim}{X})) = P_\Theta(T_n(\underset{\sim}{X}-\Theta_0) < C_n) \geq P_\Theta(X_{(n)} < \Theta_0)$$
$$= (F(\Theta_0-\Theta))^n \tag{2.6}$$

which implies (2.5).

The first question is that of attainability of the upper bound in (2.5). We shall show that the upper bound is attainable by the test based on the sample mean, provided $F(x) \to 0$ and $1-F(x) \to 0$ exponentially fast as $x \to -\infty$ and $x \to +\infty$, respectively.

THEOREM 2.2. Let ψ_n be the test of the form

$$\psi_n(\underset{\sim}{X}) = \begin{cases} 1 & \text{if } n^{1/2}(\bar{X}_n-\Theta_0) \geq u_\alpha \\ 0 & \text{if } n^{1/2}(\bar{X}_n-\Theta_0) < u_\alpha \end{cases} \tag{2.7}$$

where $u_\alpha = \Phi^{-1}(1-\alpha)$, $0 < \alpha < 1/2$, Φ is the standard normal d.f. and

$$\bar{X}_n = \frac{1}{n}\sum_{i=1}^{n} X_i.$$

Then

$$\text{(i)} \quad \lim_{\Theta \to \infty} B(\Theta, \psi_n; F) = n \tag{2.8}$$

for every d.f. satisfying

$$\lim_{x \to -\infty} \left\{-\log F(x) \left(b_1 |x|^{r_1}\right)^{-1}\right\} = 1 \tag{2.9}$$

and

$$\lim_{x \to \infty} \left\{-\log(1-F(x)) \left(b_2 |x|^{r_2}\right)^{-1}\right\} = 1 \tag{2.10}$$

for some r_1, r_2, b_1, b_2 satisfying

$$r_2 \ge r_1 \ge 1, \quad b_2 \ge b_1 > 0. \tag{2.11}$$

$$\text{(ii)} \quad \overline{\lim_{\theta\to\infty}} B(\theta, \psi_n; F) \le 1 \tag{2.12}$$

for every d.f. F satisfying

$$\lim_{x\to-\infty} \left\{-\log F(x)\left(m \log|x|\right)^{-1}\right\} = 1 \text{ for some } m > 0. \tag{2.13}$$

Proof. (i) Let F satisfy (2.9) – (2.11). Using the Markov inequality, we may write for every ε, $0 < \varepsilon < 1$, and for all sufficiently large θ,

$$\left.\begin{aligned} &E_\theta(1-\psi_n(\underset{\sim}{X})) = P_0\left(n^{1/2}\bar{X}_n < n^{1/2}(\theta_0-\theta)+u_\alpha\right) \\ &\le P_0\left(|\bar{X}_n| > \theta-\theta_0-n^{-1/2}u_\alpha\right) \\ &\le E_0\left(\exp\left\{n(1-\varepsilon)b_1|\bar{X}_n|^{r_1}\right\}\right) \\ &\quad\cdot \exp\left\{-n(1-\varepsilon)b_1\left(\theta-\theta_0-n^{-1/2}u_\alpha\right)^{r_1}\right\} \end{aligned}\right\} \tag{2.14}$$

If we are able to prove

$$E_0\left(\exp\left\{n(1-\varepsilon)b_1|\bar{X}_n|^{r_1}\right\}\right) < \infty \tag{2.15}$$

then (2.14) implies

$$\begin{aligned} &\underline{\lim}_{\theta\to\infty} B(\theta, \psi_n; F) \\ &\ge \underline{\lim}_{\theta\to\infty} \left(n(1-\varepsilon)\left(\theta-\theta_0-n^{-1/2}u_\alpha\right)^{r_1}\left(\theta-\theta_0\right)^{-r_1}\right) = n(1-\varepsilon) \end{aligned} \tag{2.16}$$

for every $\varepsilon, 0 < \varepsilon < 1$, and this further implies (2.8). Thus, it remains to prove (2.15). By Jensen's inequality,

$$\begin{aligned} E_0\left(\exp\left\{n(1-\varepsilon)b_1|\bar{X}_n|^{r_1}\right\}\right) &\le E_0\left(\exp\left\{b_1(1-\varepsilon)\sum_{i=1}^{n}|X_i|^{r_1}\right\}\right) \\ &= \left(E_0\left(\exp\left\{b_1(1-\varepsilon)|X_1|^{r_1}\right\}\right)\right)^n. \end{aligned} \tag{2.17}$$

It follows from (2.9) and (2.10) that there exists $K_\varepsilon > 0$ such that

$$1-\exp\left\{-\left(1-\tfrac{\varepsilon}{2}\right)b_2x^{r_2}\right\} \le F(x) \le 1-\exp\left\{-\left(1+\tfrac{\varepsilon}{2}\right)b_2x^{r_2}\right\} \tag{2.18}$$

for $x > K_\varepsilon$ and

$$\exp\left\{-\left(1+\tfrac{\varepsilon}{2}\right)b_1|x|^{r_1}\right\} \le F(x) \le \exp\left\{-\left(1-\tfrac{\varepsilon}{2}\right)b_1|x|^{r_1}\right\} \tag{2.19}$$

for $x < -K_\varepsilon$. Let L_1 be the smallest number of $[-K_\varepsilon, 0]$ such that

$$\exp\left\{-\left(1-\tfrac{\varepsilon}{2}\right)b_1|L_1|^{r_1}\right\} = F(L_1) \tag{2.20}$$

and L_2 be the largest number of $[0, K_\varepsilon]$ such that

$$1-\exp\left\{-\left(1-\tfrac{\varepsilon}{2}\right)b_2L_2^{r_2}\right\} = F(L_2). \tag{2.21}$$

It is easily seen that such numbers always exist. Consider the d.f.

$$G_\varepsilon(x) = \begin{cases} \exp\left\{-\left(1-\frac{\varepsilon}{2}\right)b_1|x|^{r_1}\right\} & \text{if } x < L_1 \\ F(x) & \text{if } L_1 \le x \le L_2 \\ 1-\exp\left\{-\left(1-\frac{\varepsilon}{2}\right)b_2x^{r_2}\right\} & \text{if } L_2 < x. \end{cases} \tag{2.22}$$

Then $G_\varepsilon(x)$ is continuous and

$$\begin{aligned} F(x) &\ge G_\varepsilon(x) \quad \text{if } x \ge 0 \\ F(x) &\le G_\varepsilon(x) \quad \text{if } x \le 0. \end{aligned} \tag{2.23}$$

Hence, integrating by parts and taking (2.11) into account, we get

$$\begin{aligned} E_0\left(\exp\left\{b_1(1-\varepsilon)|X_1|^{r_1}\right\}\right) &\le \int_{-\infty}^{\infty} \exp\left\{b_1(1-\varepsilon)|x|^{r_1}\right\} dG_\varepsilon(x) \\ &= \int_{L_1}^{L_2} \exp\left\{b_1(1-\varepsilon)|x|^{r_1}\right\} dF(x) \\ &+ r_2b_2\left(1-\tfrac{\varepsilon}{2}\right)\int_{L_2}^{\infty} x^{r_2-1}\exp\left\{b_1(1-\varepsilon)x^{r_1}-\left(1-\tfrac{\varepsilon}{2}\right)b_2x^{r_2}\right\}dx \\ &+ r_1b_1\left(1-\tfrac{\varepsilon}{2}\right)\int_{-\infty}^{L_1} |x|^{r_1-1}\exp\left\{-\tfrac{\varepsilon}{2}b_1|x|^{r_1}\right\}dx < \infty. \end{aligned} \tag{2.24}$$

This completes the proof of part (i).
(ii) Let F satisfy (2.13). Then

$$\begin{aligned} &E_\theta(1-\psi_n(\underset{\sim}{X})) = P_0\left(\bar{X}_n < \theta_0-\theta+n^{-1/2}u_\alpha\right) \\ &\ge P_0\left(X_1 < \theta-\theta_0-n^{-1/2}u_\alpha, \dots, X_{n-1} < \theta-\theta_0-n^{-1/2}u_\alpha,\right. \\ &\quad \left. X_n < (2n-1)\left(\theta_0-\theta+n^{-1/2}u_\alpha\right)\right) \\ &= \left(F\left(\theta-\theta_0-n^{-1/2}u_\alpha\right)\right)^{n-1} \cdot F\left((2n-1)\left(\theta_0-\theta+n^{-1/2}u_\alpha\right)\right) \end{aligned} \tag{2.25}$$

so that

$$\begin{aligned} \overline{\lim_{\theta\to\infty}} B(\theta, \psi_n; F) &\le \overline{\lim_{\theta\to\infty}} \left\{m\left(\log(2n-1)\right.\right. \\ &\left.\left. + \log\left(\theta-\theta_0-n^{-1/2}u_\alpha\right)\right)\left(m\log(\theta-\theta_0)\right)^{-1}\right\} = 1. \end{aligned} \tag{2.26}$$

The test (2.7) attains the highest possible rate of consistency for every distribution with exponentially decreasing tails. These distributions cover, among others, the normal distribution N $(0, \sigma^2)$ with unknown σ. From this point of view, the t-test cannot be better than ψ_n, even for N $(0, \sigma^2)$ with unknown σ. On the other hand, the test ψ_n is poor for long-tailed distributions satisfying (2.13) (even if the right-hand tail of F is exponential).
The following section will be devoted to the tail-behaviour of the t-test.

3. Rate of Consistency of t-Test

Let us consider the t-test of H: $\theta=\theta_0$ against K: $\theta>\theta_0$ in the form

$$\psi_n^*(\underset{\sim}{X}) = \begin{cases} 1 & \text{if } T_n(\underset{\sim}{X}-\theta_0) \ge t(n-1|1-\alpha) \\ 0 & \text{if } T_n(\underset{\sim}{X}-\theta_0) < t(n-1|1-\alpha) \end{cases} \tag{3.1}$$

where

$$\begin{gathered} T_n(\underset{\sim}{X}-\theta_0) = (n-1)^{1/2}(\bar{X}_n-\theta_0)/S_n, \\ \bar{X}_n = \frac{1}{n}\sum_{i=1}^{n} X_i, \quad S_n^2 = \frac{1}{n}\sum_{i=1}^{n}(X_i-\bar{X}_n)^2 \end{gathered} \tag{3.2}$$

and t $(n-1|1-\alpha)$ is the upper α-percentile of t-distribution with $(n-1)$ degrees of freedom. Jurečková (1980) proved that

$$\underline{\lim}_{\theta\to\infty} B(\theta, \psi_n^*; F) \ge n\left(1+t(n-1|1-\alpha)\cdot(n-1)^{-1/2}\right)^{-2} \tag{3.3}$$

provided F is normal, while

$$\overline{\lim_{\theta\to\infty}} B(\theta, \psi_n^*; F) \leq 1 \tag{3.4}$$

provided F is a long-tailed distribution satisfying (2.13). The question of interest is whether ψ_n^* is equally good for other distributions with exponential tails as for the normal distribution. This is partially answered in the following theorem.

THEOREM 3.1. Let $X_1, \ldots, X_n$ be a sample from the population with the d.f. $F(x-\Theta)$ such that F satisfies (2.9) and (2.10) with $0 < r_1 \leq r_2$, $0 < b_1 \leq b_2$. Let ψ_n^* be the t-test of (3.1) and (3.2). Then

$$\underline{\lim_{\theta\to\infty}} B(\theta, \psi_n^*; F) \geq n^{r_1/2}\left((n-1)^{-1/2} t(n-1|1-\alpha)+1\right)^{-r_1} \tag{3.5}$$

provided $0 < r_1 < 2$, and

$$\underline{\lim_{\theta\to\infty}} B(\theta, \psi_n^*; F) \geq n\left((n-1)^{-1/2} t(n-1|1-\alpha)+1\right)^{-r_1} \tag{3.6}$$

provided $r_1 \geqq 2$.

Proof. We have

$$E_\theta\left(1-\psi_n^*(\underline{X})\right) = P_0\left((n-1)^{-1/2} t(n-1|1-\alpha) S_n - \overline{X}_n > \theta-\theta_0\right)$$

$$\leq P_0\left(\left(1+(n-1)^{-1/2} t(n-1|1-\alpha)\right)\left(\frac{1}{n}\sum_{i=1}^n X_i^2\right)^{1/2} > \theta-\theta_0\right). \tag{3.7}$$

Let first $0 < r_1 < 2$. Then, using the Markov inequality, we get

$$P_0\left(\left(1+(n-1)^{-1/2} t(n-1|1-\alpha)\right)\left(\frac{1}{n}\sum_{i=1}^n X_i^2\right)^{1/2} > \theta-\theta_0\right)$$

$$\leq E_0\left(\exp\left\{n^{r_1/2}(1-\varepsilon) b_1\left(\frac{1}{n}\sum_{i=1}^n X_i^2\right)^{r_1/2}\right\}\right) \tag{3.8}$$

$$\cdot \exp\left\{-n^{-r_1/2}(1-\varepsilon) b_1 (\theta-\theta_0)^{r_1}\left(1+(n-1)^{-1/2} t(n-1|1-\alpha)\right)^{-r_1}\right\}$$

for every ε, $0 < \varepsilon < 1$, and by c_r-inequality,

$$E_0\left(\exp\left\{n^{r_1/2}(1-\varepsilon) b_1\left(\frac{1}{n}\sum_{i=1}^n X_i^2\right)^{r_1/2}\right\}\right)$$

$$\leq E_0\left(\exp\left\{n^{r_1/2}(1-\varepsilon) b_1 n^{-r_1/2}\sum_{i=1}^n |X_i|^{r_1}\right\}\right)$$

$$= \left[E_0\left(\exp\{(1-\varepsilon) b_1 |X_1|^{r_1}\}\right)\right]^n. \tag{3.9}$$

The last expression is finite by (2.24); this further implies (3.6).

Analogously, if $2 \leq r_1 \leq r_2$,

$$P_0\left(\left(1+(n-1)^{-1/2} t(n-1|1-\alpha)\right)\left(\frac{1}{n}\sum_{i=1}^n X_i^2\right)^{1/2} > \theta-\theta_0\right)$$

$$\leq \left(E_0\left(\exp\{n(1-\varepsilon) b_1 |X_1|^{r_1}\}\right)\right)^n \tag{3.10}$$

$$\cdot \exp\left\{-n(1-\varepsilon) b_1 (\theta-\theta_0)^{r_1}\left(1+(n-1)^{-1/2} t(n-1|1-\alpha)\right)^{-r_1}\right\}.$$

Regarding that the last expectation is finite by (2.24), (3.10) implies (3.6).

4. References

JUREČKOVÁ, J.
Finite-sample comparison of L-estimators of location.
Comment. Math. Univ. Carolinae 20 (1979) 507–518.

JUREČKOVÁ, J.
Rate of consistency of one-sample tests of location.
Journ. Statist. Planning and Inference 4 (1980) 249–257.

JUREČKOVÁ, J.
Tests of location and criterion of tails.
Coll. Math. Soc. J. Bolyai 32 (1982) 469–478.

Department of Probability and Statistics, Byelorussian University, Minsk, USSR

Asymptotic Robustness of Bayesian Decision Rules in Statistical Decision Theory

YURIJ S. KHARIN

Abstract

We consider the statistical classification problems when conditional probability distributions of observations are given with distortions. Robust decision rules are derived for Tukey's model with contaminating distributions and for the model with additive distortions of observations. The guaranteed risk values for robust and Bayesian decision rules are found and compared by the method of asymptotic expansions. The results are illustrated for the case of Gaussian observations.

1. Introduction

In statistical decision theory the Bayesian decision rule (BDR) is widely spread, which minimizes the risk (expected loss) of decision making. For BDR construction in statistical classification problems it is necessary to know the loss matrix, class prior probabilities and the conditional probability densities. In practice these characteristics are estimated by real data. That is why the specific prior uncertainty appears in applied classification problems: the values of the mentioned characteristics are fixed but with any possible distortions; in other words, the statistical classification model assumes any distortions. Ignoring them we receive the BDR, the risk r_0 of which is minimal for the distortionless model. However, in the situation with the real data the classification risk r of this BDR can be much more than r_0.
In this connection the following topical problems of robust statistics are considered in the paper: A) to investigate BDR's risk in the presence of distortions; B) to construct the robust decision rule the guaranteed risk value (supremum) of which is minimal; C) to find this minimal guaranteed risk value.
The review of robust statistics results was given by Rey (1978). The problems A, B were considered before in a simple case: the loss matrix (Kadane (1978)) and prior probabilities (Kadane (1978), Berger (1979)) only are subjected to distortions. A problem closed to B was considered by Randles (1978): the loss matrix and prior probabilities are exactly known and conditional densities are Gaussian, but their parameters are estimated by a sample with outliers. In Huber (1965) a special case of problem B with densities distortions was considered (two classes), but for another optimality criterion: to construct the test for which the power is maximal and the size is equal or less than the significance level.
In this paper the problems A, B, C are solved for the situation with densities distortions by the method of asymptotic expansions of risk (Kharin (1981), (1982)).

2. Mathematical Model

Let observations of L classes $\Omega_1, \dots, \Omega_L$ appear in $\Re^N$ with prior probabilities $\pi_1, \dots, \pi_L$ $(\pi_1 + \dots + \pi_L = 1)$. An observation from Ω_i is a random vector $X_i \in \Re^N$ with probability density $p_i(x)$, $x \in \Re^N$. The classification loss matrix $W = (w_{ik})$ is given; $w_{ik} \geq 0$ is the loss value when we classify an observation from Ω_i into Ω_k $(i, k = \overline{1, L})$. The density $p_i(\cdot)$ is given with distortions:

$$p_i(\cdot) \in P_i(\varepsilon_{+i}), \quad i = \overline{1, L}, \tag{1}$$

where $P_i(\varepsilon_{+i})$ is the family of admissible densities for Ω_i; $0 \leq \varepsilon_{+i} < 1$ is the distortion level for Ω_i. If $\varepsilon_{+i} = 0$, then any distortion in Ω_i is absent; $P_i(0)$ contains the single element $p_i(\cdot) = p_i^0(\cdot)$: the distortionless density. The concrete definitions of the families $\{P_i(\varepsilon_{+i})\}$ are given in the following sections.
Let $d = d(x)$ be a decision rule (DR) defined by the measurable function $d(\cdot): \Re^N \to \{1, 2, \dots, L\}$. The risk for this DR is the mean of the loss:

$$r = r(d; \{p_i\}) = \sum_{i=1}^{L} \pi_i r_i(d; p_i),$$

where

$$r_i(d; p_i) = E\{w_{i, d(X_i)}\} \text{ is}$$

the conditional risk of the classification of the observations.
By the guaranteed risk value for DR $d(\cdot)$ we mean the supremum of the risk on all admissible distributions (I):

$$r_+(d) = \sup_{\{p_i(\cdot) \in P_i(\varepsilon_{+i})\}} r(d; \{p_i\}) = \sum_{i=1}^{L} \pi_i r_{+i}(d),$$

$$r_{+i}(d) = \sup_{p_i(\cdot) \in P_i(\varepsilon_{+i})} r_i(d; p_i).$$

We say that the decision rule $d = d_*(x)$ is a robust one (RDR), if its guaranteed risk value is minimal:

$$r_+(d_*) = \inf_{d(\cdot)} r_+(d).$$

We shall characterize the asymptotic robustness of a DR $d(\cdot)$ by asymptotic expansions of its risk $r_+(d)$ on the powers of the values $\{\varepsilon_{+i}\}$.

3. Asymptotic Robustness of DR in the Case of Tukey's Model

Let the model (I) of the distortions be Tukey's model with contaminating distributions (Tukey (1960)):

$$P_i(\varepsilon_{+i}) = \{p_i(\cdot): p_i(x) = (1 - \varepsilon_i) p_i^0(x) + \varepsilon_i h_i(x), \quad 0 \leq \varepsilon_i \leq \varepsilon_{+i}, \; h_i(\cdot) \in H_i\},$$

$$H_i = \left\{ h_i(\cdot): h_i(x) \geq 0, \; \int_{R^N} h_i(x) dx = 1 \right\} \tag{2}$$

Here $h_i(\cdot)$ is a density of the contaminating distribution from the family H_i, ε_i — the coefficient of contamination influence for Ω_i. Tukey' model (2) has the following interpretation in the statistical classifcation problems. The class Ω_i of observations consists of the two subclasses: $\Omega_i = \Omega_i^0 \cup \Omega_i^h$, $\Omega_i^0 \cap \Omega_i^h = \varnothing$. Ω_i^0 is the well known (frequently observable) subclass; Ω_i^h is the non-studied (rarely observable) subclass. A random observation from Ω_i^0 has the known density $p_i^0(\cdot)$ and from Ω_i^h — an unknown density $h_i(\cdot) \in H_i$. If Ω_i is observed, then an observation from Ω_i^0 appears with probability $1-\varepsilon_i$ and from Ω_i^h — with probability ε_i.

At first we investigate the influence of the distortions (1), (2) on the BDR, which is constructed for the distortionless model $\{\Omega_i^0\}$. This rule is well known:

$$d = d_0(x) = \sum_{i=1}^{L} i\, I_{V_i^0}(x), \quad I_{V_i^0}(x) = \prod_{k=1, k\neq i}^{L} U(f_{ki}^0(x)),$$

where $U(\cdot)$ is Heaviside unity function; $V_i^0 \subset R^N$ is the region where the decision "$d=i$" is made;

$$I_{V_i^0}(x) = \{1, \text{ if } x \in V_i^0;\ 0, \text{ if } x \notin V_i^0\};$$

$$f_{ki}^0(x) = f_k^0(x) - f_i^0(x), \quad f_i^0(x) = \sum_{l=1}^{L} \pi_l p_l^0(x) w_{li}.$$

$f_{ki}^0(x)$ is the Bayesian discriminant function for classes Ω_i^0, Ω_k^0.

The BDR's risk

$$r_0 = r(d_0; \{p_i^0\}) = \sum_{i=1}^{L} \pi_i r_{0i}, \quad r_{0i} = \sum_{j=1}^{L} w_{ij} \int_{V_j^0} p_i^0(x)dx.$$

Let $w_{i+} = \max_j w_{ij}$. The following theorem holds.

Theorem 1. For conditions (1), (2) the guaranteed risk value for BDR $d_0(\cdot)$ is

$$r_+(d_0) = r_0 + \sum_{i=1}^{L} \varepsilon_{+i} \pi_i (w_{i+} - r_{0i}). \quad (3)$$

By analogy with (3) the infimum of risk can be obtained:

$$r_-(d_0) = \inf_{\{p_i(\cdot) \in P_i(\varepsilon_{+i})\}} r(d_0; \{p_i\}) = r_0 - \sum_{i=1}^{L} \pi_i \varepsilon_{+i} (r_{0i} - w_{i-}),$$

where $w_{i-} = \min_j w_{ij}$.

Corollary. If $\varepsilon_{+i} = \varepsilon_+$ is independent of i, then

$$r_+(d_0) = r_0 + \varepsilon_+(\overline{w}_+ - r_0), \quad r_-(d_0) = r_0 - \varepsilon_+(r_0 - \overline{w}_-),$$

where $\overline{w}_+ = \sum_{i=1}^{L} \pi_i w_{i+}$.

In particular, if $w_{ij} = 1 - \delta_{ij}$ (δ_{ij} is Kroneker's symbol), then the probability of classification error belongs to the interval:

$$r_0 - \varepsilon_+ r_0 \le r(d_0; \{p_i\}) \le r_0 + \varepsilon_+(1 - r_0).$$

Let

$$g_i(x) = -\sum_{j=1}^{L} \pi_j \varepsilon_{+j} w_{ji} p_j^0(x)$$

$$f_i(x) = f_i^0(x) + g_i(x), \quad f_{ki}(x) = f_k(x) - f_i(x).$$

Theorem 2. For the distortions (1), (2) the robust decision rule is given by

$$d = d_*(x) = \sum_{i=1}^{L} i\, I_{V_i^*}(x), \quad I_{V_i^*}(x) = \prod_{k=1, k\neq i}^{L} U(f_{ki}(x)). \quad (4)$$

Corollary. In the situation with equal contamination levels ($\varepsilon_{+i} = \varepsilon_+$, $i = \overline{1, L}$) the BDR is the robust decision rule.

We derive now the asymptotic expansion for $r_+(d_*)$ and coompare it with $r_+(d_0)$. Let ∇ be the gradient on $x \in R^N$; "T" is the transposition symbol; $\varepsilon_+ = \max_i \varepsilon_{+i}$.

$$A_{ijk}(x) = |\nabla f_{ij}^0(x)|^2 |\nabla f_{kj}^0(x)|^2 - (\nabla^T f_{ij}^0(x) \nabla f_{kj}^0(x))^2;$$

$$\Gamma_{ij} = \{x: f_{ij}^0(x) = 0\}$$

is the discriminant surface in R^N for the two classes Ω_i^0, Ω_j^0, $(i \neq j)$; Γ'_{ij} is that "part" of the surface Γ_{ij} which is the border for V_j^0.

Theorem 3. If $\{p_i^0(x)\}$ are differentiable and surface integrals

$$a_{lmij} = \int_{\Gamma'_{ij}} p_l^0(x) p_m^0(x) |\nabla f_{ij}^0(x)|^{-1} d\sigma_{N-1},$$

$$b_{lmijk} = \int_{\Gamma'_{ij} \cap \Gamma'_{kj}} f_i^0(x) p_l^0(x) p_m^0(x) (A_{ijk}(x))^{-1} d\sigma_{N-2}$$

are finite, then the guaranteed risk for RDR $d_*(\cdot)$ allows the asymptotic expansion

$$r_+(d_*) = r_+(d_0) - \sum_{l,m=1}^{L} \varrho_{lm} \pi_l \pi_m \varepsilon_{+l} \varepsilon_{+m} + 0(\varepsilon_+^3), \quad (5)$$

where

$$\varrho_{lm} = \sum_{i=1}^{L-1} \sum_{j=i+1}^{L} \left((w_{lj} - w_{li})(w_{mj} - w_{mi}) \left(a_{lmij}/2 - \sum_{k=j+1}^{L} b_{lmijk} \right) - \sum_{k=j+1}^{L} (w_{lj} - w_{li})(w_{mk} - w_{mi}) b_{lmijk} \right).$$

The total proof of this theorem is given in Kharin (1982). Note, that in the case with $L=2$ classes $b_{lmijk} = 0$ and the computation of expansion coefficients in (5) becomes easier.

The practical value of theorems 1, 3 lies in the fact, that (3), (5) generate the approximate formula (with the remainder term $0(\varepsilon_+^3)$) for RDR's guaranteed risk computation.

4. Asymptotic Robustness of DR for Additive Distortions of Observations

Let the observation model have the form:

$$X_i = X_i^0 + \varepsilon_{+i} Y_i, \quad i = \overline{1, L}. \quad (6)$$

Here $X_i^0 \in R^N$ is the distortionless random vector with known thrice differentiable density $p_i^0(x)$; $Y_i \in R^N$ is the random vector of distortions with unknown density $h_i(\cdot) \in H_i$; X_i^0, Y_i are independent. Let H_i be the family of densities having the moments for third order inclusive, the first and second order moments being fixed:

$$H_i = \left\{ h_i(\cdot): h_i(y) \ge 0, \int_{R^N} h_i(y)dy = 1, \ E\{Y_i\} = \mu_i, \quad \operatorname{Cov}\{Y_i, Y_i\} = \Sigma_i \right\}.$$

where $\mu_i \in R^N$ is fixed mean and Σ_i is fixed covariance matrix. The model (6) has the following interpretation. The study of the classes $\{\Omega_i\}$ (including the estimation of $\{p_i^0(\cdot)\}$) was conducted for "ideal" conditions (without distortions), but real observations are corrupted by random noises $\{y_i\}$, the statistical properties of which are partly known.

Because of additivity of the model (6) the family of densities in (1) takes the form:

$$P_i(\varepsilon_{+i}) = \left\{ p_i(\cdot): p_i(x) = \int_{R^N} p_i^0(x - \varepsilon_{+i} y) h_i(y) dy,\ h_i(\cdot) \in H_i \right\}.$$

By analogy with the section 3 we investigate the asymptotic robustness of DR for this model.

Let

$$\alpha_i = -\sum_{j=1}^{L} \int_{V_j^0} p_i^0(x) dx \text{ — column-vector},$$

$$\beta_i = \sum_{j=1}^{L} w_{ij} \int_{V_j^0} \nabla^2 p_i^0(x) dx - (N \times N) \text{ — matrix}.$$

Theorem 4. For the model (6) of distortions the guaranteed risk value allows the asymptotic expansion:

$$r_+(d_0) = r_c + \sum_{i=1}^{L} \varepsilon_{+i} \pi_i \left(\mu_i^T \alpha_i + \varepsilon_{+i} \left(\mu_i^T \beta_i \mu_i + tr(\Sigma_i \beta_i) \right) / 2 \right) + 0(\varepsilon_+^3)$$

Theorem 5. With the remainder $0(\varepsilon_+^3)$ the RDR for the observations with additive distortions has the form (4), where

$$f_j(x) = f_j^0(x) + \sum_{l=1}^{L} \pi_l w_{lj} q_l(x).$$

$$q_j(x) = \varepsilon_{+j} \left(-\mu_j^T \nabla p_j^0(x) + \varepsilon_{+j} \left(\mu_j^T \nabla^2 p_j^0(x) \mu_j + tr\left(\Sigma_j \nabla^2 p_j^0(x)\right) \right) / 2 \right)$$

The asymptotic expansion for the guaranteed risk value of this RDR is defined by (5), where the expansion coefficients are evaluated by the formulas:

5. Robust Classification of Gaussian Observations with Distortions

In statistical classifcation problems the Gaussian model is usually used as a distortionsless model for $\{\Omega_i^0\}$. For this situation we illustrate here the results of the section 3. Let us have $L = 2$ classes and $p_i(x) = n_N(x|a_i, B)$ is the given N-dimensional normal density of an observation from Ω_i^0; $a_i \in R^N$ — the mean, B — non-singular covariance matrix; $w_{ij} = 1 - \delta_{ij}$ (because of this the risk is the unconditional probability of error). It is known that the BDR is linear in the described situation:

$$d = d_0(x) = U(b^T x - \gamma_0) + 1, \quad b = B^{-1}(a_2 - a_1)$$

$$\gamma_0 = (a_2 + a_1)^T B^{-1} (a_2 - a_1)/2 + \ln(\pi_1/\pi_2)$$

and the error probability is equal to

$$r_0 = 1 - \sum_{i=1}^{2} \pi_i \Phi\left(\Delta/2 - (-1)^i \Delta^{-1} \ln(\pi_1/\pi_2)\right).$$

where $\Phi(\cdot)$ is the standard normal distribution function and $\Delta = \sqrt{(a_2 - a_1)^T B^{-1} (a_2 - a_1)}$ is Mahalanobis distance. By theorem 1

$$r_+(d_0) = r_0 + \sum_{i=1}^{2} \pi_i \varepsilon_{+i} \Phi\left(\Delta/2 - (-1)^i \Delta^{-1} \ln(\pi_1/\pi_2)\right).$$

Using theorems 2, 3 we find RDR and its guaranteed risk:

$$d = d_*(x) = U(b^T x - \gamma_*) + 1, \quad \gamma_* = \gamma_0 + \ln \frac{1 - \varepsilon_{+1}}{1 - \varepsilon_{+2}}.$$

$$r_+(d_*) = r_+(d_0) - \frac{\sqrt{\pi_1(1 - \pi_1)}\,(\varepsilon_{+2} - \varepsilon_{+1})^2}{2\sqrt{2\pi}\,\Delta \exp\left(\frac{\Delta^2}{8} + \left(\ln^2(\pi_1/\pi_2)\right)/(2\Delta^2)\right)} + 0(\varepsilon_+^3)$$

In Kharin (1982) the comparative analysis of $r'_+(d_*)$ and $r_+(d_0)$ is made and the conditions of RDR's essential superiority are established.

$$\varrho_{lm} = \mu_l^T u_{lm} \mu_m,$$

$$u_{lm} = \sum_{i=1}^{L-1} \sum_{j=i+1}^{L} \left(\frac{(w_{lj} - w_{li})(w_{mj} - w_{mi})}{2} \int_{\Gamma'_{ij}} \nabla p_l^0(x) \nabla^T p_m^0(x) \left| \nabla f_{ij}^0(x) \right|^{-1} d\sigma_{N-1} \right.$$

$$+ \sum_{k=j+1}^{L} \left((w_{lj} - w_{li})(w_{mj} - w_{mi}) + (w_{lk} - w_{li})(w_{mk} - w_{mj}) \right)$$

$$\left. \times \int_{\Gamma'_{ij} \cap \Gamma'_{kj}} f_i^0(x) \nabla p_l^0(x) \nabla^T p_m^0(x) \left(\Lambda_{ijk}(x)\right)^{-1} d\sigma_{N-2} \right).$$

6. References

BERGER, R. L.
Gamma minimax robustness of Bayes rules.
Commun. Statist.-Theor. Meth. A 8 (1979) 543–560.

HUBER, P. J.
A robust version of the probability ratio test.
Ann. Math. Stat., 36 (1965) 1753–1758.

KADANE, J. B., CHUANG, D. T.
Stable decision problems.
Ann. of Stat., 6 (1978), 1059–1110.

KHARIN, Yu. S.
About the statistical classification efficiency by using minimum contrast estimators.
Probab. Theory and Appl., XXVI (1981) 866–867.

KHARIN, Yu. S.
Stability of decision rules in pattern recognition problems.
Automatika and Remote Control, II (1982) 115–123.

RANDLES, R. H.
Generalized linear and quadratic discriminant functions using robust estimates.
JASA, 73 (1978) 564–568.

REY, W. J. J.
Robust statistical methods.
Lect. not in Math., 690 (1978).

TUKEY, J. W.
A survey of sampling from contaminated distributions.
In: Contrib. to Prob. and Statist., Stanford, 1960.

Agricultural University, Department of Mathematics, Wageningen, The Netherlands
Nederlandse Philips Bedrijven B. V., Centre of Quantitative Methods, Eindhoven, The Netherlands

Ranks, Standardized Ranks and Standardized Aligned Ranks in the Analysis of Friedman's Block Design

PAUL VAN DER LAAN, JOS DE KROON

Abstract

In this paper the use of ranks, standardized ranks and standardized aligned ranks in the analysis of Friedman's block design is elucidated. Block designs with unequal numbers of observations per cell are considered with the restriction that the design is orthogonal.

1. Introduction

A randomized block design with I blocks and J treatments and with m_{ij} (>0) observations per cell is considered. A cell is an intersection of a block and a treatment. We assume that the observations have continuous cumulative distribution functions and that the observations in different blocks are independent. In this paper we restrict ourselves to *orthogonal* designs, i.e. the number m_{ij} ($i=1, \ldots, I$ and $j=1, \ldots, J$) of observations in cell (i, j) is equal to $m_{i.} * m_{.j}/m_{..}$ where $m_{i.} = \sum_j m_{ij}$, $m_{.j} = \sum_i m_{ij}$ and $m_{..} = \sum_j \sum_j m_{ij}$ following the familiar dot notation.

In the following figure the foregoing is summarized.

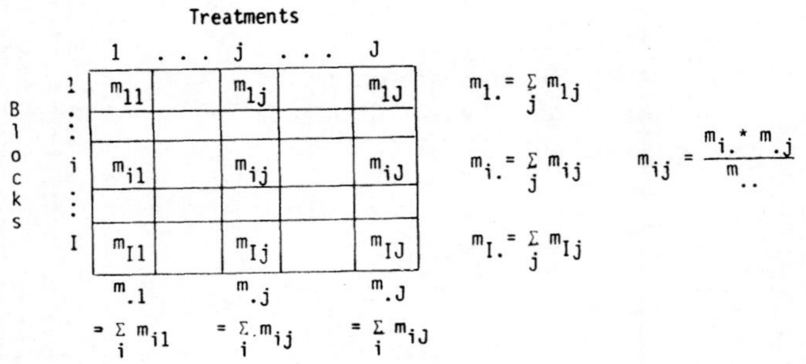

Figure 1.1.
An orthogonal randomized block design; m_{ij} represents the number of observations in cell (i, j).

We want to test the null hypothesis

H_0: no treatment effect

against shift alternatives of the form $F_{ij}(z) = F_i(x - \Delta_j)$ where F_{ij} (.) is the continuous distribution function from which the observations in the i-th block and under treatment j are drawn ($i = 1, \ldots, I$ and $j = 1, \ldots, J$) and Δ_j is a shift parameter.

2. The Use of Ranks

For testing the null hypothesis in the case that the number m_{ij} of observations per cell is equal to 1 ($i = 1, \ldots, I$; $j = 1, \ldots, J$), the distribution-free test of Friedman using *ranks* can be applied (Conover (1971)). The observations within a block are ranked in increasing order of magnitude and ranks $1, 2, \ldots, m_i = J$ are allocated ($i = 1, \ldots, I$). The test statistic Q of Friedman is defined as follows

$$Q = 12\{J(J+1)\}^{-1} \sum_i \sum_j (\bar{r}_{.j} - \bar{r}_{..})^2,$$

where r_{ij} ($i = 1, \ldots, I$; $j = 1, \ldots, J$) is the rank of the observation in cell (i, j), the combination of block i and treatment j.

Furthermore, $\bar{r}_{.j} = I^{-1} \sum_i r_{ij}$ and $\bar{r}_{..} = (IJ)^{-1} \sum_i \sum_j r_{ij}$ Tables can be found in Owen (1962) and Obeh (1977). The test statistic Q has under the null hypothesis of no The test statistic Q has under the null hypothesis of no treatment effect, asymptotically for I tends to infinity, a chi-square distribution with J-1 degrees of freedom (Hájek and Šidák (1967)).

In the case of an equal number m of observations per cell one can apply the test of Friedman for m observations per cell as we call it, with test statistic

$$Q_m = 12\{mJ(mJ+1)\}^{-1} \sum_i \sum_j \sum_l (\bar{r}_{.j.} - \bar{r}_{...})^2$$

where r_{ijl} is the rank of the l-th observation ($l = 1, \ldots, m$) in the i-th block and under treatment j ($i = 1, \ldots, I$; $j = 1, \ldots, J$), ranking each block separately, $\bar{r}_{.j.} = (Im)^{-1} \sum_i \sum_l r_{ijl}$ and $\bar{r}_{...} = (IJm)^{-1} \sum_i \sum_i \sum_l r_{ijl}$ (Conover (1071)). It is well-known that Q_m has under H_0 asymptotically, as mI tends to infinity, a chi-square distribution with J-1 degrees of freedom. In table 2.1 some critical and "almost" critical values of Q_m can be found for the levels .001; .01; .05 and .10, respectively.

Table 2.1.
Critical and "almost" critical values of Q_m for the levels .001; .01; .05 and .10, respectively.
The left column gives values C and the right column the values $P = P[Q_m \geq C]$.

J=2, m=2, I=2	
4.80	.056
2.70	.167
J=2, m=2, I=3	
7.20	.009
5.00	.037
3.20	.120
J=2, m=2, I=4	
9.60	.0015
7.35	.008
5.40	.029
3.75	.079
2.40	.179
J=2, m=2, I=5	
12.00	.0003
9.72	.0015
7.68	.007
5.88	.021
4.32	.054
3.00	.119
J=2, m=2, I=6	
12.10	.0003
10.00	.0015
8.10	.005
6.40	.015
4.90	.037
3.60	.081
2.50	.156
J=2, m=2, I=7	
12.34	.0003
10.37	.0012
8.57	.004
6.94	.011
5.49	.026
4.20	.055
3.09	.107
J=2, m=2, I=8	
10.80	.0010
9.08	.0029
7.50	.008
6.08	.018
4.80	.038
3.68	.074
2.70	.132
J=2, m=2, I=9	
11.27	.0007
9.60	.0021
8.07	.005
6.67	.013
5.40	.026

4.27	.051
3.27	.093
2.40	.156

J = 2, m = 2, I = 10	
11.76	.0006
10.14	.0016
7.26	.009
6.00	.018
4.86	.036
3.84	.065
2.94	.111

J = 2, m = 3, I = 2	
7.71	.0050
6.10	.015
4.67	.040
3.43	.090
2.38	.170

J = 2, m = 3, I = 3	
9.92	.0010
8.40	.0032
7.00	.009
5.73	.020
4.59	.041
3.57	.077
2.68	.132

J = 2, m = 3, I = 4	
10.71	.0007
9.33	.0020
6.86	.010
5.76	.020
4.76	.036
3.86	.062
3.05	.102

J = 2, m = 3, I = 5	
11.67	.0004
10.37	.0011
6.94	.009
5.95	.017
4.20*	.050
3.44	.078
2.75	.119

J = 2, m = 3, I = 6	
11.46	.0006
10.29	.0012
7.14	.008
6.22	.015
4.57	.039
3.84	.060
3.17	.090
2.57	.131

J = 2, m = 4, I = 2	
10.67	.0004
9.38	.0012
7.04	.007
6.00	.016
5.04	.029
4.17	.050
3.38	.082
2.67	.128

J = 2, m = 4, I = 3	
11.11	.0005
10.03	.0011
7.11	.008
6.25	.013
4.69	.035
4.00	.054
2.78	.114

J = 2, m = 4, I = 4	
11.02	.0007
10.08	.0012
7.52	.006
6.75	.010
4.69	.035
4.08	.050
3.00	.097
2.52	.131

J = 3, m = 2, I = 2	
9.14	.0007
8.14	.0037
7.43	.011
6.14	.041
5.57	.056
5.14	.086
4.43	.116

J = 3, m = 2, I = 3	
10.67	.0010
10.57	.0014
8.67	.009
8.00	.012
6.00	.047
5.81	.056
4.95	.082
4.67	.107

J = 3, m = 2, 1 = 4	
12.07	.0009
11.64	.0011
8.64	.009
8.36	.011
5.79	.050
5.64	.057
4.57	.097
4.50	.110

J = 3, m = 2, I = 5	
12.40	.0010
12.06	.0011
8.63	.010
8.46	.011
6.17	.046
5.89	.052
4.63	.096
4.51	.106

J = 3, m = 2, I = 6	
12.76	.0009
12.33	.0011
9.00	.009
8.71	.010
6.05	.049
5.90	.053
4.62	.100
4.43	.109

J = 3, m = 2, I = 7	
12.61	.0010
12.53	.0011
8.94	.009
8.86	.011
6.04	.047
6.00	.052
4.57	.099
4.53	.106

J = 3, m = 2, I = 8	
12.96	.0010
12.89	.0011
9.00	.009
8.82	.011
6.11	.047
6.04	.051
4.75	.095
4.61	.101

J = 3, m = 2, I = 9	
12.98	.0009
12.79	.0011
8.86	.010
8.79	.011
6.10	.048
6.00	.050
4.70	.095
4.67	.103

J = 3, m = 3, I = 2	
10.98	.0009
10.48	.0010
8.40	.009
8.13	.011
5.91	.048
5.73	.052
4.80	.095
4.58	.103

J = 3, m = 3, I = 3	
11.94	.0010
11.85	.0010
8.62	.010
8.56	.010
5.90	.048
5.81	.052
4.65	.095
4.62	.100

J = 3, m = 3, I = 4	
12.42	.0010
12.36	.0010
8.87	.010
8.82	.010
6.02	.048
5.96	.050
4.69	.097
4.62	.101

J = 3, m = 3, I = 5	
12.85	.0009
12.82	.0010
8.87	.010
8.82	.010
5.97	.049
5.92	.050
4.76	.095
4.60	.101

J = 4, m = 2, I = 2	
11.25	.0009
11.08	.0013
9.33	.009
9.25	.010
7.33	.046
7.25	.051
6.17	.094
6.08	.106

When the numbers m_{ij} of observations per cell are not equal, it is possible to apply the test of Benard and Van Elteren (1953). This test is suitable for general designs. For orthogonal designs the test statistic has the following form

$$Q_0 = 12N\left\{\sum_i m_{i.}^2 (m_{i.}+1)\right\}^{-1} \sum_j m_{.j}^{-1}\left\{r_{.j} - \frac{1}{2}\sum_i m_{ij}(m_{i.}+1)\right\}^2 .$$

* Critical at the .05 level.

The test statistic Q_0 has under the null hypothesis asymptotically, as I tends to infinity, a chi-square distribution with J-1 degrees of freedom (Benard and Van Elteren (1953)). If $m_{ij} = 1$ $(i = 1, \ldots, I;\ j = 1, \ldots, J)$ this test statistic is equivalent with Friedman's test statistic and if $m_{ij} = m$ it is easy to show that this test statistic is identical to the test statistic of Friedman's test for m observations per cell.

3. The Use of Standardized Ranks

When the numbers of observations per block are unequal, the test of Benard and Van Elteren presents difficulties. Namely, the level of the transposed response variable, the rank, depends on the number of observations in the corresponding block and therefore differs from block to block in the case that the numbers of observations per block are unequal. This objection can be avoided by using a natural generalization of Friedman's rank statistic. For this generalization of Friedman's test statistic *standardized ranks* $\tilde{r}_{ijl}$ $(i = 1, \ldots, I;\ j = 1, \ldots, J$ and $l = 1, \ldots, m_{ij})$ are defined as follows

$$\tilde{r}_{ijl} = \left\{r_{ijl} - \frac{1}{2}(m_{i.}+1)\right\}\left\{\frac{1}{12} m_{i.}(m_{i.}+1)\right\}^{-1/2}$$

Under the null hypothesis these standardized ranks have expectation zero and asymptotically, as m_i tends to infinity for $i = 1, \ldots, I$, unit variance. The "standardization factors" are derived by De Kroon and Van der Laan (1983 a). They also showed that the test statistic (for orthogonal designs)

$$Q = \sum_j m_{.j}(\tilde{r}_{.j.})^2$$

for testing against treatment effect, has under the null hypothesis asymptotically, as m_i tends to infinity for all i, a chi-square distribution with J–1 degrees of freedom. If the numbers of observations per block are equal they also showed that the proposed test statistic is equivalent with the test statistik of Benard and Van Elteren.

We shall now give a simple example in which the difference of the test procedure of Benard and Van Elteren and the proposed test procedure can be illustrated. The results of an experiment with four treatments and observations in six blocks are given. Ranking the observations per block gives the following rank results:

		Treatments 1	2	3	4	
Blocks	1	1	2	3	4	$m_{1.} = 4$
	2	1	2	3	4	$m_{2.} = 4$
	3	1	2	3	4	$m_{3.} = 4$
	4	1	2	3	4	$m_{1.} = 4$
	5	1	2	3	4	$m_{5.} = 4$
	6	3 6 8	4 5 12	1 7 10	2 9 11	$m_{6.} = 4$
Total rank sum		22	31	33	42	

We find for the test statistic of Benard and Van Elteren $Q_0 = 4.27$ and this is below the five percent critical point 7.815.

The results of the sixth block neutralize the very concordant results of the first five blocks. For the proposed test statistic based on standardized ranks we find

$\tilde{Q} = 11.15.$

This result is significant at the five percent level. The example can only be considered as an illustration for the difference of both test statistics. To obtain a better based judge it is important to make a comparison of the power functions of both tests. In table 3.1. the simulation results, based on 2 000 samples from Normal parent distributions, are presented for the following orthogonal design where the numbers of observations per cell are indicated:

		Treatments		
		1	2	3
Blocks	1	12	12	12
	2	2	2	2
	3	1	1	1

The cases considered are the null hypothesis H_0 and the alternative $\beta_j + \chi$, where χ is the standard Normal variate and β_j is the mean of treatment j (j = 1, 2, 3), with $\beta_1 = 0$, $\beta_2 = \frac{1}{2}$ and $\beta_3 = 1$.

The results for different critical values are dependent since the same samples were used to estimate the powers. The same can be said for the two cases. To get an efficient comparison the same samples were used to estimate the powers of both tests. For this design one can conclude that the power of the new test based on standardized ranks is better than the power of the test of Benard and Van Elteren, as may be expected. More Monte Carlo results can be found in De Kroon and Van der Laan (1983 a).

4. The Use of Standardized Aligned Ranks

For all these tests there is a separate ranking in each block. Thus comparisons of the reponses take place only within each block. Blocks can be made comparable by subtracting from the observations an estimate of the location of the block, for instance the median of the observations in the block. In this way we get m differences. This method of making blocks comparable is denoted by the term "aligning". We rank the m differences in increasing order of magnitude with ranks 1, 2, . . . , m . The ranks are called aligned ranks and denoted by $r^*_{ij:}$. Now we can determine *standardized aligned ranks.*

$$\tilde{r}^*_{ijl} = \left(r^*_{ijl} - \bar{r}^*_{i..}\right)\left\{m_{i.}^{-1} \sum_j \sum_l \left(r^*_{ijl} - \bar{r}^*_{i..}\right)^2\right\}^{-1/2}$$

$(i = 1, \ldots, I;\ j = 1, \ldots, J \text{ and } l = 1, \ldots, m_{ij}).$

These ranks have under the null hypothesis of no treatment effect and given ranks in the blocks, expectation zero and symptotically, as m tends to infinity for i = 1, . . . , I, unit variance. De Kroon and Van der Laan (1983 b) present a test based on these standardized aligned ranks. They also compare the power functions of the various tests. These comparisons are based on simulation experiments.

5 A Concluding Remark

From the results of the various simulation experiments which can be found in the papers of De Kroon and Van der Laan (1983 a and b) one can draw the following tentative conclusion. Standardized ranks and standardized aligned ranks may provide an improvement in power, compared with ordinary ranks, for the problem considered in this paper. Not much gain, if any, seems to be obtained by using standardized aligned ranks instead of standardized ranks.

Table 3.1.
Estimated powers of the test of Benard and Van Elteren (BE) and the test based on standardized ranks (SRT)

X	$P\left[\chi^2_2 > X\right]$	H_0 BE	H_0 SRT	BE	SRT	X	$P\left[\chi^2_2 > X\right]$	H_0 BE	H_0 SRT	BE	SRT
1.00	.607	.622	.617	.966	.967	5.80	.055	.050	.058	.543	.606
1.40	.497	.508	.505	.941	.955	6.20	.045	.039	.047	.502	.573
1.80	.407	.409	.413	.913	.932	6.60	.037	.034	.038	.464	.535
2.20	.333	.342	.336	.885	.906	7.00	.030	.026	.029	.430	.496
2.60	.273	.280	.269	.853	.885	7.40	.025	.023	.022	.397	.464
3.00	.223	.222	.225	.818	.851	7.80	.020	.018	.018	.363	.425
3.40	.183	.180	.182	.782	.825	8.20	.017	.015	.013	.328	.394
3.80	.150	.152	.149	.748	.793	8.60	.014	.011	.012	.295	.366
4.20	.122	.116	.116	.709	.760	9.00	.011	.010	.009	.264	.336
4.60	.100	.094	.095	.665	.721	9.40	.009	.007	.009	.236	.307
5.00	.082	.078	.078	.634	.689	9.80	.007	.006	.008	,209	.276
5.40	.067	.063	.066	.587	.649	10.00	.007	.004	.006	.200	.260

6. References

BENARD, A. and VAN ELTEREN, Ph.
A generalization of the method of m rankings.
Koninklijke Akademie van Wetenschappen, Amsterdam Proceed. Series A 56 (1953) 358–369.

CONOVER, W. J.
Practical nonparametric statistics.
John Wiley & Sons, New York – London, 1971.

DE KROON, J. and VAN DER LAAN, P.
A generalization of Friedman's rank statistic.
Statistica Neerlandica 37 (1983a) 1–14.

DE KROON, J. and VAN DER LAAN, P.
A comparison of the powers of a generalized Friedman's test and an aligned rank procedure based on simulation.
To be submitted for publication in Statistica Neerlandica (1983b).

HÁJEK, J. and ŠIDÁK, Z.
Theory of rank tests.
Academic Press, New York (1967).

ODEH, R. E.
Extended tables of the distribution of Friedman's S-statistic in the two-way layout.
Commun. Statist.-Simul. Comput. 86 (1977) 29–48.

OWEN, D. B.
Handbook of statistical tables.
Addison-Wesley Publ. Comp., Inc., London, 1962.

PEARSON, E. S. and HARTLEY, H. O.
Biometrika tables for statisticians, Volume I.
Cambridge University Press, 1954.

Sektion Mathematik, Friedrich-Schiller-Universität Jena, Jena, DDR

Chernoff Type Bounds of Errors in Hypothesis Testing of Diffusion Processes

FRIEDRICH LIESE

Abstract

A hypothesis testing problem of discriminating between two diffusion processes $\xi_i(t)$, $0 \leq d\xi_i = a_i(t, \xi_i)\,dt + dW(t)$, $0 \leq t \leq T$, is considered. The errors of first and second kind can be estimated in terms of Hellinger integrals. In the present paper upper bounds for the Hellinger integrals are obtained.

1. Introduction

Let $[\Omega, \mathfrak{F}]$ be a measurable space, P, Q probability measures on $[\Omega, \mathfrak{F}]$ and R a σ-finite measure dominating P, Q. The functional

$$H_s(P,Q) = \int \left(\frac{dP}{dR}\right)^s \left(\frac{dQ}{dR}\right)^{1-s} dR, \quad 0<s<1,$$

is called the Hellinger integral of order s. H_s plays an important role in probability theory in treating the problem of absolute continuity Nemetz (1974) and Liese (1976), in information theory Gallager (1968) and in statistics Chernoff (1952), Evans (1974), Hibey, Snyder, van Schuppen (1978) and Österreicher (1978). In the present paper we make use of the relation to the errors of first and second kind in hypothesis testing problem if both the hypothesis $H_0 : P$ and the alternative $H_1 : Q$ are simple.

There is a measurable partition C_1, C_2, C_3 of Ω, $C_i \in \mathfrak{F}$, $C_i \cap C_j = \emptyset$ $i \neq j$, $C_1 \cup C_2 \cup C_3 = \Omega$ with $P(C_1) = Q(C_3) = 0$ and $P(\cdot \cap C_2) \sim Q(\cdot \cap C_2)$ where $\sim$ denotes the measure theoretical equivalence. Put $Z = \frac{dP(\cdot \cap C_2)}{dQ(\cdot \cap C_2)}$.

Given a level L the likelihood ratio test is defined by the critical region

$$K = C_1 \cup (C_2 \cap \{Z \leq L\}).$$

The errors of first and second kind are given by $\alpha = P(K)$, $\beta = Q(\overline{K})$, respectively. Upper bounds for these errors in terms of Hellinger integrals are due to Chernoff (1952)

$$\alpha \leq \inf_{0<s<1} L^s H_{1-s}(P,Q), \quad \beta \leq \inf_{0<s<1} L^{-s} H_s(P,Q). \tag{1}$$

In this way the estimation of the errors of first and second kind leads to the estimation of H_s. In the present paper we investigate the Hellinger integrals of the distribution laws P_{ξ_i}, of diffusion processes

Chernoff bounds are investigated by many authors. The common method due to Evans (1974) is to introduce a new process $\eta(t)$ such that $H_s(P_{\xi_1}, P_{\xi_2})$ becomes a functional of the type $M \exp\left\{-\int_0^t V(\eta(s))\,ds\right\}$.

Then a differential equations for this function is derived. But this differential equation can not be solved in general in a closed form.

In contradiction to this method we aim to estimate $H_s(P_{\xi_1}, P_{\xi_2})$ by simpler and closed analytic expressions using only the processes ξ_i and the functionals a_i

2. Results

Let $[\Omega, \mathfrak{F}, P]$ be a complete probability space, $(\mathfrak{F}_t)_{t\geq 0}$ a nondecreasing family of sub-σ-algebras completed with the zero sets of P. Assume that $(W(t), \mathfrak{F}_t)$ is a Wiener process, i. e. $(W(t), \mathfrak{F}_t)$ is a continuous martingale with $W(0) = 0$ a. s. and

$$M((W(t) - W(s))^2 | \mathfrak{F}_s) = t - s, \qquad 0 \leqq s \leqq t < \infty.$$

Denote by C_T the space of all continuous functions $x(t)$, $0 \leq t \leq T$, with $x(0) = 0$ (if $T = \infty$ then only $0 \leq t < \infty$ is assumed). Denote by $\mathfrak{L}_t$ the σ-algebra of subsets of C_T generated by the projections up to t, $\mathfrak{L}_t = (x(s), 0 \leq s \leq t)$. Given a continuous stochastic process $\xi(t)$, $0 \leq t \leq T$ on $[\Omega, \mathfrak{F}, P]$ the distribution law which is defined on $[C_T, \mathfrak{L}_T]$ is denoted by P_ξ · $\mathfrak{R}_T$ denotes the τ-algebra of Borel sets of $[0, T]$. Assume that $a_i(t, x)$, $i = 1, 2$, $0 \leq t \leq T$, $x \in C_T$, are $\mathfrak{R}_T \otimes \mathfrak{L}_T$ measurable and that $a_i(t, \cdot)$ is $\mathfrak{L}_t$-adapted, i. e. $a_i(t, \cdot)$ is $\mathfrak{L}_t$-measurable for every fixed $0 \leq t \leq T$.

THEOREM Assume that $\xi_i(t)$, $i = 1, 2$, $0 \leq t \leq T < \infty$ are $\mathfrak{F}_t$-adapted stochastic process with

$$P\left(\left\{\int_0^T a_i^2(t, \xi_i)\,dt < \infty\right\}\right) = P\left(\left\{\int_0^T a_i^2(t, W)\,dt < \infty\right\}\right) = 1 \tag{2}$$

$$\xi_i(t) = \int_0^t a_i(u, \xi_i)\,du + W(t), \quad 0 \leq t \leq T \tag{3}$$

then

$$H_{s_1}(P_{\xi_1}, P_{\xi_2}) \leq \inf_{s_2 : s_1 < s_2 < 1} \left[M D_{s_2}(\xi_2)^{\frac{s_1}{s_2 - s_1}} \right]^{\frac{s_2 - s_1}{s_2}} \tag{4}$$

where

$$D_s(x) = \exp\left\{-\frac{1}{2} s(1-s) \int_0^T (a_1(t,x) - a_2(t,x))^2\,dt\right\}. \tag{5}$$

Corollary 1 Assume that $\xi_i(t)$, $0 \leq t < \infty$, $i = 1, 2$, are stochastic processes with (2), (3) for every $T < \infty$, then (4) is valid with $D_s(x)$ from (5), where $T = \infty$.

Corollary 2 Assume that ξ_i are from the theorem or from corollary 1 and denote by α, β the errors of first and second kind in the likelihood ratio test with given level L, then

$$\alpha \leq \inf_{0<s_1<s_2<1} L^{s_1} \left[M\left(D_{s_2}(\xi_1)\right)^{\frac{s_1}{s_2 - s_1}} \right]^{\frac{s_2 - s_1}{s_2}}$$

$$\beta \leq \inf_{0<s_1<s_2<1} L^{-s_1} \left[M\left(D_{s_2}(\xi_2)\right)^{\frac{s_1}{s_2 - s_1}} \right]^{\frac{s_2 - s_1}{s_2}}.$$

Corollary 3 Assume that ξ_i are choosen as in the preceding corollary, then

$$\alpha \leq L^{\frac{1}{4}}\left[M\exp\left\{-\frac{1}{8}\int_0^T \left(a_1(t,\xi_1)-a_2(t,\xi_1)\right)^2 dt\right\}\right]^{\frac{1}{2}}$$

$$\beta \leq L^{-\frac{1}{4}}\left[M\exp\left\{-\frac{1}{8}\int_0^T \left(a_1(t,\xi_2)-a_2(t,\xi_2)\right)^2 dt\right\}\right]^{\frac{1}{2}}.$$

In order to investigate the quality of the bounds of the errors of first and second kind given above we consider an example.

Assume that $a(t)$, $0 \leq t < T$ is a real measurable function with $\int_0^T a^2(t)dt < \infty$. Put $a_1(t, x) = a(t)$, $a_2(t, x) = 0$, then $\xi_1(t) = \int_0^t a(u)du + W(t)$ and $\xi_2(t) = W(t)$. Furthermore $P_{\xi_1} \sim P_{\xi_2}$ and

$$\frac{dP_{\xi_1}}{dP_{\xi_2}}(W) = \exp\left\{\int_0^T a(t)dW(t) - \frac{1}{2}\int_0^T a^2(t)dt\right\}$$

holds Lipster, Shiryayev (1978). The random variable $\ln Z$ where $Z = \frac{dP_{\xi_1}}{dP_{\xi_2}}$, considered on the probability space $[C_T, \mathfrak{L}_T, P_{\xi_2}]$ is a normally distributed one with mean $-\frac{1}{2}\int_0^T a^2(t)dt$ and variance $\int_0^T a^2(t)dt$. Put $A = \int_0^T a^2(t)dt$ and denote by $\Phi(x)$ the distribution function of the standard normal distribution. Given a level L we get

$$\beta = P_{\xi_2}(\{Z > L\})$$

$$= P_{\xi_2}\left(\left\{\frac{\ln Z + \frac{1}{2}A}{\sqrt{A}} > \frac{\ln L + \frac{1}{2}A}{\sqrt{A}}\right\}\right)$$

$$= 1 - \Phi\left(\frac{\ln L + \frac{1}{2}A}{\sqrt{A}}\right) \tag{6}$$

Denote by M_i the expectation with respect to P_{ξ_i} then

$$M_1\exp\{it\ln Z\} = M_2\, Z\exp\{it\ln Z\} = M_2\exp\{(1+it)\ln Z\}$$

Since $\ln Z$ is normally distributed with mean $-\frac{1}{2}A$ and variance A

$$M_2\exp\{w\ln Z\} = \exp\left\{-\frac{1}{2}Aw + \frac{1}{2}Aw^2\right\}$$

holds for every complex w. Put $w = 1 + it$ then

$$M_1\exp\{it\ln Z\} = \exp\left\{it\frac{1}{2}A - \frac{1}{2}At^2\right\}$$

That means that $\ln Z$ considered on the probability space $[C_T, \mathfrak{L}_T, P_{\xi_1}]$ is normally distributed with mean $\frac{1}{2}A$ and variance A. Consequently the error of first kind is given by

$$\alpha = P_{\xi_1}(\{Z \leq L\}) = \Phi\left(\frac{\ln L - \frac{1}{2}A}{\sqrt{A}}\right) \tag{7}$$

We now assume that there is no reason to differ between the processes ξ_1 and ξ_2. That means that the error α is not worse than β and conversely. In such a case we choose L so that $\alpha = \beta$. (6) and (7) show that we have to take $L = 1$. Hence

$$\alpha(A) = \beta(A) = \Phi\left(-\frac{1}{2}\sqrt{A}\right) \tag{8}$$

Let us now compare the exact value of α with the bound appearing in corollary 2. First of all we observe that

$$D_{s_2}(\xi_1) = \exp\left\{-\frac{1}{2}s_2(1-s_2)\int_0^T a^2(t)dt\right\}$$

is deterministic. Therefore

$$\inf_{0<s_1<s_2<1} L^{s_1}\left[M\left(D_{s_2}(\xi_1)\right)^{\frac{s_1}{s_2-s_1}}\right]^{\frac{s_2-s_1}{s_2}}$$

$$= \inf_{0<s_1<s_2<1} \exp\left\{-\frac{1}{2}s_2(1-s_2)\frac{s_1}{s_2}\int_0^T a^2(t)dt\right\}$$

$$= \exp\left\{-\frac{1}{8}A\right\}$$

In order to compare this bound with the exact value $\Phi(-\frac{1}{2}\sqrt{A})$ we remark that

$$\Phi(-x) \sim \frac{1}{x}\varphi(x) = \frac{1}{\sqrt{2\pi}}\frac{1}{x}\exp\left\{-\frac{x^2}{2}\right\} \quad x \to \infty$$

and

$$\Phi\left(-\frac{1}{2}\sqrt{A}\right) \sim \sqrt{\frac{2}{\pi}}\frac{1}{\sqrt{A}}\exp\left\{-\frac{1}{8}A\right\} \quad A \to \infty$$

This asymptotic expression shows that the exact values for the errors tend more fast to zero than the bound in corollary 2 as $A \to \infty$. In the following table some values of the upper bound and of the exact error probability are collected.

A	$\exp\left\{-\frac{1}{8}A\right\}$	$\Phi\left(-\frac{1}{2}\sqrt{A}\right)$
1	0,8825	0,3085
3	0,6873	0,1932
5	0,5353	0,1318
10	0,2865	0,0569
15	0,1534	0,0264
20	0,0825	0,0127
25	0,0439	0,0062
30	0,0235	0,0031
35	0,0126	0,0016
40	0,0067	0,0008

3. **Proofs**

First of all we remark that $T < \infty$ and (2) imply

$$P\left(\left\{\int_0^T |a_i(t,\xi_i)|dt < \infty\right\}\right) = 1$$

such that the stochastic differential equations (3) make sense. In view of Lipster, Shiryayev (1978) the relation (2) implies $P_{\xi_i} \sim P_W$ and the Randon-Nikodym derivative $X_i(x) = \frac{dP_{\xi_i}}{dP_W}(x)$, $x \in C_T$ considered as a functional of the Wiener process has the form

$$X_i(W) = \exp\left\{-\frac{1}{2}\int_0^T a_i^2(t,W)dt + \int_0^T a_i(t,W)dW(t)\right\}$$

Hence

$$X_1^s(W)X_2^{1-s}(W) = \exp\left\{-\frac{1}{2}\int_0^T \left(sa_1^2(t,W)+(1-s)a_2^2(t,W)\right)dt + \int_0^T \left(sa_1(t,W)+(1-s)a_2(t,W)\right)dW(t)\right\} \tag{9}$$

$$= D_{s,T}(W)Z_{s,T}(W) \quad \text{where}$$

$$Z_{s,t}(W) = \exp\left\{-\frac{1}{2}\int_0^t \left(sa_1(u,W)+(1-s)a_2(u,W)\right)^2du + \int_0^t \left(sa_1(u,W)+(1-s)a_2(u,W)\right)dW(u)\right\} \tag{10}$$

and

$$D_{s,t}(W) = \exp\left\{-\frac{1}{2}s(1-s)\int_0^t \left(a_1(u,W)-a_2(u,W)\right)^2du\right\}. \tag{11}$$

Suppose now $0 < s_1 < s_2 < 1$. Then in view of (9)

$$X_1^{s_1}(W)X_2^{1-s_1}(W) = D_{s_2,T}^{\frac{s_1}{s_2}}(W)Z_{s_2,T}^{\frac{s_1}{s_2}}(W)X_2^{1-\frac{s_1}{s_2}}(W)$$

Consequently

$$H_{s_1}(P_{\xi_1},P_{\xi_2}) = \int X_1^{s_1}X_2^{1-s_1}dP_W = \int Z_{s_2,T}^{\frac{s_1}{s_2}}D_{s_2,T}^{\frac{s_1}{s_2}}X_2^{1-\frac{s_1}{s_2}}dP_W$$

Applying now Hölder's inequality with $p = \frac{s_2}{s_1}$, $q = \frac{s_2}{s_2-s_1}$ we obtain

$$H_{s_1}(P_{\xi_1},P_{\xi_2}) \le \left(\int D_{s_2,T}^{\frac{s_1}{s_2-s_1}}X_2dP_W\right)^{\frac{s_2-s_1}{s_2}}\left(\int Z_{s_2,T}dP_W\right)^{\frac{s_1}{s_2}}$$

$$= \left(\int D_{s_2,T}^{\frac{s_1}{s_2-s_1}}dP_{\xi_2}\right)^{\frac{s_2-s_1}{s_2}}\left(\int Z_{s_2,T}dP_W\right)^{\frac{s_1}{s_2}}$$

$$= \left(M\, D_{s_2,T}^{\frac{s_1}{s_2-s_1}}(\xi_2)\right)^{\frac{s_2-s_1}{s_2}}\left(M\, Z_{s_2,T}(W)\right)^{\frac{s_1}{s_2}}$$

It is known Lipster, Shiryayev (1978) that $Z_{s_2,t}(W)$ forms a supermartingal with $M\, Z_{s_2,t}(W) \le 1$. Hence

$$H_{s_1}(P_{\xi_1},P_{\xi_2}) \le \left[M\, D_{s_2,T}^{\frac{s_1}{s_2-s_1}}(\xi_2)\right]^{\frac{s_2-s_1}{s_2}} \tag{12}$$

Consequently the theorem is established.

In order to prove corollary 1 we denote by $P_{\xi_i}^{(n)}$ the restriction of P_{ξ_i} to $\mathfrak{L}_n$ · $\mathfrak{L}_n$ forms an increasing sequence of sub-σ-algebras generating $\mathfrak{L} = \sigma(x(t), 0 \le t < \infty)$. It holds Vajda (1972)

$$H_s(P_{\xi_1},P_{\xi_2}) = \lim_{n\to\infty} H_s\left(P_{\xi_1}^{(n)},P_{\xi_2}^{(n)}\right) \quad \text{for every } 0<s<1. \tag{13}$$

Alternatively

$$D_{s_2,n}^{\frac{s_1}{s_2-s_1}}(\xi_2) = \exp\left\{\frac{1}{2}\frac{s_2(1-s_2)s_1}{s_1-s_2}\int_0^n \left(a_1(t,\xi_2)-a_2(t,\xi_2)\right)^2dt\right\} \xrightarrow[n\to\infty]{} \exp\left\{\frac{1}{2}\frac{s_2(1-s_2)s_1}{s_1-s_2}\int_0^\infty \left(a_1(t,\xi_2)-a_2(t,\xi_2)\right)^2dt\right\} \tag{14}$$

Substituting in (12) P_{ξ_1}, P_{ξ_2} by $P_{\xi_1}^{(n)}$, $P_{\xi_2}^{(n)}$ and using the relation (14) we get corollary 1 as $n \to \infty$, since the limit and the expectation can be changed on account of $0 \le D_{s,T} \le 1$.

In order to prove corollary 2 we notice that

$$H_{1-s}(P_{\xi_1},P_{\xi_2}) = H_s(P_{\xi_2},P_{\xi_1})$$

and therefore by (1)

$$\alpha \le \inf_{0<s_1<1} L^{s_1}H_{1-s_1}(P_{\xi_1},P_{\xi_2}) \le \inf_{0<s_1<s_2<1} L^{s_1}\left[M\left(D_{s_2}(\xi_1)\right)^{\frac{s_1}{s_2-s_1}}\right]^{\frac{s_2-s_1}{s_2}}$$

The second inequality may be proved analogously.

Corollary 3 follows immediately from corollary 2 with $s_1 = 1/4$, $s_2 = 1/2$.

4. **References**

CHERNOFF, H.
A measure of asymptotic efficiency for tests of a hypothesis based on a sum of observations.
Ann. Math. Stat. 23 (1952) 493–507.

EVANS, J. E.
Chernoff bounds on the error probability for the detection of non-Gaussian signals.
IEEE Trans. Inf. Theory 20 (1974) 569–577.

GALLAGER, R. G.
Information Theory and Reliable Communication.
Wiley, New York, 1968.

HIBEY, J. L., SNYDER, D. L., VAN SCHUPPEN, J. H.
Error probability bounds for continous-time decision problems.
IEEE Trans. Inf. Theory 24, no 5 (1978) 608–622.

LIESE, F.
Eine informationstheoretische Bedingung für die Äquivalenz unbegrenzter teilbarer Punktprozesse.
Math. Nachr. 70 (1976) 183–196.

LIPSTER, R. S., SHIRYAYEV, A. N.
Statistics of random processes.
Vol. I, II, Springer Verl. Berl., New York, 1978.

NEMETZ, T.
Equivalence orthogonality dichotomies of probability measures.
Proc. Coll. on Limit Theorems of Prob. Theor. and Stat. Keszthely 1974.

ÖSTERREICHER, F.
On the dimensioning of tests for composite hypothesis and not necessarily independent observation.
Probl. Control-Inform. Theory 7 no 5 (1978) 333–343.

VAJDA, I.
On the f-divergence and singularity of probability measures.
Periodica Math. Hung. 2 (1972) 223–234.

Department of Mathematics, Agricultural University, De Dreijen 8, 6708 BC Wageningen, the Netherlands

Power Simulation with the Same Random Sequences Under the Null Hypothesis and the Alternative

M. A. J. van MONTFORT and L. R. VERDOOREN

1. Introduction

In studies of the power of a test Monte Carlo techniques are used where the calculus to be done for getting the exact power is too complicated. The simpliest way, called crude Monte Carlo, consists of getting the fraction realized test statistics in the critical region at the alternative hypcthesis (K). In that situation simulated power (π) at the null hypcthesis (H) is an unbiased estimator of the size (α) of the test with a positive variance based on the binomial experiment.

Rothery (1982) investigates the power of a new statistic in the situation where the power curve of an old statistic is known completely, and he deals with reduction of the variance of the simulated power by simulating both statistics using the same sequence of pseudo random numbers, and compared the result with crude Monte Carlo. Also in his procedure the estimated power at H is an unbiased estimator of α with a positive variance.

In the situation where no control statistic is available variance reduction could be gained by using the simulated critical value instead of using the known critical value, see Andrews et al. (1972, p. 55). Here we give an example of the reduction of the mean square error (MSE) by estimating the power in the *same* sample as in which the critical value is estimated and dropping the knowledge of the exact critical value.

2. The MSE of the Estimated Power

Here we simulate the distribution of the test statistic under H and K simultaneously and use the α-point of the empirical distribution function at H as the critical value in order to estimate the power (π) at K. The here studied test statistic $\mathbf{x}$ could be the critical level of a test statistic with a strictly monotonically increasing cumulative distribution function (cdf). (Random variates are in bold face; their realizations are denoted by the same symbol not in bold face; $\mathbf{a} \simeq \mathbf{b}$ means that the cdf of $\mathbf{a}$ and $\mathbf{b}$ are equal.) The critical level is the smallest significance level at which H is rejected in favour of K; note that the critical level has a uniform distribution at H.

For a test statistic $\mathbf{x}$ could hold

$$\mathbf{x} \simeq \mathbf{u}^{\theta} \tag{1}$$

where $\mathbf{u}$ is uniformly distributed in (0,1), θ a positive parameter with $\theta = 1$ at H and $\theta > 1$ at K. This example is chosen because of its transparancy.

Formula (1) is closely related to the Lehmann alternative. A size α test for K results in a left sided critical region with critical value α.

For the power one gets (with $\gamma = 1/\theta$)

$$\pi = P_{\theta}(\mathbf{x} \leq \alpha) = P\left(\mathbf{u}^{\theta} \leq \alpha\right) = P\left(\mathbf{u} \leq \alpha^{1/\theta}\right) = \alpha^{\gamma}. \tag{2}$$

Note that formula (1) is equivalent to

$$-\log \mathbf{x} \simeq \theta \cdot \frac{1}{2} \chi_2^2 \tag{3}$$

and to

$$-\log(-\log \mathbf{x}) \simeq -\log \theta + \mathbf{g} \tag{4}$$

where $\mathbf{g}$ stands for the standard double exponential distribution (or Gumbel distribution or type I distribution of maxima) with

$$P(\mathbf{g} \leq g) = \exp\left(-\exp(-g)\right), \quad -\infty < g < +\infty. \tag{5}$$

In a simulation with n uniforms on (0,1) we get for the crude Monte Carlo with known critical value α:

$$\begin{aligned} n\hat{\pi} &= \#\text{ of u's with } u^{\theta} \leq \alpha \\ &= \#\text{ of u's with } u \leq \alpha^{1/\theta} = \alpha^{\gamma} \end{aligned} \tag{6}$$

(where # stands for frequency).

So $n\hat{\pi}$ has a binomial distribution with parameters n and α^{γ}

In this situation $\hat{\pi}$ is an unbiased estimator of π and

$$\mathrm{MSE}(\hat{\pi}) = \mathrm{var}(\hat{\pi}) = \alpha^{\gamma}\left(1 - \alpha^{\gamma}\right)/n. \tag{7}$$

An other estimator $\hat{\pi}$ based on $u_1, \ldots, u_n$ could be obtained in the following way. Sorting the uniforms gives $u_{(1)} \leq \ldots \leq u_{(n)}$. The critical value α is estimated by t, with

$$t = u_{(n\alpha)}. \tag{8}$$

We restrict ourselves to combinations of n and α with $n\alpha$ integer.

An estimate $\tilde{\pi}$ for the power π is given by

$$\begin{aligned} n\tilde{\pi} &= \#\text{ of } u_{(i)}^{\theta} \leq t \\ &= \#\text{ of } u_{(i)} \leq t^{\gamma} \\ &= n\alpha + \#\text{ of } u_{(i)} \text{ between t and } t^{\gamma}. \end{aligned} \tag{9}$$

We now derive the MSE of $\tilde{\pi}$.

Note that for $\theta = 1$ we find $\tilde{\pi} = \alpha$ without random fluctuations.

Note also that $(\mathbf{u}_{(n\alpha+1)}, \ldots, \mathbf{u}_{(n)})$ behaves like an ordered sample of $n - n\alpha = n(1-\alpha)$ uniforms on the interval (t, 1), so the number of $u_{(i)}$ values between t and t^{γ} has a binomial distribution with parameters $n(1-\alpha)$ and $(t^{\gamma}-t)/(1-t)$. Given t formula (9) results in

$$E(\tilde{\pi}|t) = \alpha + (1-\alpha) \cdot \left(t^{\gamma} - t\right)/(1-t). \tag{10}$$

In order to get $E(\tilde{\pi})$ we have to integrate formula (10) over t. Using the wellknown fact that $\mathbf{t} = \mathbf{u}_{(n\alpha)}$ has a beta-distribution with density

$$f(t) = \frac{\Gamma(p+q)}{\Gamma(p)\Gamma(q)} t^{p-1}(1-t)^{q-1}, \quad 0 < t < 1 \tag{11}$$

with integer valued parameters p and q with

$$p = n\alpha$$
$$q = n+1-n\alpha$$
$$p+q = n+1$$

we get

$$E(\tilde{\pi}) = \int_0^1 \left\{\alpha + (1-\alpha)\frac{t^\gamma - t}{1-t}\right\} f(t)dt$$

$$= \alpha + (1-\alpha)\int_0^1 \frac{t^\gamma - t}{1-t} \cdot f(t)dt$$

$$= \alpha + (1-\alpha)\frac{1}{q-1}\left\{\frac{\Gamma(n+1)\cdot\Gamma(p+\gamma)}{\Gamma(n+\gamma)\cdot\Gamma(p)} - p\right\}. \quad (12)$$

For the bias of $\tilde{\pi}$ we get with formula (12)

$$\text{bias}(\tilde{\pi}) = E(\tilde{\pi}) - \pi = E(\tilde{\pi}) - \alpha^\gamma. \quad (13)$$

The derivation of var $(\tilde{\pi})$ with formula (9) needs

$$\text{var}(\tilde{\pi}|t) = \frac{1}{n^2}\cdot n(1-\alpha)\cdot\frac{t^\gamma - t}{1-t}\cdot\left(1-\frac{t^\gamma - t}{1-t}\right)$$

$$= \frac{1-\alpha}{n}\cdot\frac{(t^\gamma - t)(1-t^\gamma)}{(1-t)^2} \quad (14)$$

and results in

$$\text{var}(\tilde{\pi}) = \int_0^1 \text{var}(\tilde{\pi}|t)\, f(t)dt$$

$$= \frac{(1-\alpha)\Gamma(n)}{(q-1)(q-2)\Gamma(p)}\left\{\frac{\Gamma(p+\gamma)}{\Gamma(n-1+\gamma)} - \frac{\Gamma(p+2\gamma)}{\Gamma(n-1+2\gamma)} - \frac{\Gamma(p+1)}{\Gamma(n)} + \frac{\Gamma(p+1+\gamma)}{\Gamma(n+\gamma)}\right\} \quad (15)$$

The MSE of $\tilde{\pi}$ follows by inserting formulae (13) and (15) in

$$\text{MSE}(\tilde{\pi}) = \{\text{bias}(\tilde{\pi})\}^2 + \text{var}(\tilde{\pi}). \quad (16)$$

If the efficiency of $\tilde{\pi}$ with respect to $\hat{\pi}$ is defined by

$$\text{Eff} = \frac{\text{MSE}(\hat{\pi})}{\text{MSE}(\tilde{\pi})}$$

we get with formulae (7), (13) and (15)

$$\text{Eff} = \frac{\alpha^\gamma(1-\alpha^\gamma)/n}{\{\text{bias}(\tilde{\pi})\}^2 + \text{var}(\tilde{\pi})}. \quad (17)$$

A numerical example with $\alpha = 0.05$ gives the following values of the efficiency (Eff.); the right-hand column gives the rank of the uniform used for the critical value.

Θ		1.1	1.4	3.5	5.0	($\alpha = 0.05$) $n\alpha$
power $= \alpha^{1/\Theta}$		0.066	0.118	0.425	0.549	
n =	20	4.61	1.88	1.02	0.94	1
	40	4.30	1.77	1.04	0.98	2
	100	4.11	1.70	1.06	1.02	5
	200	4.05	1.68	1.07	1.03	10
	500	4.01	1.66	1.07	1.04	25
	1 000	4.00	1.66	1.08	1.04	50
	2 000	3.99	1.65	1.08	1.04	100
	5 000	4.00	1.65	1.08	1.04	250
	10 000	3.99	1.65	1.08	1.04	500

This table shows that for small deviations from the null hypothesis H a lot can be gained by using the same random sequences at H and K including a simulated critical value. Only far from H for unrealistically small simulation sample sizes a small loss is found.

3. Concluding Remarks

In power estimation it seems to pay to replace crude Monte Carlo by simulating the power in the same sample as in which the critical value is simulated.

If one is interested in a *difference* in power, say $\pi(\Theta_2) - \pi(\Theta_1)$, the above mentioned method stresses attention to the number of uniforms between t^{1/Θ_1} and t^{1/Θ_2} with $t = u_{(n\alpha)}$, and leads to similar results.

If some testprocedures have to be compared then the use of only one random sequence and its estimated critical values compares these tests at the same critical level, being the correct critical level without random fluctuation.

4. References

ANDREWS, D. F. et al.
Robust estimates of location: survey and advances.
Princeton NJ: Princeton University Press, 1972.

ROTHERY, P.
The use of control variates in Monte Carlo estimation of power.
Applied Statistics 31 (1982), 125–129.

Academy of Agricultural Sciences of the GDR
Research Centre of Animal Production Dummerstorf-Rostock

Robustness of Two-Sample Tests for Variances

GERD NÜRNBERG

Abstract:

12 two sample-test for variances are investigated for robustness against violations of the assumed normal distribution by means of simulation. The degree of non-normality is discribed by the parameters skewness (γ_1) and kurtosis (γ_2). The real risk of first kind $\hat{\alpha}$ and the power function (at 3 points) of the 12 tests are determined for the sample sizes n = 6, 18, 42 and different pairs of (γ_1, γ_2)-values.

1. Introduction

Let $\mathbf{X}_1$ and $\mathbf{X}_2$ be independent random variables with $\mathbf{X}_1 \sim N(\mu_1, \sigma_1^2)$ and $\mathbf{X}_2 \sim N(\mu_2, \sigma_2^2)$.
By means of two independent random samples ($\mathbf{X}_{11}, \ldots, \mathbf{X}_{1n}$ and $\mathbf{X}_{21}, \ldots, \mathbf{X}_{2n}$) with $\mathbf{X}_{1j} \sim N(\mu_1, \sigma_1^2)$ and $\mathbf{X}_{2j} \sim N(\mu_2, \sigma_2^2)$ we want to test the null-hypothesis

$$H_0: \sigma_1^2 = \sigma_2^2$$

against

$$H_A: \sigma_1^2 \neq \sigma_2^2$$

A number of test statistics is available for testing variances: o. g. Bartlett (1937), Cochran (1941), Box (1953). First investigations of the effect of nonnormality were made by Box (1953), Levene (1960), Overall, Woodward (1974), Brown, Forsythe (1974) and Geng, Wang, Miller (1979).

2. Definition of Robustness

Definition 1:
A test T_α for given nominal risk of first kind α, which has a real risk of first kind α for the normal distribution is called ε-robust in a class G of distribution if

$$\max_{g \in G} |\alpha(n, g) - \alpha| \leq \varepsilon$$

for given values of α, n and ε.
The real risk of first kind α (n, g) depends on the sample size n and the distribution g$\in$G.
In this paper the class G is characterized by functions of the first four central moments $\mu_k = E(\mathbf{x} - \mu)^k$ $(k = 1, \ldots, 4)$ of the distributions:

$$E(\mathbf{x}) = \mu, \ \mathbf{V}(\mathbf{x}) = \mu_2 = \sigma^2, \ \gamma_1 = \frac{\mu_3}{\mu_2^{3/2}} \text{ and } \gamma_2 = \frac{\mu_4}{\mu_2^2} - 3$$

and so we can write G = K ($\mu, \sigma^2, \gamma_1, \gamma_2$) with skewness γ_1 and kurtosis γ_2. Since all investigated tests are invariant tests with respect to linear transformations of the observations, we can limit ourself to the class

$G = K(0, 1, \gamma_1, \gamma_2)$.

3. Generation of Random Samples with Given Distribution g$\in$G

Because the investigations of robustness (due to Definition 1) are made by means of simulation it is necessary to generate realizations of the two random samples with given first four moments.
The following three steps are carried out.

1. Generation of uniformly distributed random numbers [see Herrendörfer (1980)]
2. Transformation of uniformly distributed random numbers to normally distributed ones with mean 0 and variance 1 [Ode/Evans (1974)]
3. Using "Power Transformation" for given γ_1- and γ_2-values [Fleishman (1978)]:

$$\mathbf{y} = a + b\mathbf{x} + c\mathbf{x}^2 + d\mathbf{x}^3 \text{ with } \mathbf{x} \sim N(0,1).$$

The coefficients a, b, c, d depend on given γ_1- and γ_2-values.
Robustness of tests comparing two variances is investigated for different sample sizes n (n = 6, 18, 42) and also different values of γ_1 and γ_2.

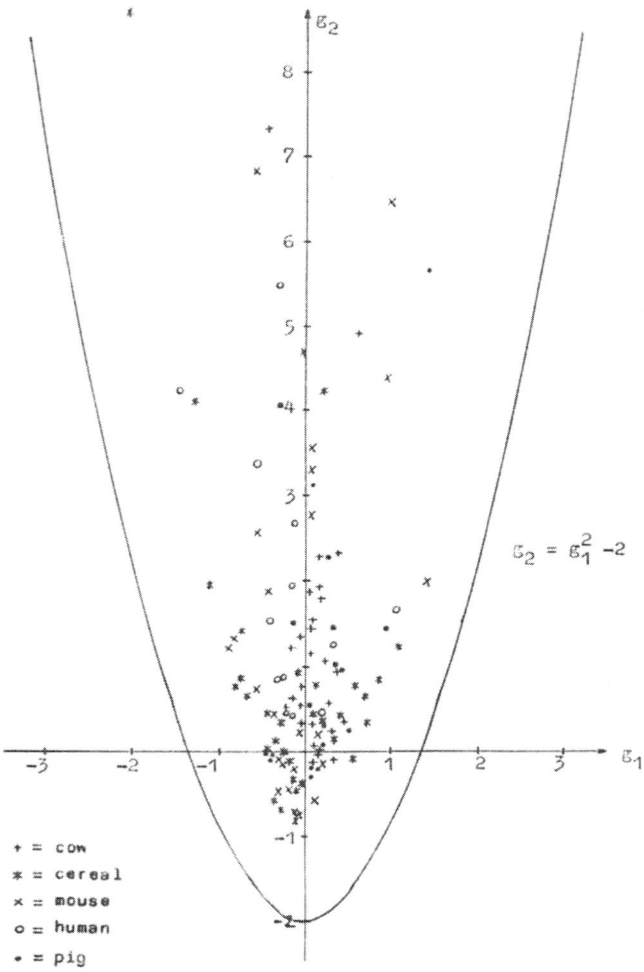

Figure 1
(g_1, g_2) of characteristics from agriculture and medicine

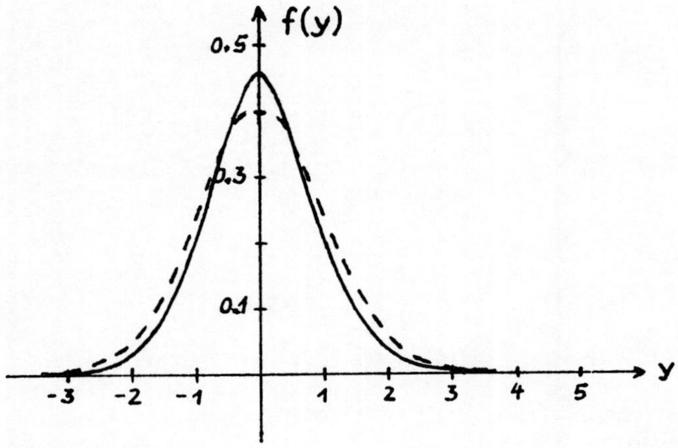

$\gamma_1 = 0;\ \gamma_2 = 1.5$

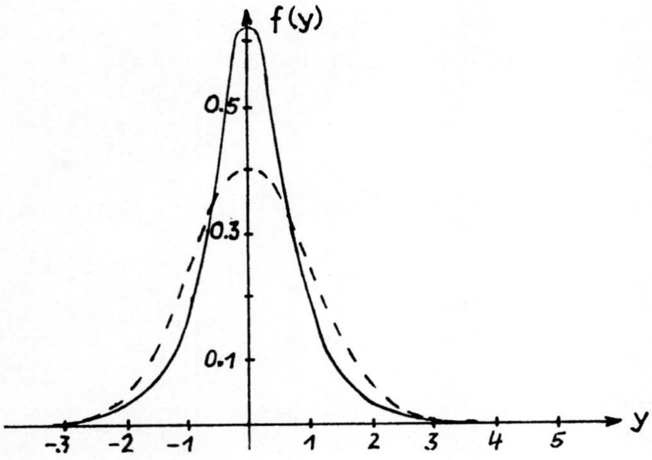

$\gamma_1 = 0;\ \gamma_2 = 7$

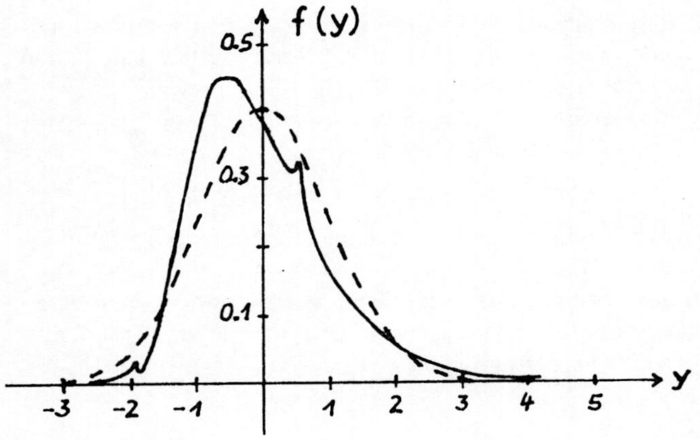

$\gamma_1 = 1;\ \gamma_2 = 1.5$

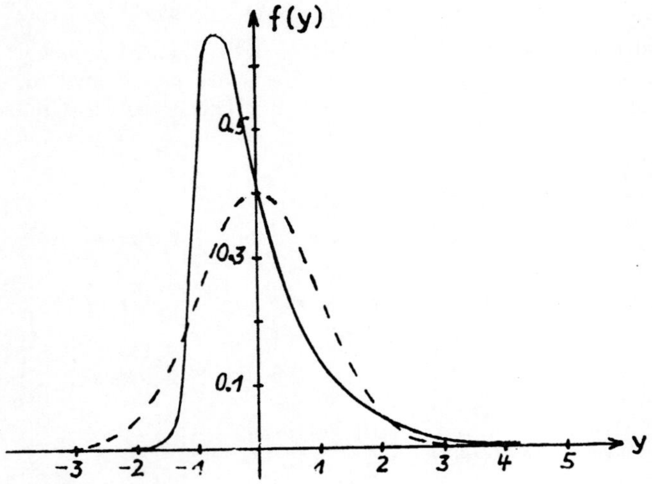

$\gamma_1 = 2;\ \gamma_2 = 7$

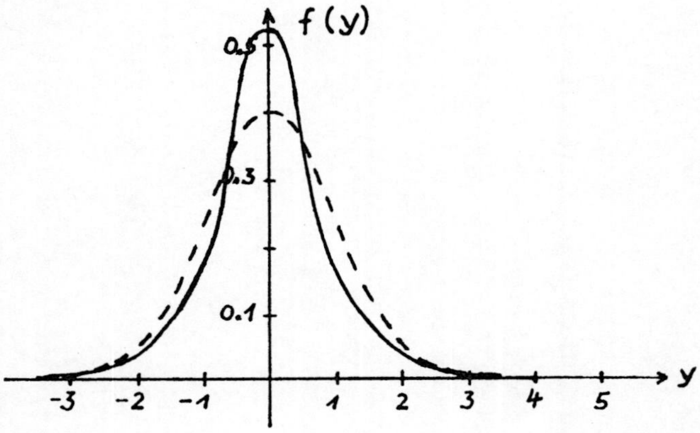

$\gamma_1 = 0;\ \gamma_2 = 3.75$

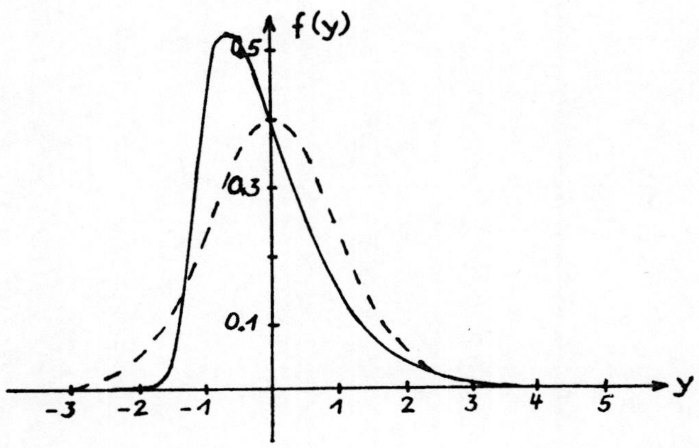

$\gamma_1 = 1.5;\ \gamma_2 = 3.75$

Figure 2

Densities of distributions generated by „Power Transformation": ——

Density of the standardized normal distribution: - - -

For choosing (γ_1, γ_2)-values we investigated the sample skewness (g_1) and sample kurtosis (g_2) of practical characteristics from agriculture and medicine.
The results are presented in Figure 1.

The selected (γ_1, γ_2)-values are:

γ_1	0	0	1	0	1.5	0	2
γ_2	0	1.5	1.5	3.75	3.75	7	7

Figure 2 shows the densities of distributions generated by "Power Transformation" for the selected (γ_1, γ_2)-values.

4. Description of the Tests

a) Bartlett-test [Bartlett (1937)]
Test statistic:

$$\mathbf{B} = -\frac{1}{C'}(n-1)\left\{\ln\frac{s_1^2}{s^2} + \ln\frac{s_2^2}{s^2}\right\} \tag{4.1}$$

with $s_i^2 = \frac{1}{n-1} \cdot \sum_{j=1}^{n} (x_{ij} - \bar{x}_{i.})^2 \quad (i = 1, 2)$,

$$s^2 = \frac{s_1^2 + s_2^2}{2}$$

and $C' = 1 + \frac{1}{2(n-1)}$.

In the special case of testing two variances and for equal sample size the Bartlett-test corresponds to the two-sided F-test!

b) Modified Bartlett-test [Box (1953)]

Test statistic:

$$B^{+} = B \cdot \left(1 + \frac{g_2'}{2}\right)^{-1}$$

with **B** from (4.1)

and

$$g_2' = \frac{2n \sum_{i=1}^{2} \sum_{j=1}^{n} (x_{ij} - \bar{x}_{i.})^4}{\left[\sum_{i=1}^{2} \sum_{j=1}^{n} (x_{ij} - \bar{x}_{i.})^2\right]^2} - 3. \qquad (4.2)$$

Approximate critical values: $\chi^2 (1; 1 - \alpha)$.

c) Modified χ^2-test [Layard (1973)]

$$\Lambda = (n-1) \sum_{i=1}^{2} \left(\ln s_i^2 - \frac{\ln s_1^2 + \ln s_2^2}{2}\right)^2 \Big/ \left[2 + \left(1 - \frac{1}{n}\right) \cdot g_2'\right]$$

with g'_2 from (4.2)

Approximate critical values: χ^2 $(1; 1 - \alpha)$.

d) Cochran-test

Test statistic:

$$G = \frac{\max_i \{s_i^2\}}{s_1^2 + s_2^2} \qquad (i = 1.2)$$

In the special case of testing two variances and for equal sample size the Cochran-test corresponds to the two-sided F-test!

e) F-test

Test statistic:

$$F = \frac{s_1^2}{s_2^2}$$

Critical values:

$$F\left(n-1, n-1, 1 - \frac{\alpha}{2}\right)$$

f) Range-test

Test statistic:

$$R = \frac{\max_i \{w_i\}}{\min_i \{w_i\}} \qquad (i = 1.2)$$

with $w_i = \max_j \{x_{ij}\} - \min_j \{x_{ij}\}$ $(j = 1, \ldots, n$.

Critical values: R (2, n—1; 1—α)

[Table in Pearson/Hartley (1966)]

g) Box-Scheffe-test (also known as Box-test or Box-Kendall-test) [Box (1953), Scheffé (1963)]

The samples are randomly divided in c groups. These groups contain m_{ij} $(i = 1, 2; j = 1, \ldots, c)$ observations

$$\left(\sum_{j=1}^{c} m_{ij} = n\right).$$

Let $Z_{ij} = \ln s_{ij}^2$ where s_{ij}^2 is an estimator of σ_i^2.

Test statistic:

$$F^{+} = \frac{2(c-1)\, c \sum_{i=1}^{2} (\bar{Z}_{i.} - \bar{Z}_{..})^2}{\sum_{i=1}^{2} \sum_{j=1}^{c} (Z_{ij} - \bar{Z}_{i.})^2}$$

Approximate critical values: F (1, 2 (c—1), 1—α)

In this study c is choosen to 2 and 3.

h) Box-Andersen-test [Box/Andersen (1955)]

Test statistic:

$$F = \frac{s_1^2}{s_2^2}$$

Approximate critical values:

$$F\left((n-1) \cdot \left(1 + \frac{g_2'}{2}\right)^{-1};\ (n-1) \cdot \left(1 + \frac{g_2'}{2}\right)^{-1};\ 1 - \frac{\alpha}{2}\right).$$

with g_2 from (4.2).

i) Jackknife-test [Miller (1968)]

$$\text{Let } \bar{x}_{i(j)} = \frac{1}{n-1} \sum_{\substack{t=1 \\ t \neq j}}^{n} x_{it} \qquad (i = 1.2;\ j = 1, \ldots, n)$$

$$s_{i(j)}^2 = \frac{1}{n-2} \sum_{\substack{t=1 \\ t \neq j}}^{n} (x_{it} - \bar{x}_{i(j)})^2$$

and $Z_{ij} = n \cdot \ln s_i^2 - (n-1) \cdot \ln s_{i(j)}^2$

Test statistic:

$$J = \frac{2(n-1)\, n \sum_{i=1}^{2} (\bar{Z}_{i.} - \bar{Z}_{..})^2}{\sum_{i=1}^{2} \sum_{j=1}^{n} (Z_{ij} - \bar{Z}_{i.})^2}$$

Approximate critical values: F (1; 2 (n—1), 1—α)

j) Levene-z-test (also known as Pfanzagl-test) [Levene (1960)]

$$\text{Let } y_{ij} = |x_{ij} - \bar{x}_{i.}| \quad i = 1.2;\ j = 1, \ldots, n$$

Test statistic:

$$F^{+} = \frac{MQ_A}{MQ_R}$$

$$\text{with } MQ_A = SQ_A \text{ and } MQ_R = \frac{SQ_R}{2(n-1)}$$

$$SQ_A = \frac{1}{n} \cdot \sum_{i=1}^{2} y_{i.}^2 - \frac{y_{..}^2}{2n}$$

$$SQ_R = \sum_{i=1}^{2} \sum_{j=1}^{n} y_{ij}^2 - \frac{1}{n} \sum_{i=1}^{2} y_{i.}^2$$

Approximate critical values: F (1, 2 (n—1), 1—α)

k) Levene-s-test

Let $y_{ij} = |x_{ij} - x_{i.}|^2$

Then the test is carried out in the same way as the Levene-z-test.

5. Results and Discussion

10,000 computer runs were carried out to evaluate the real risks of first kind α and also the power of the considered tests for each sample size n (6, 18, 42) and each selected pair (γ_1, γ_2) for nominal risks of first kind α = 0.01; 0.05; 0.10. The power is evaluated for three values λ = 1.44; 1.96; 3.24 with

$$\lambda = \frac{\sigma_1^2}{\sigma_2^2}.$$

The number of 10,000 computer runs is a result of planning the simulation experiment for the estimation of a probability in case of robustness with $\varepsilon = 0.2 \cdot \alpha$ [see Herrendörfer (1980)].

For $\alpha = 0.01$ and $\varepsilon = 0.2 \cdot \alpha$ the number of 10,000 computer runs is not large enough therefore the results show only the tendencies in the behaviour of the considered tests.

The simulation results ($\hat{\alpha}$ (n, g)) are summarized in table 2.

For $\varepsilon = 0.2 \cdot \alpha$, that means 20-%-robustness (e. g. for α = 5 %, $\hat{\alpha}$ can vary between 4 %–6 %), the Range- and F-test (and with that, the Bartlett- and Cochran-test in the case of comparing two variances and equal sample size) are not robust. The real risk of first kind $\hat{\alpha}$ of these tests increases with increasing γ_2. For $\gamma_2 = 7$, n = 42 the real risk is close to 30 % for a nominal risk α = 5 %. The skewness γ_1 seems to have only a little influence on the real risk.
The Jackknife-test isn't 20-%-robust too but the real risk $\hat{\alpha}$ is closer to the nominal than that of the F-test. Furthermore the skewness γ_1 seems to have also an influence.
The Box-Scheffé-tests (c = 2 or 3) are quite robust for all sample sizes n. The real risk $\hat{\alpha}$ is very close to the nominal risk α and in most cases conservativ.
The Levene-z-test is robust for n = 42 and $\gamma_1 = 0$. If $\gamma_1 \neq 0$, the test isn't 20-%-robust, this fact was pointed out by Miller (1968). The same behaviour we find for the modified χ^2-test. The Levene-s-test is quite robust for all investigated distributions and sample sizes n = 18 and 42.
A similar behaviour shows the Box-Andersen-test which is only in the case n = 6 nonrobust, but in all other cases the real risk is close to the nominal.
The modified Bartlett-test is 20-%-robust for n = 18 and n = 42 for all investigated distributions.
Summarizing the results we can conclude:

1. For small sample size n (n = 6) only the Box-Scheffé-test (c = 2 or 3) is 20 % robust for all investigated distributions.

2. For n = 18, and 42 the following four test are 20 % robust:
   - modified Bartlett-test
   - Box-Scheffé-test
   - Box-Andersen-test
   - Levene-s-test

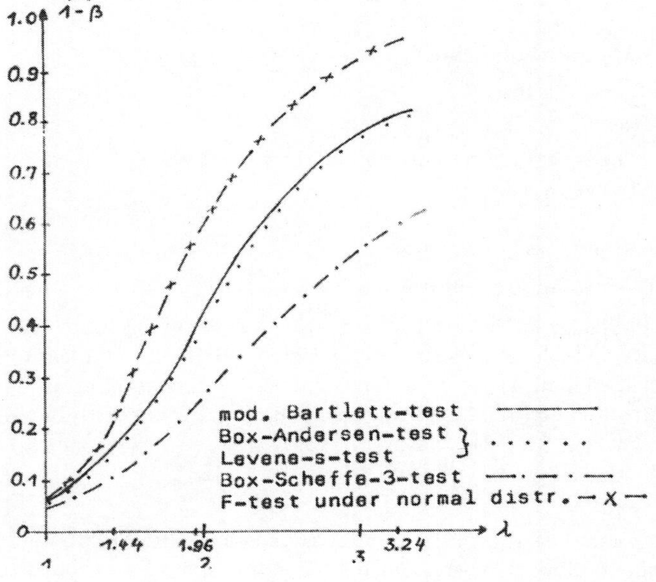

$\gamma_1 = 1.0$
$\gamma_2 = 1.5$

Now we investigate the power of these four robust test: Figure 3 contains the power of the robust tests in dependence on some (γ_1, γ_2)-values for n = 42 and $\alpha = 0.05$. The corresponding simulation results are summarized in table 3.

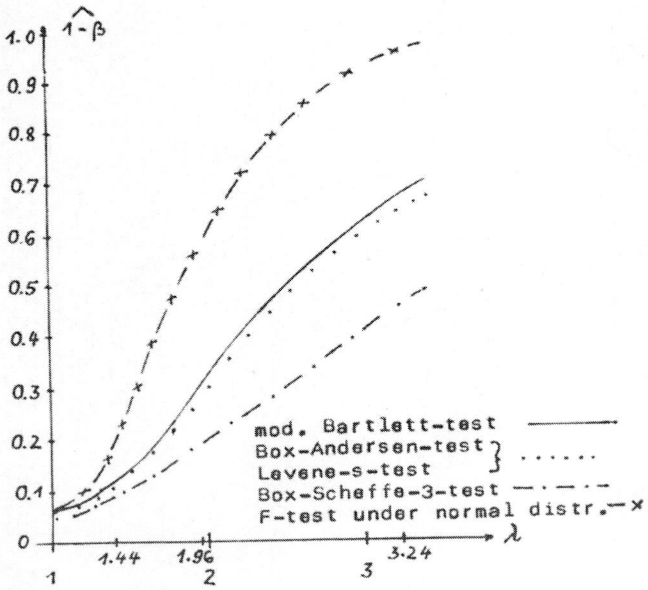

$\gamma_1 = 1.5$
$\gamma_2 = 3.75$

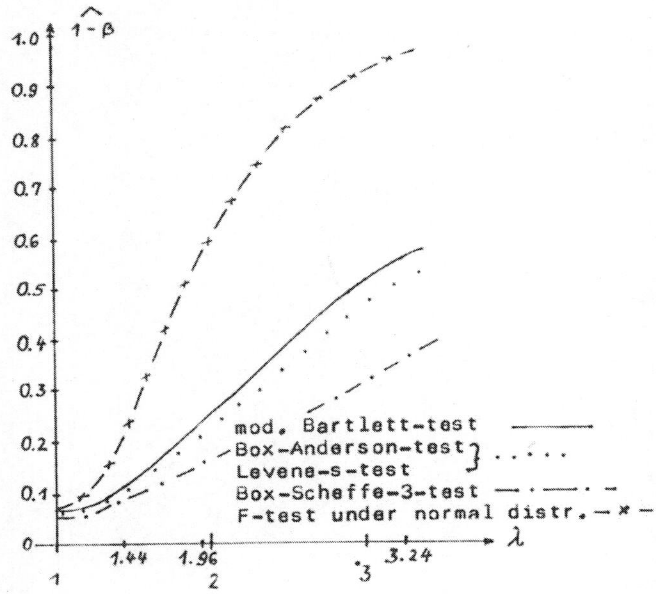

$\gamma_1 = 2.0$
$\gamma_2 = 7.0$

Figure 3
Powerfunctions of the robust tests in dependence on (γ_1, γ_2) for n = 42 and $\alpha = 0.05$

These figures show that the power of the robust tests decrease if γ_2 increase. The power of the Box-Andersen-test and the Levene-s-test are nearly the same so that only one line is drawn for both power curves. Furthermore we can see that the power of the modified Bartlett-test is slightly superior in all cases.
Only the Box-Scheffé-test (c = 3) is less powerful in all cases. To get an impression of the loss of power for increasing γ_2 the power of the F-test under n o r m a l distribution is considered in these figures too.

Figure 4 gives an impression of the power of the robust tests if the underlying distribution is the normal one. We can find that in this case the power of the modified Bartlett-test, Box-Andersen-test and Levene-s-test are similar to that of the F-test.

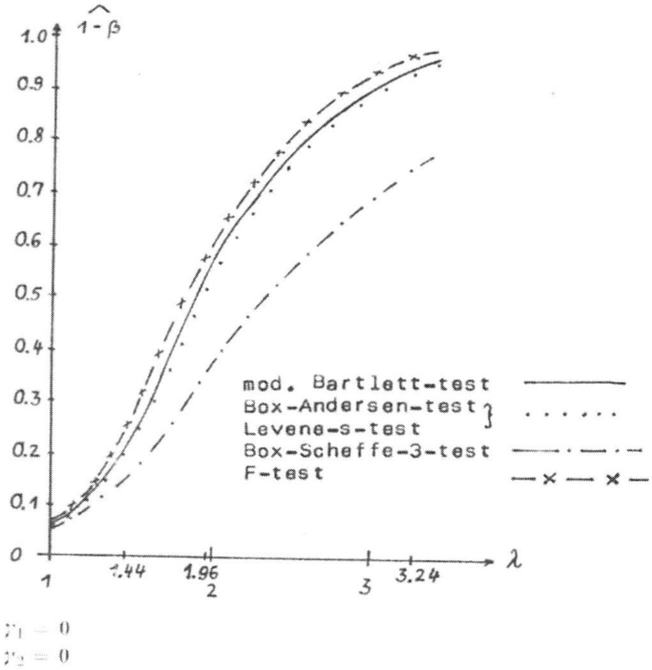

Figure 4
Powerfunctions of the robust tests under normal distribution for n = 42 and $\alpha = 0.05$

Summarizing the above results we recommend for testing equality of two variances if the underlying distributions are similar to those investigated in this study the

- modified Bartlett-test
- Box-Andersen-test
- Levene-s-test

6. Planning of Experiments

The loss of power of the robust tests for increasing γ_2 can be adjusted if we use greater sample sizes as for the F-test under normal distribution. Therefore we derived a method of planning of experiments for the Box-Andersen-test. This method is investigated by means of simulation. It is well known that for a given risk of first kind α and risk of second kind β_0 and a given

$$\lambda = \frac{\sigma_1^2}{\sigma_2^2}.$$

the sample size n' for the F-test under normal distribution is a solution of the equation

$$\lambda = F\left(n'-1, n'-1, 1-\frac{\alpha}{2}\right) \cdot F\left(n'-1, n'-1, 1-\beta_0\right). \quad (6.1)$$

Now the sample size n for the Box-Andersen-test is evaluated from n':

$$n = (n'-1)\left(1+\frac{\gamma_2}{2}\right)+1 \quad (6.2)$$

with n' from (6.1) for given α, β_0 and λ.
Table 1 shows the theoretical power $(1-\beta_0)$ and the empirical power $(\widehat{1-\beta})$ for $\alpha = 0.05$ and n = 42 using (6.1) and (6.2).

Table 1
Comparison of the theoretical and empirical power (10,000 computer runs) of the Box-Andersen-test using (6.1) and (6.2)

	$\gamma_2 = 0$			$\gamma_2 = 1.5$			$\gamma_2 = 3.75$		
λ	1.44	1.96	3.24	1.44	1.96	3.24	1.44	1.96	3.24
$1-\beta_0$	0.21	0.57	0.96	0.13	0.35	0.79	0.09	0.23	0.57
$\widehat{1-\beta}$	0.20	0.54	0.95	0.14	0.39	0.81	0.11	0.28	0.66

λ	1.44	1.96	3.24
$1-\beta_0$	0.07	0.15	0.39
$\widehat{1-\beta}$	0.09	0.21	0.52

In practice we don't know the kurtosis γ_2 of the distribution of a characteristic. Therefore we have to use an estimate g_2 of γ_2 in formula (6.2) for planning of experiment for the Box-Andersen-test.

Acknowledgement
The author wish to thank Prof. R. Verdooren for his suggestions for the improvement of the presentation of this paper.

Table 2
Results of simulation for tests comparing two variances $\hat{\alpha}(n, g) \cdot 100$ in dependence on n, α and (γ_1, γ_2)

n = 6	$\gamma_1 = 0$; $\gamma_2 = 0$		$\gamma_1 = 0$; $\gamma_2 = 1.5$		$\gamma_1 = 1.0$; $\gamma_2 = 1.5$		$\gamma_1 = 0$; $\gamma_2 = 3.75$		$\gamma_1 = 1.5$; $\gamma_2 = 3.75$	
Test \ $\alpha \cdot 100$	10	5	10	5	10	5	10	5	10	5
mod. Bartlett-	13.88	6.66	14.32	6.25	14.62	7.07	15.58	7.59	18.03	9.75
mod. χ^2-	16.30	9.11	18.42	10.01	18.14	10.28	21.65	12.86	23.23	13.97
F-	10.60	5.28	14.85	8.06	14.19	8.03	19.75	12.37	20.60	12.63
Range-	10.44	5.22	14.96	8.36	13.67	7.64	20.04	12.65	19.62	11.94
Box/Scheffé 2-	10.42	5.55	9.34	4.52	9.30	4.42	9.78	5.05	9.65	4.68
Box/Scheffé 3-	9.41	4.58	9.12	4.49	8.61	4.30	9.04	4.47	9.67	4.59
Box/Andersen-	12.38	5.72	11.57	4.89	12.38	6.05	11.67	5.91	15.01	8.07
Jackknife-	9.03	5.05	9.90	5.35	9.97	5.45	12.43	6.84	12.36	7.10
Levene-z-	13.60	7.33	13.79	6.85	15.33	8.52	14.81	7.62	19.72	11.61
Levene-s-	12.15	5.29	11.14	4.18	12.12	5.43	10.61	4.87	14.20	7.00

Table 2: continuation

n = 18	$\gamma_1 = 0$; $\gamma_2 = 0$		$\gamma_1 = 0$; $\gamma_2 = 1.5$		$\gamma_1 = 1.0$; $\gamma_2 = 1.5$		$\gamma_1 = 0$; $\gamma_2 = 3.75$		$\gamma_1 = 1.5$; $\gamma_2 = 3.75$	
Test / $\alpha \cdot 100$	10	5	10	5	10	5	10	5	10	5
mod. Bartlett-	11.56	5.53	11.41	5.37	12.32	5.90	12.53	5.86	13.60	7.01
mod. χ^2-	12.27	6.31	12.74	6.49	13.87	6.99	14.59	7.58	15.42	8.77
F-	10.39	5.30	18.04	10.94	17.86	10.54	26.12	18.07	26.37	17.99
Box/Scheffé-2-	9.93	4.99	9.35	4.61	9.39	4.80	10.10	4.92	9.66	4.69
Box/Scheffé-3-	10.23	5.16	9.63	4.87	9.47	4.58	10.71	5.09	9.62	4.67
Box/Andersen-	11.07	5.14	10.22	4.77	11.35	5.30	10.95	5.00	12.05	6.30
Jackknife-	9.86	5.31	11.59	6.36	12.60	6.66	14.12	8.21	14.54	8.54
Levene-z-	11.45	5.63	11.08	5.51	14.34	7.64	11.94	6.11	17.30	10.23
Levene-s-	11.13	4.95	10.06	4.34	11.14	5.03	10.34	4.32	11.59	5.66

Table 2: continuation

n = 42	$\gamma_1 = 0$; $\gamma_2 = 0$		$\gamma_1 = 0$; $\gamma_2 = 1.5$		$\gamma_1 = 1.0$; $\gamma_2 = 1.5$		$\gamma_1 = 0$; $\gamma_2 = 3.75$	
Test / $\alpha \cdot 100$	10	5	10	5	10	5	10	5
mod. Bartlett-	10.82	5.48	10.66	5.79	11.77	5.95	10.76	5.23
mod. χ^2-	11.15	5.74	11.30	5.36	12.43	6.53	11.60	5.93
F-	10.21	5.19	19.56	12.52	20.03	12.54	28.89	20.63
Box/Scheffé 2-	10.44	5.64	9.83	4.88	10.09	5.42	10.13	5.02
Box/Scheffé 3-	9.85	5.32	9.84	4.87	10.21	4.97	10.16	5.10
Box/Andersen-	10.61	5.29	10.78	4.49	11.40	5.68	9.92	4.84
Jackknife-	10.00	5.39	11.14	5.97	12.17	6.88	12.61	7.27
Levene-z-	10.83	5.55	10.24	5.06	14.51	8.14	11.06	5.58
Levene-s-	10.68	5.25	10.10	4.22	11.27	5.49	9.70	4.41

Table 2: continuation

n = 42	$\gamma_1 = 1.5$; $\gamma_2 = 3.75$		$\gamma_1 = 0$; $\gamma_2 = 7.0$		$\gamma_1 = 2.0$; $\gamma_2 = 7.0$	
Test / $\alpha \cdot 100$	10	5	10	5	10	5
mod. Bartlett-	12.73	6.08	10.86	5.03	12.50	5.98
mod χ^2-	13.39	6.79	12.12	5.93	13.57	6.89
F-	30.20	22.23	37.00	28.97	37.94	29.95
Box/Scheffé 2-	9.75	4.79	9.64	4.42	10.12	4.81
Box/Scheffé 3-	9.25	4.59	9.53	4.58	10.04	4.98
Box/Andersen-	11.43	5.58	9.76	4.28	11.32	5.23
Jackknife	14.35	7.94	13.87	7.82	14.90	8.75
Levene-z-	17.41	10.32	11.26	5.41	19.70	12.33
Levene-s-	11.21	5.20	9.17	3.74	10.90	4.75

Table 3

Results of simulation for the power function of the robust test $(\widehat{1-\beta}\,(n, g, \lambda)) \cdot 100$ in dependence on n, λ, α, and γ_1, γ_2

λ	1.44		1.96		3.24	
Test / $\alpha \cdot 100$	10	5	10	5	10	5
$n = 18$; $\gamma_1 = 0$, $\gamma_2 = 0$						
mod. Bartlett-	19.92	10.92	37.99	24.57	74.01	59.02
Box/Scheffé-3-	15.01	8.00	26.49	15.65	54.36	37.13
Box/Andersen-	19.13	10.36	36.53	23.13	72.21	56.48
Levene-s-	19.14	10.06	36.48	22.15	71.71	54.30
$n = 18$; $\gamma_1 = 0$, $\gamma_2 = 1.5$						
mod. Bartlett-	18.74	9.98	31.29	19.55	62.13	46.29
Box/Scheffé-3-	14.59	7.74	23.49	13.28	45.23	29.19
Box/Andersen-	17.16	9.00	29.18	17.67	58.82	42.22
Levene-s-	16.92	8.28	28.54	15.99	57.40	38.06
$n = 18$; $\gamma_1 = 1.0$, $\gamma_2 = 1.5$						
mod. Bartlett-	19.86	11.01	33.31	21.03	64.03	48.73
Box/Scheffé-3-	14.37	7.67	22.84	13.15	45.81	30.46
Box/Andersen-	18.21	10.23	31.24	19.32	60.61	45.08
Levene-s-	17.97	9.54	30.45	17.73	59.00	41.50
$n = 18$; $\gamma_1 = 0$, $\gamma_2 = 3.75$						
mod. Bartlett-	16.66	8.47	27.87	16.61	54.08	38.23
Box/Scheffé-3-	12.78	6.72	20.39	11.69	39.49	24.91
Box/Andersen-	14.45	7.23	24.88	14.10	49.17	33.65
Levene-s-	13.78	6.13	23.79	11.95	46.65	28.86

Table 3: continuation

λ	1.44		1.96		3.24	
Test / α · 100	10	5	10	5	10	5
$n = 18;\ \gamma_1 = 1.5,\ \gamma_2 = 3.75$						
mod. Bartlett	19.20	10.87	28.95	18.47	55.54	40.85
Box/Scheffé-3-	12.99	6.70	19.72	10.81	38.44	24.32
Box/Andersen-	17.18	9.48	26.13	16.21	51.06	36.67
Levene-s-	16.44	8.31	24.84	14.42	48.77	31.98
$n = 42;\ \gamma_1 = 0,\ \gamma_2 = 0$						
mod. Bartlett-	31.20	20.17	68.19	54.11	97.65	94.54
Box/Scheffé-3-	24.04	13.44	51.15	34.66	88.56	75.28
Box/Andersen-	30.84	19.78	67.69	53,46	97.48	94.13
Levene-s-	30.83	19.58	67.73	52.88	97.46	93.80
$n = 42;\ \gamma_1 = 0,\ \gamma_2 = 1.5$						
mod. Bartlett-	24.48	14.70	54.55	40.23	90.31	82.19
Box/Scheffé-3-	19.19	10.70	41.26	26.07	77.03	59.73
Box-Andersen-	23.52	13.99	53.26	38.79	89.57	80.77
Levene-s-	23.32	13.45	52.90	37.52	89.27	79.09
$n = 42;\ \gamma_1 = 1.0,\ \gamma_2 = 1.5$						
mod. Bartlett	26.33	16.64	53.82	39.93	89.86	81.64
Box/Scheffé-3-	19.72	10.55	39.25	24.71	76.84	59.75
Box-Andersen-	25.37	15.99	52.75	38.46	89.04	80.05
Levene-s-	25.25	15.40	52.23	37.40	88.68	78.59

Table 3: continuation

λ	1.44		1.96		3.24	
Test / α · 100	10	5	10	5	10	5
$n = 42;\ \gamma_1 = 0,\ \gamma_2 = 3.75$						
mod. Bartlett-	21.29	12.30	43.67	29.99	80.48	38.99
Box/Scheffé-3-	16.88	8.93	32.90	20.51	65.59	46.75
Box/Andersen-	20.32	11.29	41.76	27.96	78.41	65.56
Levene-s-	19.66	10.52	41.00	26.16	77.50	62.57
$n = 42;\ \gamma_1 = 1.5,\ \gamma_2 = 3.75$						
mod. Bartlett-	21.91	12.63	43.58	30.08	79.40	67.83
Box/Scheffé-3-	16.45	8.77	32.05	19.47	64.11	46.01
Box/Andersen-	20.59	11.66	41.67	28.14	77.08	65.22
Levene-s-	20.18	10.79	40.88	26.47	76.14	62.17
$n = 42;\ \gamma_1 = 0,\ \gamma_2 = 7.0$						
mod. Bartlett-	18.78	10.22	36.94	23.65	70.25	56.26
Box/Scheffé-3-	15.26	8.08	28.14	16.66	56.11	38.20
Box/Andersen-	17.08	8.90	34.04	21.20	66.78	52.03
Levene-s-	16.34	7.68	32.72	18.89	64.68	47.49
$n = 42;\ \gamma_1 = 2.0,\ \gamma_2 = 7.0$						
mod. Bartlett-	19.51	11.19	36.02	23,71	69,43	55.29
Box/Scheffé-3-	15.16	8.13	26.35	15.49	53.37	36.42
Box/Andersen-	17.63	10.19	32.92	21.76	65.77	51.09
Levene-s-	17.06	9.09	31.89	19.64	63.75	47.00

7. References

BARTLETT, M. S.
Properties of sufficiency and statistical tests.
Proc. Roy. Soc. A 160 (1937) 268–282.

BARTLETT, M. S., KENDALL, D. G.
The statistical analysis of variance heterogeneity and the logarithmic transformation.
J. R. S. S. 8 (1946) 128–138.

BOX, G. E. P.
Non-normality and tests on variance.
Biometrika 40 (1953) 318–335.

BOX, G. E. P., ANDERSEN, S. L.
Permutation theory in the derivation of robust criteria and the study of departures from assumptions.
J. R. S. S., Ser. B 17 (1955) 1–34.

BROWN, M. B., FORSYTHE, A. B.
Robust tests for the equality of variances.
JASA 69 (1974) 362 p. 364–367.

FLEISHMAN, A. I.
A method for simulating non-normal distributions.
Psychometrika 43 (1978) 521–532.

GAMES, P. A., WINKLER, H. B., PROBERT, D. A.
Robust tests for homogeneity of variance.
Educational and Psychological Measurement 32 (1972) 887–909.

GENG, S., WANG, W. J., MILLER, C.
Small sample size comparisons of tests for homogeneity of variances by Monte-Carlo.
Comm. Statist.-Simulation Comp. B 8 (1979) 4, 379–389.

HERRENDÖRFER, G. (ed.)
Robustheit I.
(1980) Probleme der angewandten Statistik Heft 4, AdL der DDR, Forschungszentrum für Tierproduktion Dummerstorf-Rostock.

LAYARD, M. W. J.
Robust large-sample tests for homogeneity of variances. JASA 68 (1973) 195–198.

LEVENE, H.
Robust Tests for Equality of Variances.
Contributions to probability and statistics, Stanford, University Press (1960) 278–292.

MILLER, R. G. jr.
Jackknifing Variances.
AMS 39 (1968) 567–582.

ODE, R. E., EVANS, J. O.
The percentage points of the normal distribution.
Algorithm AS 70, Appl. Stat. 23 (1974) 96–97.

OVERALL, J. F., WOODWARD, J. A.:
A simple test for heterogeneity of variance in complex factorial designs.
Psychometrika 39 (1974) 311–318.

SCHEFFÉ, H.:
Dispersiony analiz.
Gousudarstwennoe Izdatel'stvo Fiziko-Matematiceskoj Literatury, Moskva, 1963.

Academy of Agricultural Sciences of the GDR
Research Centre of Animal Production Dummerstorf-Rostock

The Influence of Different Shapes of Distributions with the Same First four Moments on Robustness

GERD NÜRNBERG and DIETER RASCH

In robustness research the robustness of a statistical procedure against non-normality often is investigated for distributions with given skewness γ_1 and kurtosis γ_2. But we know that the shape of a density function can vary even if the first four moments are fixed. The question is whether such variation may have an influence on the robustness statements.

Six distributions with $\mu = 0$, $\sigma^2 = 1$, $\gamma_1 \approx 1.5$ and $\gamma_2 \approx 3.75$ are considered:

(i) truncated standard normal (truncation points $u_u = 2.85$; $u_0 = 4.71$)
(ii) Power-transformed normal with $b = 0.865886203523$, $c = 0.221027621012$, $d = 0.027220699158$
(iii) log-normal ($\sigma^2 = 0.1786671141$)
(iv) CQ $(1, \lambda_1)$ with $\lambda_1 = 2.426$
(v) CQ $(2, \lambda_2)$ with $\lambda_2 = 1.579$
(vi) CQ $(3, \lambda_3)$ with $\lambda_3 = 0.5$

These distributions belong to three shape-types. Type I ((i), (iv), (v)) in figure 1, type II ((ii), (iii)) in figure 2 and type III ((vi)) in figure 3. We investigated the robustness of the six distributions and found different power functions.

Table 1 shows the values of the empirical power function due to $\mu = 0$ (H_0) and $\mu = d$ (H_A) for four (α, β)-combinations and $d = 0.6$; $d = 1.0$ and $d = 1.6$.

We find that the distributions of type I and II respectively show different behaviour in empirical α- and β-values.

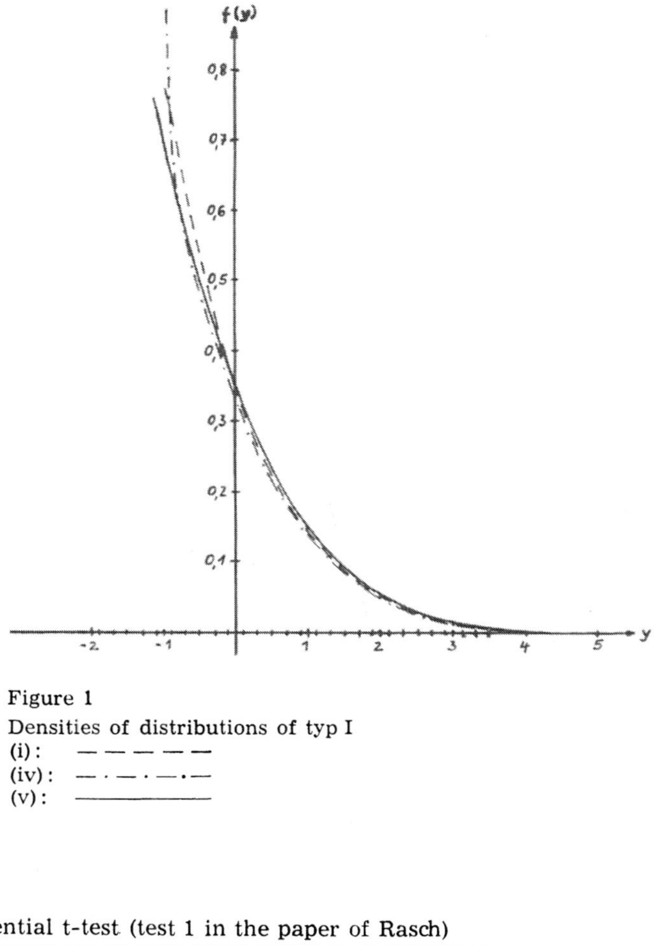

Figure 1
Densities of distributions of typ I
(i): — — — — —
(iv): — · — · — · —
(v): ————————

Table 1
Empirical values ($\alpha \cdot 100$) of $\alpha(f_\alpha)$ of the sequential t-test (test 1 in the paper of Rasch)

Distribution			I			II		III
	100α	100β	(i)	(iv)	(v)	(ii)	(iii)	(vi)
d = 0.6	5	10	10.38	10.11	9.6	7.67	7.32	9.2
	5	20	10.41	10.06	9.7	7.57	7.43	9.2
	10	20	15.91	15.68	15.1	12.63	11.99	14.6
	10	50	15.94	16.39	16.0	13.70	12.86	15.5
d = 1.0	5	10	11.06	11.13	10.7	7.93	7.09	10.0
	5	20	11.14	10.99	10.7	8.06	7.09	9.9
	10	20	16.31	16.31	16.1	12.90	11.56	15.1
	10	50	17.94	17.20	17.1	14.69	13.45	16.8
d = 1.6	5	10	10.76	11.67	10.5	7.73	6.75	9.2
	5	20	11.01	11.72	10.6	8.14	6.97	9.4
	10	20	15.58	16.50	14.8	12.22	11.32	13.6
	10	50	19.55	19.98	18.9	16.06	15.16	17.4

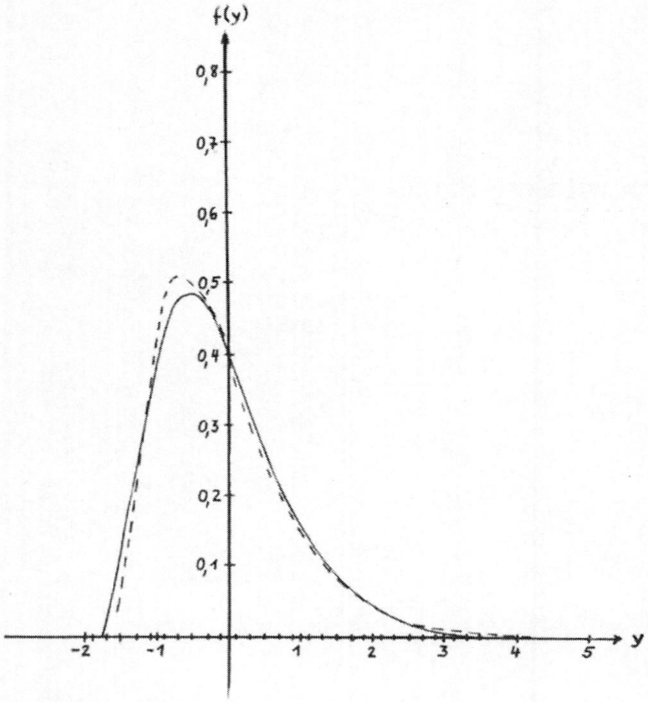

Figure 2

Densities of distributions of typ II
(ii): – – – – –
(iii): ————

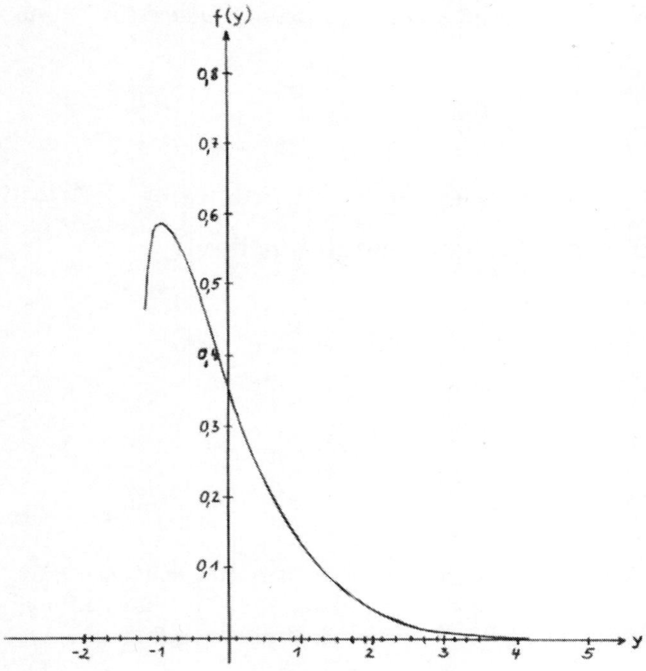

Figure 3

Densities of distribution of typ III
(vi): ————

The distribution of type III seems to have an intermediate behaviour.

The investigations of tests for comparing two variances (see paper by Nürnberg) for the distributions of type I ((i), (iv)) and type II ((ii), (iii)) led nearly to the same results for these distributions (see table 2).

So we conclude that the deviations of moments higher than the fourth for some procedures may have an influence on robustness properties and may have no influence for other procedures.

Table 2

Empirical values ($\hat{\alpha} \cdot 100$) of α of the robust two sample test of variances for different distributions with $\gamma_1 \approx 1.5$; $\gamma_2 \approx 3.75$, for $n = 42$ and $\alpha = 0.05$

test \ Distribution	Type I (i)	(iv)	Type II (ii)	(iii)
Box-Andersen-	5.94	5.79	5.58	5.25
mod. Bartlett-	6.49	6.29	6.08	5.71
Levene-s-	5.51	5.40	5.20	4.90
Box-Scheffé-3-	5.24	5.12	4.59	5.29

Sektion Mathematik, Bergakademie Freiberg, Freiberg, GDR

Robust Bayes Regression Estimation Under Weak Prior Knowledge

JÜRGEN PILZ

Abstract

We are concerned with robustifying the linear Bayes regression estimator to guard against incorrect specification of the error covariance matrix and of the first and second moments of the regression parameter. In particular, we derive a minimax Bayes estimator in case that the prior expected parameter value may vary in some subset Q of the parameter space and only upper bounds of the covariance matrices of the error and prior distribution are known. Moreover, we show that the resulting estimator has a smaller risk than the least squares estimator in some region around the centre of Q.

1. Introduction

In the present article we deal with Bayes regression estimation in case of weak and incomplete prior information. Usual conjugate Bayes estimators require exact assumptions on the type of the error and prior distribution. With qradratic loss structure, restricting consideration to linear estimators allows the analysis to be carried out in terms of the first two moments of these distributions. The computation of the linear Bayes estimator but still requires exact knowledge of the moments.
We consider the case that we have only approximate prior knowledge of the first and second moments and construct a minimax Bayes estimator. This leads to a robustification of the Bayes estimator in the sense of a minimization of the maximum possible Bayes risk which can occur in case of a misspecification of the sampling and prior distribution.
The minimax Bayes compromise criterion was used, among others, by Hodges and Lehmann (1952), Bunke (1964), Solomon (1972) and in the context of regression parameter estimation by Wind (1973), Watson (1974) and Pilz (1981). Solomon (1972) constructed a minimax linear Bayes estimator for a multivariate location parameter in case that its prior expected value is only known to belong to a prespecified rectangular solid, the sampling and prior covariances were but assumed to be known exactly. Chamberlain and Leamer (1976) considered the case of an uncertain prior covariance matrix and obtained regions where the linear Bayes estimator is constrained to lie when the prior covariance matrix may vary in some subclass of possible covariance matrices. Leamer (1982) constructed similar regions for the case that only upper and lower bounds for the prior covariance matrix (in the usual semiordering sense for positive semidefinite matrices) are known.
Another important direction in the development of robust Bayesian alternatives to classical minimax and least squares estimators roots in the construction of Bayesian Stein-type estimators as considered in Berger (1980), (1982).
The intent of our analysis is at least three-fold. First, we wish to incorporate prior knowledge in a flexible and realistic way. Secondly, we have to guard against the effects of a misspecification of the prior contents needed for the analysis. Thirdly, we aim at an improvement in efficiency (risk) over the standard least squares estimator in some parameter subregion of interest. The minimax Bayes estimator developed in the sequel meets these requirements satisfactorily.

2. The Model

In the following let denote R^r the r-dimensional Euclidean space, $\mathfrak{M}_{n\times r}$ the set of real matrices of type $n\times r$, $\mathfrak{M}_r^>$ and $\mathfrak{M}_r^{\geq}$ the sets of positive definite and positive semidefinite matrices of order r, respectively. If A, B $\in \mathfrak{M}_r^{\geq}$ then by $A\leq B$ we mean that $B-A\in\mathfrak{M}_r^{\geq}$, $A<B$ means that $B-A\in\mathfrak{M}_r^>$ and $||x||_A^2$ is shorthand for the quadratic form x^TAx with $x\in R^r$.
We consider the regression model

$$Y=X\Theta+e,\ Ee=0 \tag{1}$$

where Y is the $n\times 1$ vector of random observations, $X\in\mathfrak{M}_{n\times r}$ is the design matrix, $\Theta=(\Theta_1,\ldots,\Theta_r)^T$ is the vector of unknown regression parameters and e is the $n\times 1$ vector of random errors having expectation zero. We will make no further distributional assumptions on the error vector e but only assume knowledge of an upper bound $\Sigma_0\in\mathfrak{M}_n^>$ for the covariance matrix, i. e. the error distribution may be any member of the class

$$\mathfrak{P}_e=\{P_e: Ee=0,\ \Sigma:=\mathrm{Cov}\ e\leq\Sigma_0\}. \tag{2}$$

Suppose that we have prior knowledge about the regression parameters which can be represented by any prior distribution from the class

$$\mathfrak{P}_\Theta=\{P_\Theta:\ \mu:=E\Theta\in Q,\ \Phi:=\mathrm{Cov}\ \Theta\leq\Phi_0\} \tag{3}$$

where Q is some subset of R^r and Φ_0 is an $r\times r$ positive definite matrix which plays the role of an upper bound for the prior covariance matrix of Θ.

Assumption 1:
Let be $Q\subset R^r$ compact, convex and symmetric about some centre point $\mu_0\in R^r$.
In particular, we will consider the case in which Q is either an ellipsoid

$$Q_1=\{\mu\in R^r:\ (\mu-\mu_0)^T\Delta^{-1}(\mu-\mu_0)\leq 1\} \tag{4}$$

with centre point $\mu_0=(\mu_{01},\ldots,\mu_{0r})^T\in R^r$ and "shape" matrix $\Delta\in\mathfrak{M}_r^{\geq}$ or in which Q is a rectangular solid

$$Q_2=\{\mu\in R^r:\ |\mu_i-\mu_{0i}|\leq m_i;\ i=1,\ldots,r\} \tag{5}$$

with edges of prespecified lengths $2m_i\geq 0$ $(i=1,\ldots,r)$.
The sampling and prior information will be assumed to be stochastically independent (which is a standard assumption in Bayesian statistical inference).

Assumption 2:
The joint distribution of (Θ, e) is given by

$$P_{\Theta,e} = P_\Theta \otimes P_e \in \mathfrak{P} := \mathfrak{P}_\Theta \otimes \mathfrak{P}_e.$$

Let D be the class of all estimators of Θ. The goodness of an estimator $\hat{\Theta} \in D$ will be evaluated by a quadratic loss function

$$L(\Theta, \hat{\Theta}) = \|\Theta - \hat{\Theta}\|_U^2, \quad U \in \mathfrak{M}_r^{\geq}. \tag{6}$$

Then, for $\hat{\Theta} = \hat{\Theta}(Y) \in D$, the Bayes risk is given by

$$\varrho(P_{\Theta,e}; \hat{\Theta}) = E_\Theta E_{Y|\Theta} \|\Theta - \hat{\Theta}(Y)\|_U^2. \tag{7}$$

We wish to find an estimator $\hat{\Theta}_{MB}$ which minimizes the maximum possible Bayes risk over all distributions from $\mathfrak{P} = \mathfrak{P}_\Theta \otimes \mathfrak{P}_e$. Any estimator $\hat{\Theta}_{MB}$ which attains this minimum, i. e. for which it holds

$$\sup_{P_{\Theta,e} \in \mathfrak{P}} \varrho(P_{\Theta,e}; \hat{\Theta}_{MB}) = \inf_{\hat{\Theta} \in D} \sup_{P_{\Theta,e} \in \mathfrak{P}} \varrho(P_{\Theta,e}; \hat{\Theta}) \tag{8}$$

will be called a minimax Bayes estimator w. r. t. $\mathfrak{P}$.
If we confine ourselves to the class of linear estimators

$$D_l = \{\hat{\Theta} \in D: \hat{\Theta} = ZY + z, Z \in \mathfrak{M}_{r\times n}, z \in R^r\}$$

then the Bayes risk only depends on the first two moments of $\mathfrak{P}_\Theta$ and $\mathfrak{P}_e$, the type of these distributions does not play any role. If the moments $\mu = E\,\Theta$, $\Phi = \mathrm{Cov}\,\Theta$ and $\Sigma = \mathrm{Cov}\,e$ were known exactly then we could use the linear Bayes estimator which takes the form

$$\hat{\Theta}_B = (X^T\Sigma^{-1}X + \Phi^{-1})^{-1}(X^T\Sigma^{-1}Y + \Phi^{-1}\mu) \tag{9}$$

(see e. g. Hartigan (1969)). If we have, however, only approximate knowledge of these moments as indicated by the above classes $\mathfrak{P}_\Theta$ and $\mathfrak{P}_e$ then it will be shown in section 4 that the minimax Bayes estimator in D_l has a similar structure as $\hat{\Theta}_B$.
Before proceeding with the general problem, let us consider first the special case in which the prior expectation μ is known precisely to have the value μ_0 so that Q reduces to $Q = \{\mu_0\}$.

Theorem 1:
The estimator $\hat{\Theta}^0_{MB} = (X^T\Sigma_0^{-1}X + \Phi_0^{-1})^{-1}(X^T\Sigma_0^{-1}Y + \Phi_0^{-1}\mu_0)$ is minimax Bayes in D w. r. t. $\mathfrak{P}_0 = \{P_{\Theta,e} \in \mathfrak{P}: E\,\Theta = \mu_0\}$.
Note that the minimax Bayes optimality of $\hat{\Theta}^0_{MB}$ is not restricted to the class of linear estimators but holds under all estimators $\hat{\Theta} \in D$. This result is proved in Pilz (1983, section 6.2) for the special case in which the upper bound Σ_0 has the form $\Sigma_0 = \sigma_0^2 J_n$ with some $\sigma_0^2 > 0$, for arbitrary $\Sigma_0 \in \mathfrak{M}_n^{>}$ the proof can proceed in the same way. The proof essentially makes use of monotony relations between distribution functions and of the fact that $\hat{\Theta}^0_{MB}$ is Bayesian in D with respect to the normal distributions $P^0_\Theta = N(\mu_0, \Phi_0)$, $P^0_e = N(0, \Sigma_0)$ which are least favourable distributions within $\mathfrak{P}_0$.
Theorem 1 states that in case of known prior expectation the minimax Bayes estimator is simply the linear Bayes estimator with the unknown covariance matrices replaced by its upper bounds.

3. Reduction of the Problem

Now we deal with the search for a minimax Bayes estimator in D_l w. r. t. $\mathfrak{P} = \mathfrak{P}_\Theta \otimes \mathfrak{P}_e$ when there is uncertainty about μ and Q is no longer a single-element subset.

Lemma 1:
Let be $P_{\Theta,e} \in \mathfrak{P}$. Then $\hat{\Theta} = ZY + z \in D_l$ has Bayes risk

$$\varrho(P_{\Theta,e}; \hat{\Theta}) = \mathrm{tr}\, U[Z\Sigma Z^T + Z_0\Phi Z_0^T] + \|Z_0\mu - z\|_U^2$$

where $Z_0 = I_r - ZX$.

Proof:

$$\varrho(P_{\Theta,e}; \hat{\Theta}) = E_{\Theta,e}\|\Theta - Z(X\Theta + e) - z\|_U^2 = E_{\Theta,e}\|Z_0\Theta - (Ze + z)\|_U^2$$
$$= E_{\Theta,e}\{\Theta^T Z_0^T U Z_0 \Theta - 2\Theta^T Z_0^T U(Ze + z) + \|Ze + z\|_U^2\}$$
$$= \mathrm{tr}\, Z_0^T U Z_0(\Phi + \mu\mu^T) - 2\mu^T Z_0^T Uz + \mathrm{tr}\, Z^T UZ\Sigma + z^T Uz$$
$$= \mathrm{tr}\, UZ\Sigma Z^T + \mathrm{tr}\, UZ_0\Phi Z_0^T + \|Z_0\mu - z\|_U^2.$$

Thus, the search for a minimax Bayes estimator in D_l leads us to the problem of minimizing

$$\sup_{\mu,\Sigma,\Phi}\{\mathrm{tr}\, U[Z\Sigma Z^T + Z_0\Phi Z_0^T] + \|Z_0\mu - z\|_U^2\}$$

over all $Z \in \mathfrak{M}_{r\times n}$ and $z \in R^r$.

Lemma 2:

$$\sup_{P_{\Theta,e} \in \mathfrak{P}} \varrho(P_{\Theta,e}; ZY + z) = \mathrm{tr}\, U[Z\Sigma_0 Z^T + Z_0\Phi_0 Z_0^T] + \sup_{\mu \in Q}\|Z_0\mu - z\|_U^2.$$

This is clear since with $\Sigma \leq \Sigma_0$ and $\Phi_0 \leq \Phi_0$ it follows that

$$\mathrm{tr}\, U[Z\Sigma Z^T + Z_0\Phi Z_0^T] \leq \mathrm{tr}\, U[Z\Sigma_0 Z^T + Z_0\Phi_0 Z_0^T].$$

Lemma 3:
For every $Z \in \mathfrak{M}_{r\times n}$ and $z \in R^r$ it holds

$$\sup_{\mu \in Q}\|Z_0\mu - z\|_U^2 \geq \sup_{\mu \in Q}\|Z_0\mu - Z_0\mu_0\|_U^2.$$

Proof:
Writing $z = Z_0\mu_0 - g$ with arbitrary $g \in R^r$ it follows from the symmetry of $\bar{Q} := \{\bar{\mu} \in R^r: \bar{\mu} = \mu - \mu_0, \mu \in Q\}$ about the origin (due to Assumption 1) that

$$\sup_{\mu \in Q}\|Z_0\mu - z\|_U^2 = \sup_{\bar{\mu} \in \bar{Q}}\{\|Z_0\bar{\mu}\|_U^2 + 2g^T UZ_0\bar{\mu}\} + \|g\|_U^2$$
$$\geq \sup_{\bar{\mu} \in \bar{Q}}\|Z_0\bar{\mu}\|_U^2 = \sup_{\mu \in Q}\|Z_0(\mu - \mu_0)\|_U^2.$$

In view of lemma 3, a minimax Bayes estimator is determined by a matrix Z which minimizes

$$\sup_{P_{\Theta,e} \in \mathfrak{P}} \varrho(P_{\Theta,e}; \hat{\Theta}) = \sup_{\mu \in Q} f_Z(\mu)$$

where

$$f_Z(\mu) = \mathrm{tr}\, U[Z\Sigma_0 Z^T + Z_0\Phi_0 Z_0^T] + \|Z_0(\mu - \mu_0)\|_U^2 \tag{10}$$

and $Z_0 = I_r - ZX$ as before. However, we cannot obtain an analytical expression for this extremum. To solve the problem, we use a minimax theorem known from game theory and which, after an additional integration, permits us to perform the minimization over Z first. This minimization then can be done analytically.

Lemma 4:
Let be Q a compact Hausdorff space, F a class of real-valued and continuous functions on Q and $\tilde{F}$ the convex hull of F. Then it holds

$$\inf_{f \in \tilde{F}} \sup_{\mu \in Q} f(\mu) = \sup_{q \in Q^*} \inf_{f \in F} \int_Q f(\mu)\, q(d\mu)$$

where Q^* is the set of all probability measures defined on the σ-algebra of the Borel subsets of Q.
This result is due to Peck and Dulmage (1957).

Now, with our set $Q \subset R^r$ satisfying Assumption 1, we define for every probability measure $q \in Q^*$ the matrix

$$C_q = \int_Q (\mu - \mu_0)(\mu - \mu_0)^T q(d\mu). \tag{11}$$

Then, with f_Z from (10) we obtain

$$\int_Q f_Z(\mu) q(d\mu) = \operatorname{tr} UZ\Sigma_0 Z^T + \operatorname{tr} UZ_0(\Phi_0 + C_q) Z_0^T \tag{12}$$

and Lemma 4 reads

$$\inf_{Z \in \mathfrak{M}_{r\times n}} \sup_{\mu \in Q} f_Z(\mu) = \sup_{q \in Q^*} \inf_{Z \in \mathfrak{M}_{r\times n}} \int_Q f_Z(\mu) q(d\mu). \tag{13}$$

since $F = \{f_Z : Z \in \mathfrak{M}_{rxn}\}$ is convex.

4. The Optimal Estimator

With the preliminary considerations of section 3 we can now prove our main result.

Theorem 2:

With quadratic loss and Q according to assumption 1,

$$\hat{\Theta}_{MB} = \left(X^T \Sigma_0^{-1} X + (\Phi_0 + C_0)^{-1}\right)^{-1} \left(X^T \Sigma_0^{-1} Y + (\Phi_0 + C_0)^{-1} \mu_0\right)$$

where $C_0 = C_{q_0}$

and $q_0 = \arg \sup_{q \in Q^*} \operatorname{tr} U\left(X^T \Sigma_0^{-1} X + (\Phi_0 + C_q)^{-1}\right)^{-1}$

is the unique minimax Bayes estimator in D_1 w. r. t. $\mathfrak{P}$. The minimax Bayes risk of $\hat{\Theta}_{MB}$ is given by

$$\sup_{P_{\Theta,e} \in \mathfrak{P}} \varrho(P_{\Theta,e}; \hat{\Theta}_{MB}) = \operatorname{tr} U\left(X^T \Sigma_0^{-1} X + (\Phi_0 + C_0)^{-1}\right)^{-1}.$$

Proof:

To minimize the term on the right hand side of (13), write

$$Z = \left[X^T \Sigma_0^{-1} X + (\Phi_0 + C_q)^{-1}\right]^{-1} X^T \Sigma_0^{-1} + G$$

with arbitrary $G \in \mathfrak{M}_{r\times n}$. Observing that $Z_0 = I_r - ZX$, we obtain

$$Z\Sigma_0 Z^T + Z_0(\Phi_0 + C_q) Z_0^T = \left[X^T \Sigma_0^{-1} X + (\Phi_0 + C_q)^{-1}\right]^{-1} + GX(\Phi_0 + C_q) X^T G^T + G\Sigma_0 G^T.$$

From this it follows with (12) that

$$\int_Q f_Z(\mu) q(d\mu) \geq \operatorname{tr} U\left(X^T \Sigma_0^{-1} X + (\Phi_0 + C_q)^{-1}\right)^{-1}$$

for any $Z \in \mathfrak{M}_{r\times n}$ since $\operatorname{tr} U(GX(\Phi_0 + C_q) X^T G^T + G\Sigma_0 G^T) \geqq 0$. Thereby, equality holds if and only if $G = 0$ which implies that

$$\sup_{q \in Q^*} \inf_{Z \in \mathfrak{M}_{r\times n}} \int_Q f_Z(\mu) q(d\mu) = \sup_{q \in Q^*} \operatorname{tr} U\left(X^T \Sigma_0^{-1} X + (\Phi_0 + C_q)^{-1}\right)^{-1}.$$

Finally, the existence of a measure q_0 and thus of a matrix C_0 maximizing the trace functional is guaranteed by the fact that this functional is continuous and convex over the compact and convex set $\mathfrak{B} = \{B_q \in \mathfrak{M}_r^{>} : B_q = X^T \Sigma_0^{-1} X + (\Phi_0 + C_q)^{-1},\ q \in Q^*\}$ (see Lemma 5). The result then follows from equations (12) and (13).

Obviously, $\hat{\Theta}_{MB}$ coincides with the estimator $\hat{\Theta}^0_{MB}$ given in Theorem 1 if we have no doubts that $\mu = E\,\Theta$ is correctly specified by μ_0, i. e. if Q reduces to $Q = \{\mu_0\}$. In this case we have $C_q \equiv C_0 = 0$ and $\hat{\Theta}_{MB}$ is precisely the linear Bayes estimator w. r. t. any prior P_Θ such that $E\,\Theta = \mu_0$ and $\operatorname{Cov} \Theta = \Phi_0$ Otherwise, if we are not sure about the correctness of μ_0 and so Q has cardinality card $Q > 1$, then the minimax Bayes estimator is Bayesian (in D_1) w. r. t. any prior distribution having moments

$$E\,\Theta = \mu_0 \text{ and } \operatorname{Cov} \Theta = \Phi_0 + C_0.$$

This means that minimax Bayes estimation w. r. t. $\mathfrak{P}$ is equivalent to Bayes estimation with an enlarged prior covariance matrix, the enlargement C_0 being due to uncertainty about the first order moment $E\,\Theta$. If our prior knowledge becomes more and more diffuse, which means increasing size of Q, then $(\Phi_0 + C_0)^{-1}$ approaches the matrix of zeroes and in the limiting case $Q = R^r$ we have coincidence of $\hat{\Theta}_{MB}$ with the least squares estimator $\hat{\Theta}^0_{LS} = (X^T \Sigma_0^{-1} X)^{-1} X^T \Sigma_0^{-1} Y$ taken according to the largest possible error covariance matrix Σ_0.

5. Approximate and Particular Solutions

Lemma 4 and Theorem 2 from above accomplish the reduction of the problem of finding a minimax Bayes estimator w. r. t. $\mathfrak{P}$ to the maximization of the functional

$$T(B_q) = \operatorname{tr} UB_q^{-1} \text{ where } B_q = X^T \Sigma_0^{-1} X + (\Phi_0 + C_q)^{-1} \tag{14}$$

over the set of all matrices B_q generated by the probability measures $q \in Q^*$. In general, this maximization will have to proceed numerically, an explicit solution for a special case and rough approximations of the optimal matrix C_0 will be given below.

Lemma 5:

(i) $\mathfrak{B} = \{B_q : q \in Q^*\}$ is a compact and convex subset of $\mathfrak{M}_r^{>}$.

(ii) $T(\cdot)$ is a continuous and convex functional on $\mathfrak{B}$.

Proof:

(i) First, observe that the matrices C_q are positive semidefinite, for it holds $a^T C_q a = \int_Q (a^T \mu - a^T \mu_0)^2 q(d\mu) \geq 0$ for any vector $a \in R^r$ and any $q \in Q^*$. From this it is clear that the matrices B_q are positive definite since $X^T \Sigma_0^{-1} X \in \mathfrak{M}_r^{\geq}$ and $(\Phi_0 + C_q)^{-1} \in \mathfrak{M}_r^{>}$. The compactness and convexity of $\mathfrak{B}$ follows from the fact that Q^* is compact and convex.

(ii) The continuity of T follows immediately from the regularity of B_q and the linearity of the trace functional. The convexity of T can easily be verified by help of the well-known fact that

$$(\alpha A + (1-\alpha)B)^{-1} \leq \alpha A^{-1} + (1-\alpha)B^{-1}$$

for any two matrices $A, B \in \mathfrak{M}_r^{>}$ and any real $\alpha \in (0,1)$. From Lemma 5 and Caratheodory's Theorem we conclude that the supremum of T over $\mathfrak{B}$ and thus the optimum matrix C_0 in $\hat{\Theta}_{MB}$ will be attained for some measure q_0 which is concentrated on at most $r(r+1)/2 + 1$ extreme points of Q. This may substantially reduce the computational efforts needed in the maximization of T.

In case that Q is an ellipsoid, the shape matrix yields an upper bound for the optimal C_0.

Corollary 1:

Let be $Q = Q_1$ as given by (4). Then it holds: $C_0 \leqq \Delta$ and

$$\sup_{P_{\Theta,e} \in \mathfrak{P}} \varrho(P_{\Theta,e}; \hat{\Theta}_{MB}) \leq \operatorname{tr} U\left(X^T \Sigma_0^{-1} X + (\Phi_0 + \Delta)^{-1}\right)^{-1}$$

Proof:

Let be $\Delta^{-1/2}$ the symmetric square root of Δ^{-1} and define $\widetilde{Q}=\{\widetilde{\mu}\in R^r: \widetilde{\mu}=\Delta^{-1/2}(\mu-\mu_0),\ \mu\in Q\}$. Then it holds $\widetilde{\mu}^T\widetilde{\mu}\le 1$ or, equivalently, $\widetilde{\mu}\widetilde{\mu}^T\le I_r$ from which it follows that $(\mu-\mu_0)(\mu-\mu_0)^T\le\Delta$ for every $\mu\in Q$. Thus, $C_q=\int_Q(\mu-\mu_0)(\mu-\mu_0)^T q(d\mu)\leqq\Delta$ for all $q\in Q^*$. This, in turn, implies that

$$B_q^{-1}\le\left(X^T\Sigma_0^{-1}X+(\Phi_0+\Delta)^{-1}\right)^{-1}$$

which yields the desired bound for the maximum Bayes risk.

It is argued that by inserting Δ instead of the optimal matrix C_0 the resulting estimator

$$\hat{\Theta}^1_{MB}=\left(X^T\Sigma_0^{-1}X+(\Phi_0+\Delta)^{-1}\right)^{-1}\left(X^T\Sigma_0^{-1}Y+(\Phi_0+\Delta)^{-1}\mu_0\right)$$

is minimax Bayes with respect to an uncertainty ellipsoid Q_1' which includes the original ellipsoid Q_1.

Next we give an explicit solution for the special case in which Q is a rectangular solid and all the relevant matrices are diagonal.

Corollary 2:

Let be $Q=Q_2$ as given by (5) and assume the matrices U, $X^T\Sigma_0^{-1}X$ and Φ_0 to be diagonal. Then the estimator

$$\hat{\Theta}_{MB}=\left(X^T\Sigma_0^{-1}X+(\Phi_0+C_0)^{-1}\right)^{-1}\left(X^T\Sigma_0^{-1}Y+(\Phi_0+C_0)^{-1}\mu_0\right)$$

with

$$C_0=\mathrm{diag}\left(m_1^2,\dots,m_r^2\right)$$

is minimax Bayes in D_1.

Proof:

First it is clear from the matrix inequalities in Theobald (1975) that the optimal C_0 must also be diagonal to achieve a maximum value of T. Then it follows that the optimal C_0 must be such that its diagonal elements have maximum value. Now, for any $q\in Q^*$ the diagonal elements c_i of C_q satisfy

$$c_i=\int_Q(\mu_i-\mu_{0i})^2 q(d\mu)\le m_i^2$$

and equality is attained for any measure q^* giving weight $p_i>0$ to the corner points $\mu^{(i)}$ of Q ($i=1,\dots,s\le 2^r$) and zero weight to all the remaining points of Q. An optimal measure can then be obtained by choosing the weights p_i such that the off-diagonal elements of C will vanish, i. e.

$$c_{kl}=\sum_{i=1}^{s}p_i\left(\mu_k^{(i)}-\mu_{0k}\right)\left(\mu_l^{(i)}-\mu_{0l}\right)=0$$

for $k,l=1,\dots,r;\ k<l$.

6. Risk Comparisons

In this section we shall compare the usual risk and the minimax Bayes risk of $\hat{\Theta}_{MB}$ with that of the least squares estimator. For the sake of simplicity, let us assume that the family of possible error distributions is given by

$$\mathfrak{P}_e^0=\left\{P_e: Ee=0,\ \mathrm{Cov}\,e=\sigma^2V,\ \sigma^2\in(0,\sigma_0^2]\right\}\qquad(15)$$

with some given constant $\sigma_0^2>0$ and given matrix $V\in\mathfrak{M}_n^>$, i. e. the error covariance matrix is known up to a multiple. Further, let us assume that X is of full rank. Then the LSE for Θ and the covariance matrix are given by

$$\hat{\Theta}_{LS}=\left(X^TV^{-1}X\right)^{-1}X^TV^{-1}Y,\quad \mathrm{Cov}\,\hat{\Theta}_{LS}=\sigma^2\left(X^TV^{-1}X\right)^{-1}.$$

Let denote $B_0=X^T\Sigma_0^{-1}X+(\Phi_0+C_0)^{-1}$ the optimal matrix B_q computed with $C_q=C_0$, $\Sigma_0=\sigma_0^2V$ being the upper bound for the error covariance matrix, and define

$$B:=\sigma_0^2B_0=X^TV^{-1}X+\sigma_0^2\left(\Phi_0+C_0\right)^{-1}\qquad(16)$$

so that the minimax Bayes estimator takes the form

$$\hat{\Theta}_{MB}=B^{-1}\left(X^TV^{-1}Y+\sigma_0^2\left(\Phi_0+C_0\right)^{-1}\mu_0\right).$$

Clearly, $\hat{\Theta}_{MB}$ is a biased estimator with bias given by

$$b(\Theta)=\sigma_0^2B^{-1}\left(\Phi_0+C_0\right)^{-1}(\mu_0-\Theta)\qquad(17)$$

and covariance matrix

$$\mathrm{Cov}\,\hat{\Theta}_{MB}=\sigma^2B^{-1}X^TV^{-1}XB^{-1}.\qquad(18)$$

Lemma 6:

(i) $\mathrm{Cov}\,\hat{\Theta}_{MB}<\mathrm{Cov}\,\hat{\Theta}_{LS}$

(ii) With $H=\sigma_0^4\left(\Phi_0+C_0\right)^{-1}B^{-1}UB^{-1}\left(\Phi_0+C_0\right)^{-1}$, $\hat{\Theta}_{MB}$ has risk

$$R\left(\Theta,\sigma;\hat{\Theta}_{MB}\right)=\sigma^2\,\mathrm{tr}\,UB^{-1}X^TV^{-1}XB^{-1}+\|\Theta-\mu_0\|_H^2.$$

Proof:

(i) By direct computation we have $B\,(X^T V^{-1} X)^{-1}\,B=X^T V^{-1} X+A$ where $A=2\sigma_0^2(\Phi_0+C_0)^{-1}+\sigma_0^4[(\Phi_0+C_0)X^T V^{-1} X(\Phi_0+C_0)]^{-1}$ is positive definite. Thus, it follows that

$$\mathrm{Cov}\,\hat{\Theta}_{MB}=\sigma^2\left[B\left(X^TV^{-1}X\right)^{-1}B\right]^{-1}<\sigma^2\left(X^TV^{-1}X\right)^{-1}=\mathrm{Cov}\,\hat{\Theta}_{LS}.$$

(ii)
$$\begin{aligned}\mathbf{R}(\boldsymbol{\Theta},\boldsymbol{\sigma};\hat{\boldsymbol{\Theta}}_{MB})&=E_{Y|\Theta}\|\Theta-\hat{\Theta}_{MB}(Y)\|_U^2\\&=\mathrm{tr}\,U\left(\mathrm{Cov}\,\hat{\Theta}_{MB}\right)+\|b(\Theta)\|_U^2\\&=\sigma^2\,\mathrm{tr}\,UB^{-1}X^TV^{-1}XB^{-1}+\|\Theta-\mu_0\|_H^2.\end{aligned}$$

The following result demonstrates that there exists a non-empty subregion in the parameter space for which the minimax Bayes estimator has smaller risk than the least squares estimator.

Theorem 3:

With quadratic loss (6) and a full rank matrix X, $\hat{\Theta}_{MB}$ has smaller risk than $\hat{\Theta}_{LS}$ for all parameters $(\Theta,\sigma^2)\in R^r\times(0,\sigma_0^2)$ for which it holds

$$\|\Theta-\mu_0\|_H^2\le\sigma^2\,\mathrm{tr}\,U\left[\left(X^TV^{-1}X\right)^{-1}-B^{-1}X^TV^{-1}XB^{-1}\right].$$

Proof:

Observing that $\hat{\Theta}_{LS}$ has risk $R(\Theta,\sigma;\hat{\Theta}_{LS})=\mathrm{tr}\,U(\mathrm{Cov}\,\hat{\Theta}_{LS})$ we obtain

$$R\left(\Theta,\sigma;\hat{\Theta}_{LS}\right)-R\left(\Theta,\sigma;\hat{\Theta}_{MB}\right)=\mathrm{tr}\,U\left(\mathrm{Cov}\,\hat{\Theta}_{LS}-\mathrm{Cov}\,\hat{\Theta}_{MB}\right)-\|\Theta-\mu_0\|_H^2$$

This yields the result by virtue of the fact that

$$\mathrm{tr}\,U\left(\mathrm{Cov}\,\hat{\Theta}_{LS}-\mathrm{Cov}\,\hat{\Theta}_{MB}\right)=\sigma^2\,\mathrm{tr}\,U\left(\left(X^TV^{-1}X\right)^{-1}-B^{-1}X^TV^{-1}XB^{-1}\right)$$

is positive due to the positive definite character of the difference $\mathrm{Cov}\,\hat{\Theta}_{LS}-\mathrm{Cov}\,\hat{\Theta}_{MB}$.

Obviously, the subregion of parameter values Θ for which $\widehat{\Theta}_{MB}$ leads to an improvement in risk over the LSE is an ellipsoid which is centered at the symmetry point μ_0 of the uncertainty region Q associated with the possible values of the prior expectation $\mu = E\Theta$. The size of this ellipsoid and the magnitude of the improvement depend on the precision of our prior knowledge expressed by the matrix $(\Phi + C_0)^{-1}$ and on the quality of the upper bound σ_0^2 for the variance of observation.

When comparing the corresponding minimax Bayes risks, it turns out that the risk of $\widehat{\Theta}_{MB}$ is bounded from above and below, respectively, by that of the LSE and of the minimax Bayes estimator $\widehat{\Theta}^{\circ}_{MB}$ in case of $Q = \{\mu_0\}$.

Corollary 3:

Let be $\mathfrak{P} = \mathfrak{P}_\Theta \otimes \mathfrak{P}^0_e$. Then it holds

$$\sigma_0^2 \operatorname{tr} U\left(X^T V^{-1} X + \sigma_0^2 \Phi_0^{-1}\right)^{-1} \leq \sup_{P_{\Theta,e} \in \mathfrak{P}} \varrho\left(P_{\Theta,e}; \widehat{\Theta}_{MB}\right) \leq \sigma_0^2 \operatorname{tr} U\left(X^T V^{-1} X\right)^{-1}.$$

This follows from theorem 2 observing that

$$\sup_{P_{\Theta,e} \in \mathfrak{P}} \varrho\left(P_{\Theta,e}; \widehat{\Theta}_{MB}\right) = \operatorname{tr} U B_0^{-1} = \sigma_0^2 \operatorname{tr} U B^{-1}$$

and $X^T V^{-1} X \leq B \leq X^T V^{-1} X + \sigma_0^2 \Phi_0^{-1}$.

In corollary 3, the upper bound refers to the case of non-informative prior knowledge implying that $(\Phi_0 + C_0)^{-1} = 0$ and the lower bound refers to the case of exact knowledge of the prior expectation implying that $C_0 = 0$. In the intermediate nonextreme situations, sharper lower bounds for the minimax Bayes risk of $\widehat{\Theta}_{MB}$ can be obtained by inserting B_q (with a particular choice of the measure $q \in Q^*$) instead of B_0. On the other hand, sharper upper bounds can be obtained using the particular shape of the region Q as demonstrated e. g. in Corollary 1 for the case of an ellipsoid.

7. References

BERGER, J. O.
A robust generalized Bayes estimator and confidence region for a multivariate normal mean.
Ann. Statist. **8** (1980), 716—761.

BERGER, J. O.
Selecting a minimax estimator of a multivariate normal mean.
Ann. Statist. **10** (1982), 81—92.

BUNKE, O.
Bedingte Strategien in der Spieltheorie: Existenzsätze und Anwendung auf statistische Entscheidungsprobleme.
Transact. Third Prague Conf. on Information Theory, Statist. Decision Functions, Random Processes, Prague, 1964, 35—43.

CHAMBERLAIN, G. and LEAMER, E. E.
Matrix weighted averages and posterior bounds.
J. Roy. Statist. Soc., Ser. B **38** (1976), 73—84.

HARTIGAN, J. A.
Linear Bayesian methods.
J. Roy. Statist. Soc., Ser. B **31** (1969), 446—454.

HODGES, J. L. and LEHMANN, E. L.
The use of previous experience in reaching statistical decisions.
Ann. Math. Statist. **23** (1952), 396—407.

LEAMER, E. E.
Sets of posterior means with bounded variance priors.
Econometrica **50** (1982), 725—736.

PECK, J. E. L. and DULMAGE, A. L.
Games on a compact set.
Canad. J. Math. **9** (1957), 450—458.

PILZ, J.
Robust Bayes and minimax-Bayes estimation and design in linear regression.
Math. Operationsforsch. Stat., Ser. Statist. **12** (1981), 163—177.

PILZ, J.
Bayesian estimation and experimental design in linear regression models.
Teubner-Verlag, Leipzig, 1983.

SOLOMON, D. L.
Λ-Minimax estimation of a multivariate location parameter.
J. Amer. Statist. Assoc. **67** (1972), 641—646.

THEOBALD, C. M.
An inequality with application to multivariate analysis.
Biometrika **62** (1975), 461—466.

WATSON, S. R.
On Bayesian inference with incompletely specified prior distribution.
Biometrika **61** (1974), 193—196.

WIND, S. L.
An empirical Bayes approach to multiple linear regression.
Ann. Statist **1** (1973), 93—103.

Inst. Math., Acad. Sciences GDR Berlin

Distribution of the Maximal Gap in a Sample and its Application for Outlier Detection

RICHARD PINCUS

Abstract

Regardless of the underlying distribution the maximal gap is asymptotically stochastically larger than a Gumbel-distributed variable. An asymptotic expression for the distribution of the maximal gap is given and is compared with the outcome of simulation studies for truncated normal distributions.

1. Introduction

Given $n \geq 3$ independent observations $x_1, \ldots, x_n$ the normalized maximal gap in the sample

$$M_n = \max_{i=2..n} \frac{x_{(i)} - x_{(i-1)}}{x_{(n)} - x_{(1)}}$$

forms a reasonable test statistic for detecting an unknown number of possible outliers.
The distribution of M_n depends on the underlying distribution of $x_1, \ldots, x_n$, of course, and gets a simple form if the observations are uniformly distributed. It will be shown that M_n or rather a transformation $Z_n = nM_n - \log n$ is asymptotically 'stochastically minimal' if the underlying distribution is just uniform.

2. The Distribution of the Maximal Gap Under Uniform Distributions

By invariance it is evident that the distribution of M_n does not depend on location and scale.
In the following $(1-jx)_+$ means $\max(1-jx;\ 0)$, $0 \leq x \leq 1$.
Proposition 1:
If $x_1, \ldots, x_n$ are uniform distributed, then the distribution of M_n is given by

$$P(M_n \leq x) = \sum_{j=0}^{n-1} (-1)^j \binom{n-1}{j} (1-jx)_+^{n-2}. \tag{1}$$

Proof:
M_n has the same distribution as

$$M'_n = \max_{i=2..n} x'_{(i)} - x'_{(i-1)}$$

where $x'_{(2)}, \ldots, x'_{(n-1)}$ are independently uniform $(0,1)$-distributed and $x'_{(1)} = 0$, $x'_{(n)} = 1$.
Validity of (1) is easy to see for $n = 3$. Now we have

$$P(M'_{n+1} \leq x) = \int_{1-x}^{1} P\left(M'_n \leq \frac{x}{x'}\right) \cdot (n-1)x'^{n-2} dx' \tag{2}$$

Substituting (1) in (2) gives the result.
The distribution (1) was found by Fisher (1929) as the distribution of $y_{(n)} / \Sigma y_i$ for the exponential distribution, see Barnett and Lewis (1977), p. 79.
For the transformed variable

$$Z_n = nM_n - \log n$$

we get from (1), substituting $x = (z + \log n)/n$:
Proposition 2:
A non degenerated limit distribution of Z_n exists and has the form

$$P(Z_n \leq z) \xrightarrow[n]{} e^{-e^{-z}}, \quad -\infty < z < \infty \tag{3}$$

The limit distribution is the so called extreme value distribution of first kind, or Gumbel distribution.
Figure 1 presents the exact distribution function of Z_n for selected n at the 5-, 10-, 20- and 30-percent points of limit distribution, and shows that (3) forms a satisfactory approximation if $n \geq 20$.

Figure 1
$P(nM - \log n \leq z)$

n \ z	2.97	2.25	1.50	1.03
5	.997	.953	.784	.580
10	.977	.930	.804	.654
20	.967	.921	.809	.683
30	.963	.917	.810	.692
50	.960	.912	.809	.698
∞	.950	.900	.800	.700

3. Asymptotic Distribution of the Maximal Gap Under Non Uniform Distribution

Let f be a density with $f > 0$ on a finite interval, and let $x_1, \ldots, x_n$ independently distributed according to that distribution.
Without loss of generality we may assume that the support is the interval $(0, 1)$.
A rough approximation of the density by a step function with values f_j on $\left(\frac{j-1}{K}, \frac{j}{K}\right)$, $j = 1, \ldots, K$, gives by (3) for the maximal gap in the j-th interval, provided it contains n_j observations

$$P(n_j M_{(j)} K \leq y) \longrightarrow e^{-e^{-(y - \log n_j)}}, \quad y > 0.$$

By the law of large numbers we can further approximate n_j, by nf_j/K, thus getting

$$P(nM_{(j)} \leq x) \longrightarrow e^{-e^{-(xf_j - \log(nf_j/K))}} \tag{4}$$

Since $nM \leq x$ iff $nM_{(j)} \leq x$ for all j, (4) gives taking the product of all expressions

$$P(nM \leq x) \xrightarrow[n]{} e^{-n \sum \frac{f_j}{K} e^{-xf_j}} \tag{5}$$

If we know choose K sufficiently large and form the limit in the exponent of (5) we get

Proposition 3:

Under the above condition on the boundedness of the distribution we have asymptotically

$$P(nM \leq x) \xrightarrow[n]{} e^{-n\int e^{-xf(t)}F(dt)}, \quad x>0. \tag{6}$$

Remark: Convergence $P_n \to Q_n$ is used in this section in the sense that for any sequence of intervals, A_n say, Q_n $(A_n) \to c$ implies P_n $(A_n) \to c$.
An interesting property of the right side of (6) is, that it attains it minimum iff f forms a rectangular distribution, i. e. $f \equiv 1$.

In that case (6) can be written as $e^{-e^{(-x-\log n)}}$ and is equivalent to (3).

Proposition 4:

We have

$$e^{-n\int e^{-xf(t)}F(dt)} \geq e^{-e^{-(x-\log n)}}, \quad x>0, \tag{7}$$

with equality iff $f(t) \equiv 1$ (a. e.), $0 \leq t \leq 1$.
Proof:
If we return to (5) then replacing f_j/K by p_j, we have $\Sigma p_i = 1$, and see that $\Sigma p_i \; e^{-xKp_j}$, $x \geq \frac{4}{K}$, is maximized subject to $\Sigma p_j = 1$, $p_j \geq 0$, iff $p_1 = \ldots = p_K = \frac{1}{K}$.
This immediately implies the assertion.
An interpretation of inequality (7) is that the Maximal Gap test forms a consistent Goodness-of-Fit test for uniform distributions.
On the other hand the Maximal Gap test indicates outliers in the presence of an underlying non-uniform distribution with a probability larger than α, even if there are not outlier.
If one has information on the underlying distribution one can use (6) to find the asymptotically true significance point.

4. Monte-Carlo Results

Monte-Carlo studies showed, however, that the approximation (6) is not very accurate for moderate sample sizes like $n = 30$ and even $n = 100$, so that higher order approximations could be useful.

Figures 2 and 3 show results of simulation with standard normal distributions truncated at $\pm\sqrt{3}$. For sample sizes $n = 30$ and $n = 100$, respectively, 1000 repetitions were done. As to expect, for truncation points ± 1, i. e. for a distribution which is 'nearer' to a uniform one, the approximation is better, while for truncation at $\pm\sqrt{5}$ it is worse.

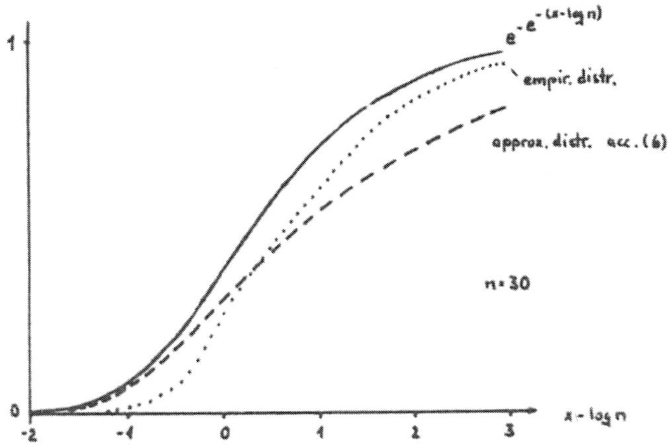

Figure 2

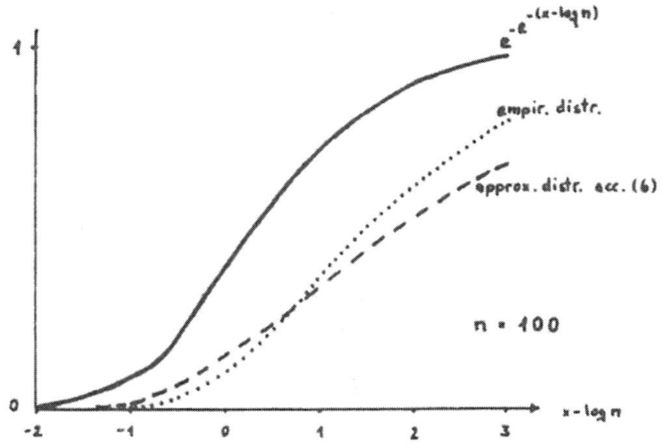

Figure 3

5. References

BARNETT, V. and LEWIS, T.
Outliers in Statistical Data.
Wiley, New York, 1978.
FISHER, R. A.
Tests of significance in harmonic analysis.
Proc. Roy. Soc. A, 125, (1929) 54–59.

Department of Statistics, University of Connecticut Storrs, Connecticut 06268 USA

Robustness of the Two-Sample T-Test

HARRY O. POSTEN

Abstract

In the literature, one finds evidence that the two-sample t-test is robust with respect to departures from normality, and departures from homogeneity of variance (at least when sample sizes are equal or nearly equal). This evidence, presented in various articles, is usually based on an approximate approach without error analysis or on a simulation approach that is of limited extent. The present paper takes a closer and more extensive look at the quality of this procedure under departures from the primary assumptions of normality and of equal variances. The results presented are a synthesis of several previous papers by the author and colleagues, with particular emphasis on the use of a broad Monte Carlo approach to the assessment of robustness.

1. Introduction

In robustness research, there are two directions one may take. One may attempt to quantify or measure the degree of robustness inherent in a standard statistical procedure, or one may attempt to develop a new alternative procedure which, in some sense, is more robust than the standard procedure. In recent years, much of the robustness literature has been concerned with the development of such new procedures. However, significant contributions can still be made in the study of the robustness of standard procedures since, even for the most familiar procedures, there exists vagueness concerning the conditions under which the procedure is robust and under which it is nonrobust. For example, one finds general evidence in the literature that the two-sample t-test is fairly robust with respect to departures from normality and also with respect to departures from homogeneity of variance (at least when sample sizes are equal or nearly equal). On the other hand, one can also find evidence that the two-sample t-test may not be robust under certain conditions. Bradley (1980) provided results from a simulation study (30,000 generated values of the two-sample t-statistic for samples from several pairs of populations and sample sizes) which suggested that dramatically different shapes for the two populations could produce significant nonrobustness in the Type 1 error probability. Also, studies by Hyrenius (1950) and Zachrisson (1959) for the one-sample t-test hint at the possibility of nonrobustness of the two-sample t-test under two types of practical conditions: the condition of samples from a compound population (occurring when a population is a mixture of two or more distinct populations), and the condition of the samples being stratified samples from two or more populations (occurring when conditions change during the selection of the sample). Despite this negative evidence, there is sufficient support in the literature (see Hatch and Posten (1966) for a survey of robustness research for the one- and two-sample t-tests) to indicate that under simple random sampling from populations which do not differ strongly in shape and which are not extreme departures from normality, the two-sample t-test is not very sensitive to nonnormality. Also, the evidence indicates that with equal or nearly equal sample sizes this test is robust with respect to departures from homogeneity of variance.

The present paper is concerned with a synthesis of several recent studies which in a detailed manner provide an assessment of the robustness level of the two-sample t-test under common practical conditions. In some cases, the results are restricted to the two-tailed test or to equal sample sizes, but on the whole they provide answers to the question "What level of robustness does the two-sample t-test have?" for important practical cases.

2. Robustness Under Heterogeneity of Variance

A recent paper, Posten, Yeh and Owen (1982), studied the change in the true significance level $\alpha(\lambda)$ of the two-sample double-tailed t-test when the populations are normal but the ratio $\lambda = \sigma_1^2/\sigma_2^2$ varies from the assumed value $\lambda = 1$. The results of this theoretical study indicate an extremely strong level of robustness under departures from equal variances when the sample sizes are equal. This level of robustness is probably stronger than most people realize. Table 1 provides these results in terms of the concept of "total robustness at a given robustness level". Specifically, the t-test is considered to be totally robust at level ε if, no matter what the value of λ, one makes no more an error than ε in assuming the significance level to be $\alpha(1)$, the value under the condition of equal variances. Mathematically, this means that as λ ranges over $(0, \infty)$, $|\alpha(\lambda) - \alpha(1)|$ varies only within a range bounded by ε. For example, from table 1, with $n_1 = n_2 = 20$ and $\alpha(1) = .05$, the maximum error one can make in assuming the significance level to be 0.05 is 0.0072. Thus, the true significance level will be no more than 0.0072 from the assumed level of 0.05, no matter how much σ_1^2 varies from σ_2^2. Further, an error of a magnitude near the value 0.0072 will occur only when σ_1^2 is very much larger or smaller than σ_2^2 and table 1 may therefore be used to conservatively determine the degree of robustness of the equal sample size t-test under violations of the assumption of equal variances. From table 1, it is clear that this test is quite robust when sample sizes are equal.

The question of what happens to the robustness of the two-sample t-test when sample sizes are unequal is discussed in the same paper. The results are given in table 2 in terms of maximal regions of robustness. A "maximal region of robustness" of level ε is the region of λ-values over which the true significance level, $\alpha(\lambda)$, deviates from the assumed value, $\alpha(1)$, by no more than ε. If this range of λ is wide in a practical sense, then the t-test is robust at this level ε. The maximal regions of robustness are given in table 2 for equal sample sizes and for sample sizes that vary 10 % and 20 % from equality. Table 2 indicates that the sizes of maximal regions of robustness

Table 1
Minimum value of ε for which the t-test is totally robust at level ε ($n_1 = n_2 = n$)

	nominal value $\alpha = \alpha(l)$			nominal value $\alpha = \alpha(l)$	
	$\alpha = 0.05$	$\alpha = 0.01$		$\alpha = 0.05$	$\alpha = 0.01$
n	ε	ε	n	ε	ε
2	.0954	.0539	15	.0098	.0052
3	.0589	.0341	20	.0072	.0038
4	.0419	.0241	25	.0057	.0030
5	.0324	.0184	30	.0048	.0025
6	.0263	.0148	50	.0028	.0015
7	.0222	.0124	100	.0014	.0007
8	.0191	.0106	500	.0003	.0001
9	.0168	.0093	1000	.0001	.0001
10	.0150	.0082	∞	.0000	.0000

Table 2
Maximal regions of robustness of level ε for the twotailed t-test (nominal significance level $\alpha(l) = 0.05$)

$\varepsilon = 0.03$

Equal Sample Sizes			10 % Sample Size Change			20 % Sample Size Change		
n_1	n_2	λ-range	n_1	n_2	λ-range	n_1	n_2	λ-range
5	5	0.02—85.63				4	6	0.00—2.89
10	10	0.00—∞	9	11	0.00— 8.33	8	12	0.00—3.06
15	15	0.00—∞	14	16*	0.00— ∞	12	18	0.00—3.17
20	20	0.00—∞	18	22	0.00— 17.58	16	24	0.00—3.25
25	25	0.00—∞	23	27*	0.00— ∞	20	30	0.02—3.30
30	30	0.00—∞	27	33	0.00— 36.95	24	36	0.03—3.33
40	40	0.00—∞	36	44	0.00—104.22	32	48	0.06—3.38
50	50	0.00—∞	45	55	0.00— ∞	40	60	0.05—3.42

$\varepsilon = 0.02$

n_1	n_2	λ-range	n_1	n_2	λ-range	n_1	n_2	λ-range
5	5	0.09—12.33				4	6	0.00—2.14
10	10	0.00—∞	9	11	0.00—4.02	8	12	0.21—2.17
15	15	0.00—∞	14	16*	0.00—9.31	12	18	0.26—2.20
20	20	0.00—∞	18	22	0.00—4.95	16	24	0.28—2.21
25	25	0.00—∞	23	27*	0.00—3.90	20	30	0.29—2.22
30	30	0.00—∞	27	33	0.00—5.55	24	36	0.30—2.23
40	40	0.00—∞	36	44	0.00—5.96	32	48	0.31—2.24
50	50	0.00—∞	45	55	0.00—6.26	40	60	0.31—2.25

* = sample size change nearest to 10 % change from equality but not greater

reduce dramatically as sample sizes vary significantly from equality. Thus, the t-test tends to lose its strong degree of robustness rapidly as the sample sizes become unequal. When each sample size varies by 10 % from a condition of equal sample sizes, the t-test still has a respectable amount of robustness with respect to the Type 1 error probability. However, when the sample sizes reach a 20 % difference from equality, one might wish to be more cautious with the use of the t-test. To an important degree, the loss of robustness when sample sizes are unequal is in the range where $\lambda > 1$, that is, when the larger variance is associated with the smaller sample size. The level of robustness for the unequal sample size test can, therefore, be significantly improved if one knows beforehand which population has this smaller variance. In this case, the smaller sample size may be assigned to the population with the smaller variance. The range of λ is the restricted to (0, 1] and table 2 can be used with the righthand entries all replaced by 1. The result is that the t-test becomes somewhat robust with respect to the Type 1 error probability, even when sample sizes are somewhat unequal, as long as the smaller sample is taken from the population having the smaller variance. The original paper also contains results for $\alpha(l) = .01$ with similar results.

3. Robustness Under Nonnormality

To precisely determine the degree of robustness of the two-sample t-test over a wide range of practical non-normal distributions is a difficult problem. An exact theoretical approach is impractical because of its mathematical intractability, an approximate approach would lack accuracy assurances, and a simulation approach requires an exhorbitant amount of computer time to achieve respectable precision over an extensive practical range of distributions. A simulation approach, however, can be made practical by using a computer artifice to speed up sample generation and by using low priority computer time to reduce computer costs.

Such a simulation study was provided by Posten (1978). The intent of that study was to accurately quantify the degree of robustness of the two-sample t-test for a range of sample sizes over a wide range of practical distributions. The Pearson family of distributions was chosen because it appeared to have best withstood the test of time, in terms of representing practical data. The range of coverage was for both negative and positive skewness over $0 \leq \beta_1 \leq 2.0$ and $1.4 \leq \beta_2 \leq 7.8$, where $\beta_1 = \mu_3^2/\sigma^6$ and $\beta_2 = \mu_4/\sigma^4$. This seems to be a wide range of coverage for practical distributions if one judges by the range of reported values of β_1 and β_2 in, for example, Scheffe (1959) and Pearson and Please (1975). The decision on the fineness of the grid covering this region was conservatively made and the final coverage was for $\beta_1 = 0\ (0.4)\ 2.0$ and

$\beta_2 = 1.4$ (0.4) 7.8, excluding the impossible distributions in this range. The study was constructed over sample sizes $n_1 = n_2 = n$ with $n = 5$ (5) 30.

The basic strategy of the study was to generate a sufficiently large number of observations for each of the 87 Pearson distributions to provide very good accuracy in the significance level and power evaluations for each sample size. The key to bringing the simulation study within a practical range was a computer artifice used for producing a large enough number of generated values from each of these distributions, to provide 100,000 values of t for each sample size $n = 5$ (5) 25. For practical programming reasons, it was decided to settle for 83,000 generated t-values in the case $n = 30$.

Random variable generation on a computer is usually performed by using a linear congruential method to generate uniformly distributed random values and then transforming these to values that follow the required distribution. This transformation can be rapidly effected by use of the inverse distribution function. In the case of the Pearson distributions, these inverse functions are not available. Therefore, the transformations for the Pearson distributions were accomplished by numerically tabulating the distribution function within the computer and obtaining the transformed values by interpolation from this interval table. The numerical tabulation was over many values (2000—6000 values, depending on the distribution) of the argument. This approach ordinarily would provide a relatively slow transformation method, because for each generated uniform value it is necessary to search for the tabulated interval which includes it before performing the transformation by interpolation. Since searching is a relatively slow computer process, a significant reduction of generation time can be achieved if the search is eliminated. To accomplish this, uniform numbers were first generated in blocks of 10,000. Each block was ordered and stored on tape in ascending order. This strategy allowed each uniform value to be trans formed without searching for the proper interval which included it, since each succeeding uniform value is very close to the preceding value. The transformed values are then no longer random but may easily be restored to the random order of the original uniform values by storing each transformed value in a storage position corresponding to the original occurrence position of the uniform values. The tape, then, has a block of 10,000 uniform values ordered in ascending order followed by a block of 10,000 integers which identify the original order position for each of the 10,000 uniform values. Five hundred such pairs of blocks were ordered on tape, enough to provide at least 100,000 double-samples for each value of $n = 5(5)25$. These ordered uniform values were then used to efficiently generate the five million observations required for each Pearson distribution in the study.

Each stage of the Monte Carlo study was conducted with extreme care in order to provide quality assurances on the accuracy of the results. Since the study was identical in form for all Pearson distributions, with the only programming variation being in the actual numerical evaluation of each Pearson distribution, the overall quality of this study can be judged by viewing the results for the normal distribution member of the Pearson family. For the normal distribution, the results (36 results in all) can be compared with known correct probabilities for this distribution. The latter are given in table 3, where each entry is the evaluated probability using 100,000 generated t-values (except for $n = 30$, as previously indicated). Since these values are actually the Monte Carlo evaluations of 0.05 for the first two columns, 0.50 for the next two columns and 0.95 for the last two columns, table 3 provides strong quality assurances for this study. To two decimal places, all round off exactly to the correct probability.

The original paper provides results for both one- and two-tailed tests. However, only the two-tailed results will be presented here (the one-tailed results are similar). Table 4 provides the results for the significance level. Rounding off to two decimals, it is evident from the table that for all sample sizes and Pearson distributions studied, the significance level is in the range 0.03 to 0.06. Further, for only a few extreme distributions, when $n = 5$, does the significance level round off to 0.03, 0.04 or 0.06. All other significance levels round off to 0.05. Thus, there is extremely little variation in the significance level over this extensive range of distributions. This means that one of the arguments favoring the choice of a nonparametric procedure for this problem, that the significance level of a nonparametric procedure is always exactly the nominal level, is of little consequence since for the equal sample t-test (over the range of distributions studied), the significance level is very near the nominal level. Arguments favoring a nonparametric approach for this problem should therefore be based only on power considerations. Results for the power evaluated at noncentrality parameter values corresponding to powers of 0.50 and 0.95 under normal conditions are provided in tables 5 and 6. A review of these tables shows that for all sample sizes and all distributions studied, the range of the power levels is only 0.444 to 0.566 for the 50 % normal power level and 0.934 to 0.983 for the 95 % normal power level. In each of these cases, the more extreme probabilities occur with the more extreme distributions and, as sample size increases, the robustness level improves further. If n is at least 15, the two power levels to two decimals are in the range 0.48—0.53 and 0.94—0.96. The latter result along with the results for the significance level indicate a very strong level of robustness for the equal sample t-test, probably stronger than most users realize.

4. Comparisons with the Wilcoxon Test

The foregoing suggests that, at least for the equal sample case, the two-sample t-test should be quite useful as a solution to the two-sample location problem. However, considerable theoretical support, particularly asymptotic efficiency levels, recommends the Wilcoxon test as a viable general alternative to the t-test. The prominent nonparametricist G. E. Noether, in fact, recommends its use most of the time in lieu of the t-test.

Unfortunately, there are comparatively few studies in the literature on the small sample power of the Wilcoxon test and the existing papers are, for the most part, concerned with Wilcoxon power for normal parent populations. A recent paper by the author, Posten (1982), however, using the broad simulation approach discussed above, provides an evaluation of Wilcoxon power over the same sample sizes and Pearson distributions discussed in the previous section. Again, the objective was an accurate assessment of power with the two noncentrality values used being the exact values required to produce 50 % and 95 % power for the t-test under normality. These values of the shift parameter are the values used in the foregoing t-test study. Thus, this Wilcoxon study is identical in form to the t-test study, thereby allowing direct comparisons of power for the two procedures. However, because the length of computer time required for the

Table 3
Monte Carlo probabilities for the two-sample two-tailed t-test (normal case, significance level = 0.05)

	Significance level		50 % Power		95 % Power	
n	Upper Tail Test	Double Tail Test	Upper Tail Test	Double Tail Test	Upper Tail Test	Double Tail Test
5	0.0503	0.0510	0.4999	0.4997	0.9499	0.9501
10	0.0497	0.0495	0.4984	0.4986	0.9507	0.9507
15	0.0507	0.0486	0.4999	0.5005	0.9511	0.9512
20	0.0511	0.0495	0.5012	0.5012	0.9510	0.9511
25	0.0507	0.0495	0.4991	0.4986	0.9508	0.9506
30	0.0483	0.0491	0.4989	0.4987	0.9495	0.9495

Monte Carlo evaluation of Wilcoxon power is significantly greater than for the t-test, only two million values were generated for each of the 87 Pearson distributions involved. This still provided 40,000 generated values of the Wilcoxon statistic, U, for each sample size $n = 5(5)25$. For $n = 30$, 33,200 values of U were generated each time.

The original paper provides results for both one- and two-tailed tests but for the present purpose only the two-tailed results will be discussed (the one-tailed results are not very different). As it turns out, the power of the Wilcoxon test for a fixed shift value is more variable over the Pearson family than the power of the t-test. In many cases, particularly for the 50 % (normal t-test power) shift value, this variability is favorable to the Wilcoxon test since it results in higher power for this test. Tables 7 and 8 provide the Wilcoxon Monte Carlo results. For purposes of comparing Wilcoxon and t-test power, these tables provide the difference in power of the t-test and the Wilcoxon test (t-test power minus Wilcoxon test power).

It is evident from the large number of negative entries in tables 7 and 8 that, except for sample size $n = 5$, these results support, to a large degree, the recommendations of some nonparametricists that the Wilcoxon test be used generally rather than the t-test. For $n = 5$, the tables indicate that the power function of the t-test is dominant over that of the Wilcoxon test over essentially the entire range of the Pearson family presented. Other than this, the Wilcoxon function is dominant over a substantial part of the Pearson family if one is concerned with the 50 % (normal) shift value of the power curve. The region of dominance is considerably reduced if one considers the power curve region around the 95 % (normal) shift value. In fact, the situation often changes, with the t-test power increasing (around the 95 % shift value) to being nearly equal or superior to the Wilcoxon power. The general pattern of the Wilcoxon power curve, if one stays away from a modest sized region of the Pearson family near the normal distribution, seems to be approximately the following: the power curve of the Wilcoxon test starts off (near the null value) more sharply than the t-test and is superior to the t-test power curve until the shift parameter reaches values associated with higher probabilities, at which point the t-test power may be nearly equal or superior to the Wilcoxon power. This is not a perfect picture since it depends upon the region of the Pearson system involved. The reader may review the pattern of the tables for particulars.

As previously indicated, the results of this study, to a large degree, would appear to support the nonparametric recommendation to use the Wilcoxon test as a general solution to the two-sample location problem. A note of caution, however, is probably justified. The range of distributional coverage provided by the Pearson distributions of this study is extensive. It is conceivable that in many application areas the range of distributions one is likely to meet are of considerably less extent. Consequently, depending upon the area of application, one may wish to consider a more restricted region of the Pearson family. In particular, it seems reasonable to at least eliminate distributions near the left boundary and lower extreme left boundary of the Pearson plane, where the U-shape and J-shape distributions reside. The latter distributions seem more likely to announce their presence in the sample data. Despite the support provided by this study for the Wilcoxon test, the results could also provide to some degree an argument for general use of the t-test, particularly if U- and J-shaped distributions are disregarded. Suppose one asks what is the region in the Pearson system where the power of the t-test at the 50 % value of the shift parameter is not less than 0.05 of the Wilcoxon power and for the 95 % value of the shift parameter is not less than 0.03 of the Wilcoxon power? This would be a region in which the power function of the t-test is either superior of "not bad at all" compared to the Wilcoxon power. An inspection of tables 7 and 8 shows that this region varies with the sample sizes but in general is a fairly substantial region. A similar argument could also be applied to the Wilcoxon test since the truth of the matter is that over a substantial part of the Pearson plane the power of the two tests does not differ dramatically.

5. Conclusions

It would seem that despite the informativeness of the foregoing studies, these results do not definitely make a choice between the equal sample t-test and Wilcoxon test for the general two-sample location problem. The results tend to at least partially support a Wilcoxon choice in a situation when the population differ only by a shift parameter. However, several authors, Boneau (1962), Van der Vaart (1960) and Glazer (1963), have indicated that for normal populations and equal sample sizes, the t-test is superior to the Wilcoxon test in robustness for departures from the assumption of equal variances. It would seem, therefore, that further studies of the effects of variance heterogeneity on the two tests would be needed over an extensive practical family of nonnormal distributions before a single procedure might be specified as a somewhat general choice for the two-sample location problem.

6. References

BONEAU, C. A.
A comparison of the power of the U and t-tests.
Psychological Review 69 (1962) 245—256.

BRADLEY, J. V.
Nonrobustness in classical tests on means and variances: a large scale study.
Bulletin of the Psychonomic Society 15 (1980) 275—278.

GLAZER, H.
Comparison of the Student two-sample t-test and the Wilcoxon test for normal distributions with equal variances.
Ph. D. Dissertation, Boston University, 1963.

HATCH, L. O. and POSTEN, H. O.
Robustness of the Student procedure: a survey.
Res Rep. No. 24, Dept. of Statistics, Univ. of Conn. (1966) (currently available for a fee as AD 643 494 from National Technical Information Service, Springfield, Va. 22161 USA).

HYRENIUS, H.
Distribution of 'Student'-Fisher's t in samples from compound normal functions.
Biometrika 37 (1950) 429—442.

PEARSON, E. S. and PLEASE, N. W.
Relation between the shape of population distribution and the robustness of four simple test statistics.
Biometrika 62 (1975) 223—241.

POSTEN, H. O.
The robustness of the two-sample t-test over the Pearson system.
J. of Statistical Computation and Simulation 6 (1978) 295—311.

POSTEN, H. O.
Two-sample Wilcoxon power over the Pearson system and comparisons with the t-test.
J. of Statistical Computation and Simulation 16 (1982) 1—18.

POSTEN, H. O., YEH, H. C. and OWEN, D. B.
Robustness of the two-sample t-test under violations of the homogeneity of variance assumption.
Communication in Statistics Theory and Methods 11 (1982) 109—126.

SCHEFFE, H.
The Analysis of Variance.
John Wiley and Sons, New York, 1959.

Van der LAAN, P. and WEIMA. J.
Experimental comparison of the powers of the two-sample tests of Wilcoxon and Student under logistic parent distributions.
J. of Statistical Computation and Simulation 8 (1978) 133—144.

Van der VAART, H. R.
On the robustness of Wilcoxon's two-sample test.
In Qualitative Methods in Pharmacology (ed. De Jonge), North Holland Pub. Co. (1960) 140—158.

ZACHRISSON, U.
The distribution of "Student's" t in samples from individual nonnormal populations.
Publication of Statistical Institute, No. 6, University of Gothenburg, Sweden, 1959.

Table 4
Monte Carlo Results for Significance Level (Two-Sample t-Test, Double Tail Test, $\alpha = .05$)

β_1 β_2	1.4	1.8	2.2	2.6	3.0	3.4	3.8	4.2	4.6	5.0	5.4	5.8	6.2	6.6	7.0	7.4	7.8
0.0	.0563	.0555	.0540	.0523	.0510	.0498	.0491	.0484	.0479	.0474	.0471	.0468	.0466	.0464	.0463	.0461	.0460
0.4		.0514	.0520	.0514	.0509	.0500	.0493	.0488	.0482	.0478	.0475	.0473	.0470	.0468	.0465	.0464	.0463
0.8			.0454	.0480	.0488	.0489	.0488	.0486	.0483	.0481	.0477	.0474	.0471	.0469	.0466	.0466	.0465
1.2				.0404	.0438	.0459	.0467	.0472	.0472	.0473	.0473	.0472	.0471	.0469	.0466	.0465	.0463
1.6	n = 5				.0348	.0398	.0427	.0445	.0455	.0460	.0463	.0463	.0464	.0465	.0465	.0465	.0464
2.0						.0307	.0353	.0393	.0416	.0434	.0445	.0451	.0454	.0456	.0458	.0459	.0460
0.0	.0515	.0509	.0501	.0498	.0495	.0489	.0485	.0481	.0478	.0474	.0472	.0469	.0468	.0466	.0465	.0463	.0463
0.4		.0516	.0507	.0497	.0492	.0486	.0482	.0480	.0477	.0475	.0474	.0472	.0470	.0470	.0468	.0466	.0465
0.8			.0508	.0502	.0492	.0484	.0476	.0476	.0474	.0472	.0470	.0469	.0468	.0467	.0466	.0466	.0465
1.2				.0497	.0495	.0486	.0476	.0471	.0468	.0468	.0466	.0465	.0466	.0464	.0463	.0463	.0463
1.6	n = 10				.0477	.0480	.0478	.0472	.0465	.0463	.0459	.0460	.0460	.0460	.0459	.0459	.0460
2.0						.0450	.0463	.0469	.0464	.0462	.0458	.0457	.0455	.0455	.0454	.0455	.0456
0.0	.0509	.0494	.0492	.0491	.0486	.0484	.0482	.0479	.0477	.0476	.0475	.0473	.0472	.0471	.0470	.0469	.0468
0.4		.0505	.0501	.0492	.0487	.0483	.0481	.0477	.0477	.0474	.0473	.0472	.0470	.0470	.0470	.0468	.0467
0.8			.0508	.0501	.0490	.0486	.0481	.0477	.0474	.0472	.0471	.0470	.0470	.0468	.0467	.0467	.0466
1.2				.0507	.0500	.0487	.0480	.0479	.0475	.0472	.0468	.0468	.0468	.0469	.0468	.0466	.0467
1.6	n = 15				.0503	.0497	.0486	.0481	.0472	.0471	.0467	.0468	.0466	.0464	.0466	.0465	.0464
2.0						.0494	.0491	.0482	.0477	.0472	.0467	.0464	.0462	.0462	.0462	.0461	.0460
0.0	.0510	.0504	.0499	.0498	.0495	.0492	.0491	.0490	.0489	.0487	.0486	.0485	.0484	.0483	.0483	.0481	.0480
0.4		.0509	.0504	.0499	.0496	.0494	.0492	.0489	.0487	.0486	.0484	.0482	.0481	.0481	.0481	.0480	.0480
0.8			.0500	.0502	.0499	.0494	.0492	.0492	.0491	.0487	.0486	.0484	.0482	.0481	.0479	.0478	.0477
1.2				.0499	.0500	.0498	.0493	.0491	.0487	.0487	.0485	.0486	.0484	.0482	.0481	.0480	.0479
1.6	n = 20				.0494	.0495	.0493	.0490	.0487	.0486	.0485	.0484	.0483	.0481	.0481	.0481	.0479
2.0						.0494	.0487	.0488	.0486	.0485	.0482	.0482	.0483	.0482	.0480	.0478	.0478
0.0	.0505	.0503	.0503	.0499	.0495	.0496	.0494	.0493	.0491	.0488	.0487	.0486	.0485	.0484	.0483	.0484	.0483
0.4		.0502	.0496	.0491	.0492	.0494	.0493	.0493	.0491	.0491	.0490	.0489	.0486	.0485	.0483	.0482	.0482
0.8			.0500	.0497	.0490	.0490	.0487	.0488	.0487	.0486	.0485	.0484	.0485	.0485	.0485	.0483	.0483
1.2				.0493	.0493	.0494	.0490	.0488	.0486	.0485	.0485	.0485	.0482	.0481	.0482	.0482	.0482
1.6	n = 25				.0494	.0490	.0491	.0489	.0489	.0487	.0483	.0480	.0480	.0482	.0482	.0480	.0479
2.0						.0495	.0488	.0489	.0487	.0486	.0484	.0483	.0481	.0478	.0477	.0478	.0479
0.0	.0496	.0494	.0495	.0494	.0491	.0491	.0492	.0490	.0490	.0488	.0488	.0487	.0487	.0486	.0486	.0485	.0484
0.4		.0498	.0495	.0495	.0493	.0494	.0494	.0491	.0491	.0489	.0487	.0487	.0485	.0484	.0483	.0482	.0481
0.8			.0495	.0492	.0493	.0492	.0488	.0489	.0490	.0488	.0487	.0488	.0487	.0487	.0484	.0483	.0482
1.2				.0492	.0492	.0492	.0489	.0486	.0488	.0485	.0487	.0486	.0484	.0484	.0485	.0484	.0485
1.6	n = 30				.0490	.0490	.0489	.0485	.0486	.0482	.0481	.0481	.0480	.0480	.0481	.0483	.0481
2.0						.0490	.0492	.0484	.0480	.0480	.0480	.0478	.0476	.0478	.0479	.0477	.0477

Table 5
Monte Carlo Results for .50 Normal Power N.C.P. (Two-Sample t-Test, Double Tail Test, $\alpha = .05$)

β_1 β_2	1.4	1.8	2.2	2.6	3.0	3.4	3.8	4.2	4.6	5.0	5.4	5.8	6.2	6.6	7.0	7.4	7.8
0.0	.4444	.4599	.4756	.4881	.4997	.5099	.5182	.5256	.5306	.5352	.5393	.5426	.5454	.5480	.5500	.5521	.5538
0.4		.4582	.4764	.4909	.5014	.5112	.5181	.5247	.5304	.5350	.5391	.5425	.5453	.5480	.5503	.5525	.5543
0.8			.4660	.4887	.5040	.5146	.5222	.5281	.5326	.5369	.5407	.5440	.5467	.5490	.5510	.5533	.5550
1.2				.4721	.5001	.5161	.5257	.5322	.5377	.5412	.5447	.5471	.5495	.5519	.5540	.5557	.5576
1.6	n = 5				.4789	.5100	.5268	.5364	.5423	.5466	.5494	.5525	.5545	.5565	.5580	.5594	.5609
2.0						.4870	.5202	.5371	.5463	.5515	.5561	.5588	.5603	.5619	.5636	.5645	.5659
0.0	.4714	.4791	.4862	.4929	.4986	.5044	.5095	.5137	.5172	.5202	.5229	.5251	.5274	.5292	.5308	.5323	.5337
0.4		.4809	.4863	.4930	.4988	.5035	.5080	.5125	.5161	.5196	.5224	.5250	.5271	.5291	.5310	.5327	.5342
0.8			.4876	.4933	.4994	.5050	.5091	.5136	.5171	.5201	.5227	.5251	.5274	.5295	.5312	.5331	.5346
1.2				.4953	.5007	.5054	.5111	.5151	.5188	.5218	.5247	.5271	.5292	.5311	.5331	.5346	.5359
1.6	n = 10				.5009	.5059	.5119	.5171	.5211	.5249	.5275	.5295	.5313	.5333	.5352	.5366	.5378
2.0						.5069	.5109	.5173	.5227	.5268	.5304	.5332	.5352	.5366	.5377	.5388	.5407
0.0	.4835	.4882	.4938	.4969	.5005	.5044	.5080	.5112	.5138	.5164	.5188	.5206	.5225	.5240	.5254	.5266	.5277
0.4		.4882	.4922	.4973	.5019	.5053	.5087	.5118	.5140	.5162	.5186	.5204	.5224	.5235	.5249	.5262	.5273
0.8			.4928	.4970	.5023	.5063	.5098	.5125	.5153	.5181	.5198	.5217	.5234	.5246	.5261	.5272	.5284
1.2				.4973	.5014	.5060	.5109	.5145	.5172	.5192	.5213	.5230	.5248	.5259	.5273	.5288	.5302
1.6	n = 15				.5020	.5067	.5098	.5141	.5179	.5210	.5233	.5249	.5261	.5279	.5292	.5304	.5313
2.0						.5058	.5106	.5139	.5173	.5209	.5242	,5270	.5286	.5300	.5309	.5321	.5332
0.0	.4884	.4922	.4955	.4987	.5012	.5041	.5068	.5089	.5110	.5133	.5151	.5162	.5174	.5187	.5197	.5206	.5215
0.4		.4923	.4955	.4985	.5015	.5041	.5065	.5092	.5114	.5135	.5155	.5171	.5184	.5198	.5210	.5219	.5226
0.8			.4958	.4988	.5011	.5043	.5070	.5089	.5114	.5135	.5158	.5178	.5195	.5206	.5222	.5232	.5239
1.2				.4996	.5025	.5050	.5074	.5099	.5122	.5144	.5159	.5180	.5196	.5210	.5224	.5234	.5242
1.6	n = 20				.5034	.5053	.5084	.5111	.5135	.5155	.5171	.5190	.5208	.5220	.5232	.5244	.5257
2.0						.5058	.5087	.5116	.5138	.5170	.5187	.5206	.5222	.5235	.5248	.5259	.5267
0.0	.4912	.4926	.4943	.4967	.4986	.5010	.5032	.5055	.5075	.5090	.5106	.5120	.5131	.5143	.5153	.5163	.5171
0.4		.4927	.4945	.4981	.5006	.5030	.5048	.5065	.5083	.5099	.5111	.5124	.5136	.5147	.5155	.5163	.5171
0.8			.4947	.4958	.4997	.5021	.5044	.5072	.5089	.5101	.5117	.5127	.5139	.5153	.5162	.5174	.5186
1.2				.4978	.4987	.5015	.5041	.5064	.5091	.5109	.5130	.5139	.5153	.5163	.5174	.5184	.5192
1.6	n = 25				.5001	.5018	.5037	.5065	.5089	.5112	.5133	.5147	.5160	.5175	.5182	.5191	.5199
2.0						.5025	.5046	.5061	.5091	.5113	.5135	.5147	.5163	.5177	.5193	.5204	.5210
0.0	.4910	.4938	.4957	.4976	.4987	.5011	.5027	.5035	.5049	.5065	.5076	.5087	.5096	.5106	.5117	.5125	.5134
0.4		.4941	.4953	.4975	.4994	.5009	.5031	.5047	.5064	.5076	.5085	.5098	.5108	.5119	.5129	.5135	.5143
0.8			.4961	.4971	.4997	.5010	.5026	.5048	.5068	.5079	.5087	.5101	.5114	.5126	.5135	.5141	.5149
1.2				.4981	.4990	.5013	.5028	.5043	.5068	.5084	.5094	.5105	.5114	.5129	.5141	.5150	.5161
1.6	n = 30				.5003	.5018	.5040	.5050	.5064	.5087	.5103	.5115	.5122	.5135	.5149	.5157	.5164
2.0						.5026	.5042	.5066	.5072	.5086	.5108	.5127	.5134	.5147	.5158	.5167	.5180

Table 6
Monte Carlo Results for .95 Normal Power N.C.P. (Two-Sample t-Test, Double Tail Test, $\alpha = .05$)

β_1 β_2	1.4	1.8	2.2	2.6	3.0	3.4	3.8	4.2	4.6	5.0	5.4	5.8	6.2	6.6	7.0	7.4	7.8
0.0	.9828	.9713	.9631	.9560	.9501	.9453	.9425	.9407	.9391	.9383	.9378	.9370	.9364	.9361	.9358	.9356	.9356
0.4		.9740	.9636	.9555	.9495	.9461	.9440	.9422	.9408	.9399	.9392	.9386	.9380	.9377	.9374	.9370	.9367
0.8			.9657	.9562	.9491	.9441	.9416	.9403	.9395	.9392	.9388	.9385	.9383	.9379	.9375	.9372	.9369
1.2				.9578	.9493	.9430	.9386	.9372	.9364	.9362	.9363	.9368	.9369	.9368	.9368	.9366	.9365
1.6	n = 5				.9491	.9434	.9372	.9332	.9322	.9327	.9334	.9340	.9344	.9349	.9351	.9355	.9356
2.0						9407	.9382	.9323	.9283	.9282	.9290	.9302	.9311	.9318	.9325	.9332	.9336
0.0	.9636	.9601	.9568	.9536	.9507	.9480	.9459	.9443	.9433	.9425	.9418	.9411	.9406	.9401	.9398	.9395	.9394
0.4		.9595	.9564	.9536	.9506	.9481	.9463	.9449	.9440	.9427	.9421	.9414	.9408	.9403	.9398	.9396	.9393
0.8			.9559	.9526	.9499	.9474	.9451	.9439	.9428	.9422	.9417	.9410	.9406	.9401	.9399	.9397	.9397
1.2				.9524	.9498	.9469	.9444	.9426	.9413	.9406	.9400	.9398	.9398	.9396	.9395	.9394	.9392
1.6	n = 10				.9495	.9467	.9437	.9417	.9399	.9387	.9386	.9385	.9382	.9382	.9384	.9385	.9383
2.0						.9464	.9439	.9411	.9393	.9379	.9365	.9365	.9366	.9367	.9367	.9370	.9369
0.0	.9594	.9575	.9549	.9531	.9512	.9494	.9482	.9471	.9464	.9455	.9447	.9441	.9436	.9431	.9427	.9423	.9422
0.4		.9574	.9554	.9531	.9511	.9494	.9478	.9467	.9457	.9450	.9444	.9440	.9435	.9431	.9427	.9425	.9425
0.8			.9550	.9532	.9512	.9491	.9475	.9460	.9450	.9447	.9436	.9432	.9428	.9425	.9425	.9422	.9420
1.2				.9533	.9510	.9488	.9474	.9457	.9441	.9435	.9431	.9427	.9425	.9422	.9420	.9416	.9415
1.6	n = 15				.9508	.9488	.9472	.9453	.9440	.9427	.9421	.9415	.9412	.9411	.9409	.9405	.9405
2.0						.9488	.9471	.9452	.9435	.9425	.9413	.9406	.9402	.9399	.9397	.9395	.9395
0.0	.9571	.9555	.9538	.9523	.9511	.9493	.9482	.9471	.9460	.9452	.9446	.9442	.9439	.9435	.9432	.9430	.9426
0.4		.9553	.9534	.9517	.9506	.9494	.9483	.9472	.9463	.9458	.9449	.9446	.9441	.9436	.9433	.9428	.9427
0.8			.9539	.9518	.9503	.9491	.9481	.9471	.9461	.9453	.9447	.9441	.9438	.9436	.9433	.9431	.9429
1.2				.9522	.9503	.9492	.9476	.9467	.9459	.9451	.9443	.9439	.9435	.9433	.9430	.9427	.9426
1.6	n = 20				.9506	.9491	.9475	.9464	.9454	.9447	.9437	.9434	.9429	.9425	.9424	.9424	.9421
2.0						.9490	.9476	.9461	.9450	.9442	.9435	.9425	.9421	.9418	.9415	.9415	.9413
0.0	.9555	.9546	.9536	.9522	.9506	.9496	.9484	.9474	.9465	.9460	.9453	.9450	.9447	.9443	.9440	.9438	.9437
0.4		.9538	.9529	.9521	.9508	.9500	.9490	.9482	.9473	.9465	.9462	.9455	.9449	.9444	.9441	.9440	.9437
0.8			.9525	.9519	.9505	.9499	.9486	.9480	.9473	.9466	.9459	.9454	.9450	.9447	.9444	.9443	.9440
1.2				.9515	.9502	.9492	.9487	.9478	.9469	.9462	.9458	,9451	.9446	.9443	.9442	.9441	.9440
1.6	n = 25				.9502	.9489	.9481	.9472	.9460	.9455	.9452	.9446	.9442	.9441	.9438	.9436	.9434
2.0						.9486	.9471	.9466	.9457	.9448	.9444	.9440	.9434	.9431	.9430	.9429	.9428
0.0	.9539	.9530	.9519	.9509	.9495	.9487	.9478	.9471	.9463	.9453	.9448	.9445	.9441	.9438	.9434	.9431	.9430
0.4		.9531	.9520	.9504	.9492	.9485	.9472	.9462	.9456	.9454	.9451	.9448	.9444	.9441	.9436	.9432	.9429
0.8			.9522	.9503	.9493	.9486	.9472	.9464	.9457	.9451	.9447	.9446	.9441	.9439	.9435	.9432	.9431
1.2				.9512	.9498	.9486	.9475	.9466	.9455	.9448	.9445	.9437	.9433	.9431	.9430	.9430	.9428
1.6	n = 30				.9503	.9493	.9475	.9467	.9455	.9449	.9440	.9434	.9430	.9427	.9425	.9420	.9417
2.0						.9494	.9483	.9466	.9455	.9447	.9438	.9433	.9428	.9423	.9420	.9416	.9417

Table 7
Monte Carlo Results for the Difference — t-Test Minus Wilcoxon Test Power
($\alpha = .05$, Double Tail Test, .50 Normal Power N.C.P.)

β_1 \ β_2	1.4	1.8	2.2	2.6	3.0	3.4	3.8	4.2	4.6	5.0	5.4	5.8	6.2	6.6	7.0	7.4	7.8
0.0	0.0513	0.0322	0.0292	0.0275	0.0264	0.0258	0.0248	0.0254	0.0244	0.0236	0.0235	0.0230	0.0229	0.0226	0.0223	0.0222	0.0219
0.4		0.0510	0.0311	0.0255	0.0227	0.0238	0.0227	0.0234	0.0239	0.0237	0.0236	0.0232	0.0229	0.0223	0.0219	0.0216	0.0215
0.8			0.0496	0.0341	0.0270	0.0214	0.0198	0.0199	0.0206	0.0211	0.0215	0.0219	0.0219	0.0220	0.0213	0.0217	0.0209
1.2				0.0468	0.0379	0.0299	0.0235	0.0192	0.0184	0.0176	0.0182	0.0181	0.0191	0.0199	0.0203	0.0205	0.0207
1.6	n = 5				0.0438	0.0403	0.0334	0.0273	0.0219	0.0189	0.0164	0.0165	0.0162	0.0168	0.0167	0.0174	0.0183
2.0						0.0428	0.0449	0.0383	0.0314	0.0254	0.0216	0.0195	0.0164	0.0152	0.0152	0.0153	0.0163
0.0	0.0112	0.0413	0.0418	0.0315	0.0194	0.0097	0.0033	—0.0023	—0.0071	—0.0111	—0.0140	—0.0167	—0.0184	—0.0203	—0.0221	—0.0234	—0.0246
0.4		—0.0432	—0.0095	0.0032	0.0051	0.0034	0.0010	—0.0034	—0.0069	—0.0107	—0.0137	—0.0161	—0.0183	—0.0203	—0.0218	—0.0233	—0.0244
0.8			—0.0911	—0.0538	—0.0348	—0.0252	—0.0190	—0.0155	—0.0147	—0.0156	—0.0169	—0.0190	—0.0201	—0.0216	—0.0231	—0.0248	—0.0259
1.2				—0.1288	—0.0897	—0.0684	—0.0540	—0.0440	—0.0351	—0.0309	—0.0278	—0.0265	—0.0259	—0.0263	—0.0274	—0.0285	—0.0292
1.6	n = 10				—0.1620	—0.1212	—0.0976	—0.0808	—0.0678	—0.0576	—0.0492	—0.0430	—0.0392	—0.0362	—0.0345	—0.0339	—0.0343
2.0						—0.1893	—0.1495	—0.1238	—0.1052	—0.0905	—0.0786	—0.0693	—0.0619	—0.0549	—0.0500	—0.0470	—0.0440
0.0	—0.0192	0.0430	0.0487	0.0353	0.0208	0.0095	0.0003	—0.0077	—0.0139	—0.0191	—0.0228	—0.0266	—0.0301	—0.0325	—0.0348	—0.0367	—0.0381
0.4		—0.1070	—0.0325	—0.0042	0.0048	0.0032	—0.0022	—0.0068	—0.0123	—0.0176	—0.0217	—0.0252	—0.0286	—0.0318	—0.0344	—0.0367	—0.0390
0.8			—0.1800	—0.0995	—0.0582	—0.0369	—0.0260	—0.0218	—0.0213	—0.0221	—0.0258	—0.0279	—0.0310	—0.0341	—0.0358	—0.0376	—0.0397
1.2				—0.2340	—0.1580	—0.1127	—0.0812	—0.0624	—0.0493	—0.0423	—0.0396	—0.0388	—0.0382	—0.0393	—0.0402	—0.0411	—0.0428
1.6	n = 15				—0.2786	—0.2050	—0.1611	—0.1258	—0.1005	—0.0816	—0.0694	—0.0614	—0.0564	—0.0534	—0.0514	—0.0501	—0.0505
2.0						—0.3127	—0.2456	—0.1998	—0.1659	—0.1376	—0.1158	—0.0978	—0.0869	—0.0778	—0.0716	—0.0679	—0.0650
0.0	—0.0530	0.0410	0.0493	0.0373	0.0212	0.0075	—0.0033	—0.0122	—0.0191	—0.0244	—0.0288	—0.0333	—0.0369	—0.0393	—0.0419	—0.0446	—0.0469
0.4		—0.1587	—0.0517	—0.0118	0.0004	—0.0003	—0.0068	—0.0125	—0.0186	—0.0247	—0.0293	—0.0328	—0.0366	—0.0396	—0.0422	—0.0448	—0.0470
0.8			—0.2423	—0.1352	—0.0801	—0.0499	—0.0339	—0.0300	—0.0290	—0.0310	—0.0331	—0.0360	—0.0384	—0.0424	—0.0439	—0.0464	—0.0483
1.2				—0.3030	—0.2053	—0.1439	—0.1044	—0.0781	—0.0622	—0.0515	—0.0485	—0.0474	—0.0476	—0.0483	—0.0495	—0.0512	—0.0533
1.6	n = 20				—0.3473	—0.2614	—0.2010	—0.1562	—0.1236	—0.1003	—0.0840	—0.0728	—0.0655	—0.0621	—0.0609	—0.0602	—0.0597
2.0						—0.3809	—0.3061	—0.2485	—0.2040	—0.1671	—0.1404	—0.1189	—0.1024	—0.0906	—0.0825	—0.0776	—0.0740
0.0	—0.0762	0.0336	0.0463	0.0323	0.0156	0.0011	—0.0098	—0.0190	—0.0261	—0.0321	—0.0371	—0.0415	—0.0448	—0.0479	—0.0505	—0.0530	—0.0551
0.4		—0.1983	—0.0708	—0.0196	—0.0049	—0.0050	—0.0109	—0.0172	—0.0237	—0.0302	—0.0360	—0.0402	—0.0442	—0.0474	—0.0509	—0.0535	—0.0555
0.8			—0.2905	—0.1634	—0.0949	—0.0582	—0.0404	—0.0342	—0.0339	—0.0364	—0.0397	—0.0433	—0.0462	—0.0490	—0.0518	—0.0538	—0.0555
1.2				—0.3527	—0.2415	—0.1663	—0.1182	—0.0870	—0.0687	—0.0598	—0.0546	—0.0540	—0.0550	—0.0558	—0.0573	—0.0587	—0.0600
1.6	n = 25				—0.3965	—0.3031	—0.2317	—0.1767	—0.1382	—0.1105	—0.0910	—0.0797	—0.0744	—0.0699	—0.0685	—0.0680	—0.0680
2.0						—0.4271	—0.3502	—0.2868	—0.2313	—0.1887	—0.1555	—0.1314	—0.1124	—0.1012	—0.0914	—0.0856	—0.0825
0.0	—0.0990	0.0353	0.0489	0.0350	0.0169	0.0032	—0.0092	—0.0199	—0.0280	—0.0342	—0.0400	—0.0444	—0.0487	—0.0525	—0.0555	—0.0581	—0.0604
0.4		—0.2332	—0.0827	—0.0240	—0.0056	—0.0068	—0.0131	—0.0194	—0.0261	—0.0325	—0.0380	—0.0428	—0.0469	—0.0508	—0.0539	—0.0570	—0.0594
0.8			—0.3271	—0.1877	—0.1069	—0.0645	—0.0443	—0.0368	—0.0357	—0.0385	—0.0424	—0.0458	—0.0486	—0.0524	—0.0557	—0.0585	—0.0609
1.2				—0.3901	—0.2719	—0.1875	—0.1305	—0.0942	—0.0731	—0.0629	—0.0587	—0.0575	—0.0585	—0.0594	—0.0608	—0.0626	—0.0643
1.6	n = 30				—0.4285	—0.3344	—0.2554	—0.1954	—0.1519	—0.1206	—0.0981	—0.0869	—0.0799	—0.0761	—0.0741	—0.0736	—0.0741
						—0.4520	—0.3831	—0.3125	—0.2535	—0.2068	—0.1684	—0.1415	—0.1221	—0.1079	—0.0989	—0.0938	—0.0903

Table 8

Monte Carlo Results for the Difference — t-Test Minus Wilcoxon Test Power
($\alpha = .05$, Double Tail Test, .95 Normal Power N.C.P.)

β_1 \ β_2	1.4	1.8	2.2	2.6	3.0	3.4	3.8	4.2	4.6	5.0	5.4	5.8	6.2	6.6	7.0	7.4	7.8
0.0	-0.0087	0.0237	0.0211	0.0207	0.0226	0.0233	0.0247	0.0253	0.0255	0.0261	0.0265	0.0264	0.0265	0.0267	0.0266	0.0267	0.0266
0.4		0.0297	0.0438	0.0362	0.0327	0.0298	0.0285	0.0278	0.0277	0.0279	0.0283	0.0287	0.0286	0.0290	0.0292	0.0292	0.0294
0.8			0.0805	0.0674	0.0527	0.0436	0.0380	0.0342	0.0320	0.0307	0.0298	0.0294	0.0292	0.0290	0.0289	0.0290	0.0288
1.2				0.1277	0.0893	0.0674	0.0540	0.0465	0.0407	0.0369	0.0347	0.0333	0.0318	0.0311	0.0306	0.0301	0.0298
1.6	n = 5				0.1651	0.1108	0.0813	0.0641	0.0533	0.0472	0.0425	0.0396	0.0369	0.0351	0.0336	0.0327	0.0319
2.0						0.1929	0.1297	0.0950	0.0739	0.0606	0.0525	0.0475	0.0437	0.0406	0.0384	0.0364	0.0351
0.0	0.1000	0.0459	0.0279	0.0191	0.0127	0.0067	0.0021	-0.0017	-0.0046	-0.0069	-0.0090	-0.0108	-0.0123	-0.0135	-0.0145	-0.0155	-0.0161
0.4		0.0998	0.0470	0.0264	0.0152	0.0079	0.0024	-0.0015	-0.0042	-0.0069	-0.0088	-0.0108	-0.0124	-0.0138	-0.0152	-0.0160	-0.0167
0.8			0.0975	0.0459	0.0237	0.0114	0.0025	-0.0031	-0.0065	-0.0088	-0.0103	-0.0123	-0.0138	-0.0152	-0.0160	-0.0165	-0.0172
1.2				0.0907	0.0454	0.0222	0.0087	-0.0002	-0.0068	-0.0103	-0.0127	-0.0143	-0.0152	-0.0162	-0.0172	-0.0180	-0.0189
1.6	n = 10				0.0848	0.0431	0.0198	0.0062	-0.0025	-0.0093	-0.0129	-0.0155	-0.0173	-0.0184	-0.0190	-0.0194	-0.0199
2.0						0.0762	0.0397	0.0180	0.0046	-0.0039	-0.0106	-0.0147	-0.0178	-0.0199	-0.0211	-0.0214	-0.0223
0.0	0.0833	0.0470	0.0311	0.0207	0.0129	0.0056	0.0004	-0.0040	-0.0077	-0.0107	-0.0129	-0.0151	-0.0169	-0.0183	-0.0198	-0.0209	-0.0220
0.4		0.0682	0.0363	0.0205	0.0113	0.0035	-0.0016	-0.0056	-0.0092	-0.0118	-0.0141	-0.0159	-0.0177	-0.0194	-0.0206	-0.0215	-0.0223
0.8			0.0536	0.0261	0.0116	0.0024	-0.0035	-0.0080	-0.0112	-0.0134	-0.0160	-0.0179	-0.0190	-0.0201	-0.0211	-0.0223	-0.0230
1.2				0.0388	0.0170	0.0036	-0.0049	-0.0100	-0.0143	-0.0170	-0.0185	-0.0196	-0.0209	-0.0221	-0.0231	-0.0236	-0.0243
1.6	n = 15				0.0247	0.0086	-0.0021	-0.0104	-0.0161	-0.0194	-0.0219	-0.0233	-0.0249	-0.0251	-0.0259	-0.0267	-0.0272
2.0						0.0132	0.0012	-0.0080	-0.0149	-0.0203	-0.0239	-0.0263	-0.0276	-0.0285	-0.0295	-0.0299	-0.0304
0.0	0.0664	0.0444	0.0318	0.0209	0.0111	0.0033	-0.0024	-0.0070	-0.0109	-0.0142	-0.0166	-0.0189	-0.0204	-0.0218	-0.0229	-0.0239	-0.0248
0.4		0.0407	0.0242	0.0134	0.0073	0.0021	-0.0027	-0.0074	-0.0110	-0.0138	-0.0163	-0.0183	-0.0201	-0.0218	-0.0233	-0.0245	-0.0254
0.8			0.0222	0.0094	0.0016	-0.0040	-0.0087	-0.0110	-0.0136	-0.0159	-0.0178	-0.0196	-0.0212	-0.0224	-0.0237	-0.0245	-0.0256
1.2				0.0060	-0.0027	-0.0081	-0.0131	-0.0163	-0.0186	-0.0204	-0.0221	-0.0227	-0.0240	-0.0245	-0.0255	-0.0266	-0.0270
1.6	n = 20				-0.0078	-0.0121	-0.0173	-0.0209	-0.0233	-0.0248	-0.0262	-0.0272	-0.0279	-0.0287	-0.0292	-0.0295	-0.0300
2.0						-0.0190	-0.0207	-0.0238	-0.0272	-0.0290	-0.0305	-0.0318	-0.0326	-0.0326	-0.0330	-0.0331	-0.0335
0.0	0.0501	0.0425	0.0328	0.0215	0.0107	0.0022	-0.0043	-0.0092	-0.0133	-0.0162	-0.0189	-0.0207	-0.0229	-0.0245	-0.0258	-0.0270	-0.0280
0.4		0.0215	0.0166	0.0110	0.0052	0.0002	-0.0047	-0.0093	-0.0124	-0.0156	-0.0176	-0.0205	-0.0225	-0.0240	-0.0256	-0.0268	-0.0279
0.8			0.0009	-0.0011	-0.0048	-0.0070	-0.0102	-0.0129	-0.0152	-0.0174	-0.0200	-0.0214	-0.0228	-0.0244	-0.0257	-0.0268	-0.0282
1.2				-0.0151	-0.0144	-0.0160	-0.0174	-0.0187	-0.0209	-0.0219	-0.0228	-0.0243	-0.0253	-0.0265	-0.0271	-0.0278	-0.0289
1.6	n = 25				-0.0258	-0.0247	-0.0251	-0.0262	-0.0277	-0.0285	-0.0284	-0.0291	-0.0294	-0.0299	-0.0304	-0.0310	-0.0313
2.0						-0.0343	-0.0329	-0.0325	-0.0334	-0.0343	-0.0346	-0.0346	-0.0349	-0.0348	-0.0346	-0.0346	-0.0347
0.0	0.0380	0.0391	0.0319	0.0213	0.0109	0.0020	-0.0048	-0.0098	-0.0146	-0.0181	-0.0207	-0.0230	-0.0247	-0.0261	-0.0273	-0.0287	-0.0295
0.4		0.0063	0.0093	0.0072	0.0034	-0.0013	-0.0065	-0.0109	-0.0146	-0.0173	-0.0197	-0.0221	-0.0241	-0.0255	-0.0270	-0.0283	-0.0293
0.8			-0.0141	-0.0110	-0.0109	-0.0110	-0.0129	-0.0151	-0.0167	-0.0191	-0.0213	-0.0228	-0.0248	-0.0259	-0.0276	-0.0287	-0.0295
1.2				-0.0277	-0.0238	-0.0228	-0.0226	-0.0227	-0.0235	-0.0246	-0.0252	-0.0265	-0.0279	-0.0288	-0.0295	-0.0299	-0.0308
1.6	n = 30				-0.0360	-0.0335	-0.0319	-0.0311	-0.0312	-0.0312	-0.0314	-0.0315	-0.0319	-0.0326	-0.0327	-0.0340	-0.0343
2.0						-0.0417	-0.0394	-0.0390	-0.0378	-0.0375	-0.0374	-0.0374	-0.0374	-0.0374	-0.0371	-0.0372	-0.0367

7*

Academy of Agricultural Sciences of the GDR,
Research Centre of Animal Production Dummerstorf-Rostock

Robustness of Three Sequential One-Sample Tests Against Non-Normality

DIETER RASCH

Abstract

By simulation it was shown that only one of three sequential tests of the hypothesis $H_0: \mu = \mu_0$ against $H_A: (\mu - \mu_0)^2 = \sigma^2 d^2$ is robust with respect to the first kind risk α but not with respect to the second kind risk β. Another test is robust with respect to β but not with respect to α.

1. The Test Statistics

The sequence $\{y_1, y_2, \ldots,\}$ with identically and independently distributed components with mean μ and variance σ^2 $(0 < \sigma^2 < \infty)$ was used to test the hypothesis

$H_0 : \mu = \mu_0$

against

$H_A : (\mu - \mu_0)^2 = \sigma^2 d^2 \qquad (d > 0).$

We consider three tests developed for normally distributed $\mathbf{y}_i$ following Wald ($\mathbf{T}_1$ (n), 1947), Bartlett ($\mathbf{T}_2$ (n), 1946) and Mann ($\mathbf{T}_3$ (n), 1980) respectively. The test statistics $\mathbf{T}_j$ (n) (j = 1, 2, 3) are

$$\mathbf{T}_1(n) = -\frac{nd^2}{2} + \ln H\left(\frac{n}{2}; \frac{1}{2}; \frac{nd^2 \mathbf{t}_n^2}{2(n-1+\mathbf{t}_n^2)}\right). \tag{1}$$

$$\mathbf{T}_2(n) = -\frac{n-1}{2}\left\{\ln\frac{n-1}{n}\mathbf{s}_n^2 - \ln\left[\frac{n-1}{n}\mathbf{s}_n^2 + (\bar{\mathbf{y}}_n - d)^2\right]\right\}. \tag{2}$$

$$\mathbf{T}_3(n) = \frac{1}{2}\left[\sum_{i=1}^{n}\mathbf{y}_i^2\right]^{\frac{n-1}{2}}\left\{\left[\sum_{i=1}^{n}(\mathbf{y}_i+d)^2\right]^{-\frac{n-1}{2}} + \left[\sum_{i=1}^{n}(\mathbf{y}_i-d)^2\right]^{-\frac{n-1}{2}}\right\} \tag{3}$$

In (1) to (3) we used

$$\mathbf{t}_n^2 = \frac{n}{\mathbf{s}_n^2}(\bar{\mathbf{y}}_n - \mu_0)^2,$$

$$\bar{\mathbf{y}}_n = \frac{1}{n}\sum_{i=1}^{n}\mathbf{y}_i,$$

$$\mathbf{s}_n^2 = \frac{1}{n-1}\sum_{i=1}^{n}(\mathbf{y}_i - \bar{\mathbf{y}})^2$$

and the confluent hypergeometric function H (a; b; c). Usually the decision rule used for the sequential test is;

Accept H_0 if $\mathbf{T}_j(n) \le b = \ln B \qquad (j = 1, 2)$

or if $\mathbf{T}_3(n) \le B$.

Reject H_0 if $\mathbf{T}_j(n) \ge a = \ln A \qquad (j = 1, 2)$

or if $\mathbf{T}_3(n) \ge A$.

Take a further observation $\mathbf{y}_{n+1}$, if

$b < \mathbf{T}_j(n) < a \qquad (j = 1, 2)$

or if $B < \mathbf{T}_3(n) < A$

with $A = \frac{1-\beta}{\alpha}$ and $B = \frac{\beta}{1-\alpha}$

Using this rule, test T_1 has approximately the strength (α, β).

2. The Simulation Experiment

The simulation experiment consists of two parts. Without loss of generality we put $\mu_0 = 0$ and $\sigma^2 = 1$.

Part I

We investigated the behaviour of the three tests under the normal assumption and for truncated normal distributions. 10 000 samples were generated for the normal distribution and one truncated distribution for each of the 264 possible combinations of

(α, β) = (0.05; 0.10), (0.05; 0.20), (0.10; 0.20) (0.10; 0.50)
d = 0.6; 1.0; 1.6
T (n) = T_1 (n), T_2 (n); T_3 (n)
μ = 0 (0.2) d + 0.2

and for the other truncated distributions and the distributions of part II only for $\mu = 0$ and $\mu = d$ (72 combinations per distribution). (See Table 1) Pseudo-random numbers (p. r. n) from N (μ, 1) are generated by the combination ZZGD/NVØ1 (Feige et. al. 1984).

Part II

P.r.n. from N(μ, 1) were generated for the 72 parameter combinations mentioned in part I. These p.r.n. were transformed into p.r.n. with given skewness γ_1 and kurtosis γ_2 by the power transformation $z = -c + by + cy^2 + dy^3$ (Fleishman 1978). For both parts we calculated for each μ-value the relative frequency of the 10 000 samples in which H_0 was accepted and used this for $\mu = 0$ as an estimate of α and for $\mu = d$ as an estimate of $1-\beta$. Further, we calculated the average n of the sizes of the 10 000 samples and the variance of these sample sizes s_n^2. We used n as an estimate of E(**n**). We also determined n_{min} and n_{max} and the frequency distribution of the sample sizes.

3. Results

(i) With four exceptions the empirical α-values are, under normal conditions lower than the nominal ones.
Under normal conditions the empirical $(1-\beta)$-values are with the exception of T_3; $(\alpha, \beta) = (0.05, 0.10)$ higher than the nominal ones, so that all tests are in most cases conservative, as could be expected from theory in the case of test 1.

(ii) With respect to α test 2 is robust for all investigated alternative distributions (it is worst in the normal case), tests 1 and 3 are non-robust for small d if γ_1 and γ_2 both differ greatly from zero.

(iii) With respect to β test 1 is always robust, test 2 is robust only if the values of d are not to small and test 3 is robust only for d = 1.6.

(iv) The ASN is low for test 2 if d is small and is also good for test 1 if d is large. Test 3 always needs larger average sample sizes than either test 1 or test 2. For median d-values, the sample sizes do not differ too much between the three tests.

4. **Proposal**

Lim and Fung (1982) investigated sequential t-test based on M-estimators under three non-normal (long-tailed) distributions. All simulated tests where either in the first or in the second kind risk non-robust. We therefore propose the following test statistic

$$\mathbf{T} = \begin{cases} \mathbf{T}_1, & \text{if} \quad u < \frac{1}{2} \\ \mathbf{T}_2, & \text{if} \quad u \geq \frac{1}{2} \end{cases}$$

where **u** is an in (0,1) uniformely distributed pseudorandom number. The properties of this test **T** will be investigated for some more d-values.

Table 1
Distributions used in the simulation experiment and their parameters (u, v standarized truncation points, b, c, d parameters of the power transformation)

No of distribution	Truncation points u	v		γ_1	γ_2
1	$-\infty$	∞		0	0
2	0.5	3.0		1.0057	0.5915
3	−1.5	3.0		0.3480	−0.3488
4	2.85	4.71		1.505	3.75
	Parameters of power transformation b	c	d	γ_1	γ_2
5	0.748020807992	0	0.077872716101	0	3.75
6	0.63044672784	0	0.11069674204	0	7.00
7	0.953076897706	0.163194276264	0.006597369744	1.00	1.50
8	0.865886203523	0.221027621012	0.027220699158	1.50	3.75

Table 2
Percentage $10^2 f_\alpha$ of false rejection of H_0 for 8 distributions and test 1

d	$10^2\alpha$	$10^2\beta$	Number of distribution in table 1: 1	2	3	4	5	6	7	8
0.6	5	10	4.19	7.52	4.74	10.38	3.16	2.64	6.12	7.67
		20	4.34	7.44	4.93	10.41	3.23	2.95	6.17	7.57
	10	20	8.85	12.99	9.33	15.91	6.94	6.46	10.52	12.63
		50	9.94	13.79	10.27	15.94	8.30	8.06	11.65	13.70
1.0	5	10	3.90	8.49	4.47	11.06	2.76	2.13	6.40	7.93
		20	4.15	8.67	4.70	11.14	3.03	2.42	6.36	8.06
	10	20	7.90	13.33	9.00	16.31	6.60	5.79	10.65	12.90
		50	10.39	14.64	11.20	17.94	8.78	7.77	12.60	14.69
1.6	5	10	3.87	7.77	4.26	10.76	2.32	2.24	6.04	7.73
		20	4.27	8.01	4.58	11.01	2.60	2.43	6.43	8.14
	10	20	7.68	12.47	8.51	15.58	5.58	5.13	10.51	12.22
		50	11.82	16.33	12.72	19.55	9.70	8.55	14.35	16.06

Table 3
Percentage $10^2 f_\beta$ of false acception of H_0 for 8 distributions and test 1

d	$10^2\alpha$	$10^2\beta$	Number of distribution in table 1: 1	2	3	4	5	6	7	8
0.6	5	10	7.54	3.26	6.09	1.68	8.00	7.86	4.14	2.37
		20	15.54	8.95	13.78	5.05	14.64	14.76	10.46	7.24
	10	20	15.72	9.43	14.20	5.45	14.34	14.15	10.85	7.66
		50	39.90	34.30	40.79	27.91	34.35	31.97	36.22	30.56
1.0	5	10	5.95	0.35	3.42	0	7.16	7.22	1.22	0.34
		20	11.77	1.52	8.01	0.06	11.90	11.73	3.73	1.27
	10	20	11.58	1.68	8.30	0.07	11.52	11.09	3.94	1.36
		50	30.13	13.13	28.59	0.21	24.25	22.50	18.93	11.23
1.6	5	10	4.30	0	0.61	0	6.83	6.42	0.02	0.01
		20	7.59	0	1.78	0	9.53	9.27	0.12	0.04
	10	20	7.60	0	2.01	0	9.38	8.87	0.13	0.04
		50	15.01	0.17	10.58	0.08	13.98	12.64	2.37	0.44

Table 4
Percentage $10^2 f_\alpha$ of false rejection of H_0 for 8 distributions and test 2

d	$10^2\alpha$	$10^2\beta$	1	2	3	4	5	6	7	8
			Number of distribution in table 1							
0.6	5	10	4.25	1.75	3.44	0.98	3.26	2.51	1.62	1.17
		20	4.64	1.69	3.59	1.11	3.33	2.68	1.78	1.36
	10	20	8.92	3.95	7.11	2.62	6.78	5.94	4.25	3.26
		50	11.15	5.46	9.32	3.96	9.06	8.03	6.26	4.41
1.0	5	10	2.61	1.14	2.00	0.61	1.26	0.77	1.14	0.80
		20	2.65	1.20	2.28	0.61	1.39	0.84	1.31	0.88
	10	20	5.21	2.73	4.92	1.91	3.01	2.12	3.14	1.94
		50	7.21	4.20	7.00	3.18	4.36	3.34	4.50	2.99
1.6	5	10	0.51	0.42	0.67	0.25	0.21	0.18	0.37	0.28
		20	0.58	0.45	0.71	0.26	0.25	0.19	0.42	0.29
	10	20	1.37	1.08	1.42	0.58	0.78	0.42	1.01	0.77
		50	2.37	2.00	2.50	1.29	1.31	0.88	1.84	1.36

Table 5
Percentage $10^2 f_\beta$ of false acception of H_0 for 8 distributions and test 2

d	$10^2\alpha$	$10^2\beta$	1	2	3	4	5	6	7	8
			Number of distribution in table 1							
0.6	5	10	8.37	19.18	11.48	22.85	6.85	5.71	15.48	18.38
		20	15.46	27.80	19.45	31.30	13.15	11.53	24.01	27.01
	10	20	15.51	28.15	19.64	31.58	13.11	11.46	24.10	27.30
		50	32.27	44.53	36.28	46.32	28.56	26.52	39.93	42.27
1.0	5	10	4.45	10.37	6.66	9.55	2.78	1.93	8.07	8.07
		20	8.90	16.70	11.56	14.13	5.96	4.85	13.60	12.90
	10	20	9.17	17.11	11.97	14.55	6.10	4.97	13.94	13.29
		50	20.55	26.81	22.86	24.94	16.04	14.44	24.18	24.32
1.6	5	10	11.19	0.19	1.60	0.01	0.54	0.35	0.76	0.13
		20	12.55	0.85	3.21	0.04	1.31	1.03	1.58	0.53
	10	20	12.65	0.94	3.36	0.06	1.33	1.06	1.67	0.61
		50	8.09	5.34	8.77	1.46	5.75	5.03	6.04	3.77

Table 6
Percentage $10^2 f_\alpha$ of false rejection of H_0 for 8 distributions and test 3

d	$10^2\alpha$	$10^2\beta$	1	2	3	4	5	6	7	8
			Number of distribution in table 1							
0.6	5	10	4.80	8.80	5.26	12.60	3.27	2.41	7.03	6.73
		20	4,90	9.18	5.41	12.54	3.33	3.53	7.15	7.02
	10	20	9.17	14.47	9.32	18.17	6.60	5.43	11.70	12.01
		50	10.77	15.82	11.68	19.10	8.60	7.03	13.30	13:15
1.0	5	10	2.61	5.43	3.46	4.41	1.05	0.58	3.22	4.01
		20	2.80	5.74	3.70	4.63	1.12	0.65	3.54	3.23
	10	20	5.50	8.51	6.53	7.69	2.37	1.69	6.11	5.78
		50	7.23	10.28	8.45	9.42	3.64	2.64	8.40	7.89
1.6	5	10	0.43	0.09	0.66	0.11	0.13	0.06	0.34	0.45
		20	0.48	0.11	0.69	0.16	0.18	0.08	0.39	0.63
	10	20	1.12	0.48	1.47	0.39	0.43	0.23	0.73	0.78
		50	1.80	0.93	2.33	0.66	0.79	0.49	1.22	1.02

Table 7
Percentage $10^2 f_\beta$ of false acception of H_0 for 8 distributions and test 3

d	$10^2\alpha$	$10^2\beta$	1	2	3	4	5	6	7	8
			Number of distribution in table 1							
0.6	5	10	10.36	19.47	12.60	25.36	9.81	8.86	17.46	16.79
		20	18.90	29.87	21.68	34.84	17.74	16.80	27.21	22.75
	10	20	18.91	30.09	21.74	35.35	17.67	16.57	27.20	27.15
		50	40.44	49.59	43.46	53.64	37.96	36.37	47.72	45.12
1.0	5	10	6.56	12.72	8.72	12.36	3.77	3.02	9.80	10.25
		20	12.44	19.85	14.88	18.69	8.70	7.28	16.19	15.80
	10	20	12.77	20.34	15.31	19.25	8.86	7.36	16.58	16.03
		50	27.84	34.33	30.88	33.71	23.04	21.12	31.45	29.50
1.6	5	10	1.75	0.59	2.14	0.04	0.82	0.61	0.94	0.99
		20	4.19	2.01	4.76	0.34	2.30	1.82	2.48	2.12
	10	20	4.47	2.17	5.05	0.36	2.48	1.95	2.78	2.55
		50	14.01	10.99	13.71	6.58	10.71	9.29	11.74	10.79

5. **References**

BARTLETT, M. S.
The large sample theory of sequential tests.
Proc. Comb. Phil. Soc. 42, (1946) 239–244.

FEIGE, K.-D. et. al.
Results of comparisons between different random number generators.
in: Rasch, D., and Tiku, M. L.: Robustness of statistical methods and nonparametric statistics; VEB Deutscher Verlag der Wissenschaften Berlin 1984. 80–84.

FLEISHMAN, A. L.
A method for simulating non-normal distributions.
Psychometrika 43 (1978) 521–532.

GUIARD, V. (ed.)
Robustheit II.
Probleme der angewandten Statistik.
FZ für Tierproduktion Dummerstorf-Rostock, Heft 5 (1981).

HERRENDÖRFER, G. (ed.)
Robustheit I.
Probleme der angewandten Statistik.
FZ für Tierproduktion Dummerstorf-Rostock, Heft 4 (1980).

LIM, T. K., and FUNG, K. Y.
Sequential t-tests based on M estimation.
Statist. Comp. Simul. 14 (1982) 271–281.

MANN, I. E.
A simple sequential criterion for testing a normal mean with unknown variance.
Commun. Statist.-Theor. Meth. A 9 (1980) 1619–1624.

NÜRNBERG, G.
Beiträge zur Versuchsplanung für die Schätzung von Varianzkomponenten und Robustheitsuntersuchungen zum Vergleich zweier Tests.
Probleme der angewandten Statistik Heft 6, FZ für Tierproduktion Dummerstorf-Rostock der AdL der DDR (1982).

RASCH, D., and HERRENDÖRFER, G.
Robustheit statistischer Verfahren.
Rostock, Math. Kolloq. 17 (1981) 87–104.

RASCH, D., HERRENDÖRFER, G. (ed.)
Robustheit III.
Probleme der angewandten Statistik.
FZ für Tierproduktion Dummerstorf-Rostock, Heft 7, (1982).

RASCH, D., and TEUSCHER, F.
Das System der gestutzten Normalverteilungen.
7. Sitzungsbericht der Interessengemeinschaft Mathematische Statistik MG DDR, Berlin (1982).

RASCH, D.
First results on robustness of the one-sample sequential t-test.
Trans 9th Prague Conference on Inform. Theory and Mathem. Statist. 1983, 133–140.

WALD, A. (1947)
Sequential Analysis.
New York: Wiley.

Academy of Agricultural Sciences of the GDR

Research Centre of Animal Production Dummerstorf-Rostock

A Test for Exponential Regression and its Robustness

DIETER RASCH AND ERHARD SCHIMKE

Abstract

We consider the regression model $\mathbf{y}_i = \alpha + \beta \exp(\gamma x_i) + \mathbf{e}_i$ ($\gamma < 0$) with i. i. d. error variables $\mathbf{e}_i$ with $E(\mathbf{e}_i) = 0$, $V(\mathbf{e}_i) = \sigma^2$. By simulation experiments we investigate the possibility of using the elements of the sequence $V_A(n)$ ($n = 4, \ldots$) as approximations of the variance $V(\hat{\boldsymbol{\vartheta}})$ of the least squares estimator $\hat{\boldsymbol{\vartheta}}$ of $\vartheta' = (\alpha, \beta, \gamma) = (\vartheta_1, \vartheta_2, \vartheta_3)$. Here $V_A(n)$ is equal to

$$V_A(n) = \left[\sum_{i=1}^{n} g'(x_i, \vartheta)\, g(x_i, \vartheta)\right]^{-1} \text{ with}$$

$$g(x_i, \vartheta) = \left(g_1(x_i, \vartheta), g_2(x_i, \vartheta), g_3(x_i, \vartheta)\right),$$

$$g_j(x_i, \vartheta) = \frac{\partial}{\partial \vartheta_j} f(x_i, \vartheta) \text{ and } f(x_i, \vartheta) = \alpha + \beta \exp(\gamma x_i).$$

We find that approximations for the estimation problem are already good for $n = 4$ and very good from $n = 6$ on. Tests and confidence intervals in respect of γ can easily be constructed with sufficient accuracy from $n = 4$ on. We also investigate the robustness of the proposed test against non-normality.

1. Introduction

We consider the exponential regression model of the form

$$\mathbf{y}_i = \alpha + \beta e^{\gamma x_i} + \mathbf{e}_i = f(x_i, \vartheta) + \mathbf{e}_i \tag{1}$$

$$\left(i = 1, \ldots, n; \quad (\alpha, \beta, \gamma) \in \Omega \subset R^3, V(\mathbf{e}_i) = \sigma^2 > 0\right)$$

and limit ourselves to the parameters $\beta < 0$, $\gamma < 0$ ($\Omega = R^1 \times R^- \times R^-$), a case which often arises in describing growth or production functions. The $\mathbf{e}_i$ may be i. i. d. random variables, in the main part of the work we consider the normal case. It is well known that the realisation $\hat{\vartheta}' = (a, b, c)$ of the least squares estimatior $\hat{\boldsymbol{\vartheta}} = (\mathbf{a}, \mathbf{b}, \mathbf{c})$ is obtained by first computing c iteratively from

$$h(c) = \left(\hat{F} - \bar{y}\hat{A}\right)\left(\hat{D} - \frac{1}{n}\hat{A}\hat{B}\right) - \left(\hat{C} - \frac{1}{n}\hat{A}^2\right)\left(\hat{G} - \bar{y}\hat{B}\right) = 0 \tag{2}$$

and then calculating a and b from

$$a = \bar{y} - \frac{b}{n}\hat{A} \tag{3}$$

and

$$b = \frac{\hat{G} - \bar{y}\hat{B}}{\hat{D} - \frac{1}{n}\hat{A}\hat{B}} \tag{4}$$

respectively.

We use the following abbreviations

$$A = \sum_{i=1}^{n} e^{\gamma x_i},\ B = \sum_{i=1}^{n} x_i e^{\gamma x_i},\ C = \sum_{i=1}^{n} e^{2\gamma x_i},\ D = \sum_{i=1}^{n} x_i e^{2\gamma x_i},$$

$$E = \sum_{i=1}^{n} x_i^2 e^{2\gamma x_i},\ F = \sum_{i=1}^{n} y_i e^{\gamma x_i},\ G = \sum_{i=1}^{n} x_i y_i e^{\gamma x_i}$$

If we replace γ by its least squares estimate c we write $\hat{A}, \hat{B}, \ldots, \hat{G}$.

It is well known that $\hat{\boldsymbol{\vartheta}}$ is consistent (Malinvaud (1970)) and that

$$\sqrt{n}\left(\hat{\boldsymbol{\vartheta}} - \vartheta\right) \tag{5}$$

is asymptotically $N[0_3, \sigma^2 n V_A(n)]$ where

$$V_A(n) = \begin{pmatrix} n & A & \beta B \\ A & C & \beta D \\ \beta B & \beta D & \beta^2 E \end{pmatrix}^{-1}$$

$$= \frac{1}{\Delta}\begin{pmatrix} CE - D^2 & BD - AE & \frac{1}{\beta}(AD - BC) \\ BD - AE & nE - B^2 & \frac{1}{\beta}(AB - nD) \\ \frac{1}{\beta}(AD - BC) & \frac{1}{\beta}(AB - nD) & \frac{1}{\beta^2}(nC - A^2) \end{pmatrix}$$

(Jennrich 1969),

with $\Delta = n\,(CE - D^2) + 2ABD - A^2E - B^2C$.

It is the aim of this paper to find by simulation experiments a lower bound n_0 for n such that, for $n > n_0$,

$$E\left(\hat{\boldsymbol{\vartheta}}\right) \approx \vartheta \text{ and } V\left(\hat{\boldsymbol{\vartheta}}\right) \approx V_A(n)\sigma^2$$

The problem of testing hypothesis and constructing confidence intervals with respect to γ was discussed by Bates and Watts (1981), Beale (1960), Bismarck et al. (1974), Brandt (1973), Broemling and Hartley (1968), Gallant (1975 a, b), Halperin (1963), Halperin and Mantel (1963), Hartley (1964), Maritz (1962), Milliken and Graybill (1970), Milliken (1978), Schmidt (1979, 1980, 1982, 1983), Sundararary (1978), Williams (1962) and in most cases solved approximately or asymptotically.

Exact methods for constructing confidence intervals (also used to obtain tests) are often difficult or lead to difficult interpretations.

From theorems of Gallant (1975 b) and Schmidt (1980) we find that

$$\mathbf{z} = \frac{(\mathbf{c} - \gamma_0)^2 \Delta \beta^2}{(nC - A^2)\mathbf{s}^2} \tag{7}$$

is asymptotically (for $\gamma = \gamma_0$ central) chi-square distributed with one d. f. if

$$s^2 = \frac{1}{n-3} \sum_{i=1}^{n} \left(y_i - a - b e^{c x_i}\right)^2, \quad (n \geq 4) \tag{8}$$

is used as an estimator of σ^2.

So with $\mathfrak{y}' = (y_1, \ldots, y_n)$

$$k(\mathfrak{y}) = \begin{cases} 1, & \text{if } z > \chi^2(1, 1-\alpha^*) \\ 0, & \text{otherwise} \end{cases} \tag{9}$$

will be an asymptotical α^*-test for the null hypothesis

$$H_0: \gamma = \gamma_0 \quad \text{against} \quad H_A: \gamma \neq \gamma_0 \tag{10}$$

Furthermore we know from Gallant (1975 b) that, for a normal error distribution,

$$\mathbf{z} = \mathbf{F} + \varepsilon,$$

so that **F** under H_0 is F-distributed with one and $n-3$ d. f. and ε converges in probability to zero. In section 3 we investigate the small sample properties of a modification of (7) which, we conjectured, has an approximate t-distribution (central under H_0 and noncentral otherwise). The conjecture was based on the asymptotic properties of (7) and on the results of section 2. In section 4 we discuss the robustness of the test based on the statistic of section 3.

In the simulation experiments we used the program ZZGD to generate pseudo-random numbers from a uniform distribution in (0; 1) and transformed them by the program NV01 into normally distributed random numbers (see Feige et al. (1984) on page 30 ff. for details).

2. Small Sample Distribution of a, b, c and s^2

The simulation experiment was based on a parameter configuration of a real growth process. Rasch (1970) fitted an exponential regression function to growth data of cattle. For the wither height for instance the following estimates for $0 \leq x_i \leq 60$ month were found: $a = 133$, $b = -56$, $c = -0.068$, $s^2 = 1.04$. Without loss of generality we used $\alpha = 0$ to save computing time and we put $\sigma^2 = 1$. The simulation experiment for this section was done in two parts. In part I we considered 48 (β, γ)-combinations of $\gamma = -0.10(0.01)-0.03$ and $\beta = -80(10)-30$, $n = 14$ and equidistant points $x_i \in \langle 0; 65 \rangle$. This part led to the conclusion that the influence of (β, γ) is not so great as expected, and therefore we used only 12 combinations (β, γ) for further investigations. Part I was realized with $N = 1000$ samples for the 48 parameter combinations.

Part II was realized with $N = 5000$ samples for the 12 parameter combinations of $\gamma = -0.09(0.02)-0.03$ and $\beta = -70(20)-30$ fore $n = 4$ and $n = 6$ equidistant $x_i \in \langle 0; 0.65 \rangle$ respectively. For both parts in each of the N samples we added a N(0; 1) pseudorandom number to each of the n values

$$\beta e^{\gamma x_i} \quad (i = 1, \ldots, n)$$

and estimated $\alpha = 0$, β, γ and $\sigma^2 = 1$ by (3), (4), (2) and (8) respectively. For the N samples we calculated the mean, variance, skewness and kurtosis of the empirical distribution of a, b, c and s^2, and we also calculated the correlation coefficient $r_{s,c}$ between the residual standard deviation s and the estimate c of γ. Furthermore we computed the empirical covariances of the vector (a, b, c) and the determinant of the empirical covariance matrix of this vector.

Tables 1, 2 and 3 contain the means and variances of the estimates of α, β and γ from $N = 5000$ samples for $n = 4$ and 6 and from $N = 1000$ samples for $n = 14$ for the (β, γ)-values used in part II of the experiment. More information for $n = 14$ and results concerning the covariances and the determinant can be found in Rasch and Schimke (1982).

We see that **a** and **b** are a little biased especially for $\gamma = -0.03$, but the relative bias (for instance for $\alpha = 130$) is small. The bias of **c** is negligible and decreases with increasing n. We can thus state that **c** is nearly unbiased for $n \geq 10$. Empirical and asymptotic variances of **a, b** and **c** are also in good agreement for $n = 4$; agreement is worst for a and best for c. The correlation coefficients between c- and s-values (s square root of s^2 in (8)) lie between -0.013 and 0.026 for $n = 4$ and between -0.016 and 0.044 for $n = 6$ so it seems that **s** and **c** are uncorrelated and also independent.

Goodness of fit tests were performed to test the hypothesis H_{0_γ} that **c** is normally distributed and $H_{0_{\sigma^2}}$ that $\frac{(n-3)s^2}{\sigma^2}$ is chi-square distributed with $n-3$ d. f.

The results are contained in table 4. The null hypothesis $H_{0_{\sigma^2}}$ was at the 5 %-level in no case rejected. In table 4 we have $x_j = \frac{(f_j - \varphi_j)^2}{\varphi_j}$ with empirical (f_j) and theoretical (φ_j) absolute frequency in class j. The empirical skewness and kurtosis values are in good agreement with the acceptance of the corresponding distributional hypothesis, as can be seen from table 5 taking into account that a random variable **z** with chi-square distribution with v d. f. has mean v, variance 2 v, skewness $\gamma_1 = 2\sqrt{\frac{2}{v}}$ and kurtosis $\gamma_2 = \frac{12}{v}$.

Summarizing it seems that most of the assumptions needed for a statistic

$$\mathbf{t} = \frac{(\mathbf{c} - \gamma_0)\,\mathbf{b}\sqrt{\hat{\Delta}}}{\mathbf{s}\sqrt{n\hat{\mathbf{C}} - \hat{\mathbf{A}}^2}} \tag{11}$$

with $\hat{\Delta} = n\left(\hat{\mathbf{C}}\hat{\mathbf{E}} - \hat{\mathbf{D}}^2\right) + 2\hat{\mathbf{A}}\hat{\mathbf{B}}\hat{\mathbf{D}} - \hat{\mathbf{A}}^2\hat{\mathbf{E}} - \hat{\mathbf{B}}^2\hat{\mathbf{C}}$

to have a Student distribution with $n-3$ d. f. also for finite n if $\gamma = \gamma_0$ are fulfilled. But we do not know whether **c** is independent of **C** and $\hat{\mathbf{A}}$, and other difficulties may arise from estimating $V_A(\mathbf{c})$, so we had to investigate the properties of a test based on (11).

3. Properties of Testing $H_0: \gamma = \gamma_0$ by the Statistic (11)

The null hypothesis

$$H_0 \; : \gamma = \gamma_0$$

wes tested against

$$H_A : \gamma \neq \gamma_0$$

in $N = 10\,000$ samples in a simulation experiment performed in an analogous manner to that dealt with in chapter 2 for the twelve (β, γ)-combinations of part II, with $n = 10$ and for four (β, γ)-combinations and $n = 4$ after we found that $n = 4$ is acceptable in the normal case (table 14). For $n = 9$ we also performed a simulation with $\sigma^2 > 1$ and found that the restriction on $\sigma^2 = 1$ is without loss of generality (table 15).

Tables 6 to 10 give the relative frequencies of rejection of H_0 under the null hypothesis and for some values of the noncentrality parameter for first kind risks $\alpha^* = 0.01$, $\alpha^* = 0.05$ and $\alpha^* = 0.10$. The results are easy to interpret. The test based on the statistic (11) can be used if $n \geq 4$. The properties of the test of H_0 by (11) with the percentiles of Students distribution with $n-3$ d. f. are nearly independent of (β, γ). For $n \geq 4$ this test is approximately an α^*-test if $\alpha^* \geq 0.05$, and it seems that it as also almost uniformly most powerful unbiased (in this connection see figure 1).

So we found that the test

$$\mathbf{k}(t) = \begin{cases} 1, & \text{if } \mathbf{t} > t\left(n - 3.1 - \frac{\alpha^*}{2}\right) \\ 0, & \text{otherwise} \end{cases}$$

is approximately a uniformly most powerful unbiased α^*-test ($\alpha^* = 0.05$). The properties for $\alpha^* = 0.01$ are also not bad. Therefore, and because **c** is nearly unbiased for γ,

$$\mathbf{c} \pm t\left(n - 3.1 - \frac{\alpha^*}{2}\right) \frac{s\sqrt{n\hat{\mathbf{C}} - \hat{\mathbf{A}}^2}}{\mathbf{b}\sqrt{\hat{\Delta}}}$$

are the limits of an approximate $(1 - \alpha^*)$ confidence interval for γ.

4. Robustness of the Test Investigated in Chapter 3

The t-test based on test statistic (11) seems to be fairly acceptable from n larger than or equal to 4. To investigate the behaviour of the test for non-normal distributions we generated six distributions by Fleishman's power transformation as described in detail by Guiard (1984) with the same sample size $n = 4$. We characterized the non-normal distributions by their skewness γ_1 and kurtosis γ_2, and simulated 10 000 runs of t-tests for each of the four extreme combinations of $\beta = -30$ and $\beta = -70$ with $\gamma = -0.03$ and $\gamma = -0.09$ to save computing time and for each of the distributions characterized by (γ_1, γ_2) as follows

γ_1	0	0	0	1	1,5	2
γ_2	1,5	4	7	1,5	4	7

Table 11 contains the percentages of rejecting H_0 if it is true for three values of α^*. The behaviour of the empirical power function can be seen in tables 12 and 13 for two (β, γ)-pairs. We find that with only few exceptions the t-test based on (11) is ε-robust with $\varepsilon = 0.2\,\alpha^*$ and that the empirical power function is nearly the same for all distributions, including the normal one.

5. Definition of the Robustness of a Test and Planning the Sample Size of a Simulation Study

Following Rasch and Herrendörfer (1981) we define the robustness of a test as follows

Definition:

Let $\varphi(\mathfrak{y})$ be a test for the hypothesis $\vartheta = \vartheta_0$ based on a sample $\mathfrak{y}$ and let $\varphi(\mathfrak{y})$ be an α^*-test if the distribution g of the components of $\mathfrak{y}$ is a element of G. Then $\varphi(\mathfrak{y})$ is called $\varepsilon(\alpha^*)$-robust in the class $H \supset G$ of distributions if

$$\operatorname*{Max}_{h \in H} \left|\alpha^* - \alpha^h\right| \leq \varepsilon(\alpha^*)$$

where α^h is the size of the test for a given element of H. In an analogue way robustness with respect to the whole power function can be defined.

We usually use $\varepsilon(\alpha^* = 0.2\,\alpha^*$ (20 %-robustness).

To determine the size of a simulation experiment we need a precision requirement. We will estimate a probability P by a confidence interval with coefficient $1 - \alpha_c$ in such a way that the half expected width of this interval is not larger than δ that is $E(\mathbf{d}) \leq \delta$, Then with the normal fractile the sample size needed is given by

$$N = \frac{u^2_{1-\frac{\alpha_c}{2}} P(1-P)}{\delta^2}$$

We will estimate by simulation the first kind risk α^* of a test ($\alpha^* < 1/2$). The sample size increases if α^* increases to $1/2$. So we propose to use for P in the above formula the conservative value $P = \alpha^* + \varepsilon(\alpha^*) = \alpha^* + \varepsilon$ and we are on the safe side. If we choose $\alpha^* = 0.05$ we obtain the following sample sizes

	δ					
ε	0.001	0.003	0.004	0.005	0.007	0.010
0.005	199 668	22 186	12 480	7 987	5 547	1 997
0.010	216 667	24 075	13 542	8 667	6 019	2 167

For $\varepsilon(\alpha^*) = \varepsilon = 0.2\,\alpha^* = 0.01$ ($\alpha^* = 0.05$) we found $N = 10\,000$ runs as a reasonable choice and so 10 000 runs were performed in each paper of the research group of Dummerstorf-Rostock.

Conclusions:

The test statistic $\mathbf{t} = \sqrt{}\,\mathbf{z}$ is a 20 %-robust (approximately) α^*-test of $H_0 : \gamma = \gamma_0$ (against $H_0 : \gamma \neq \gamma_0$) in the Fleishman-system ($1 \leq \sigma^2 \leq 16$ if $0 \leq x \leq 65$).

Acknowledgement:

The authors thank Mr. Rieman (Computing Center of the Wilhelm-Pieck-Universität Rostock) for helping to perform the simulations on an ES 1040.

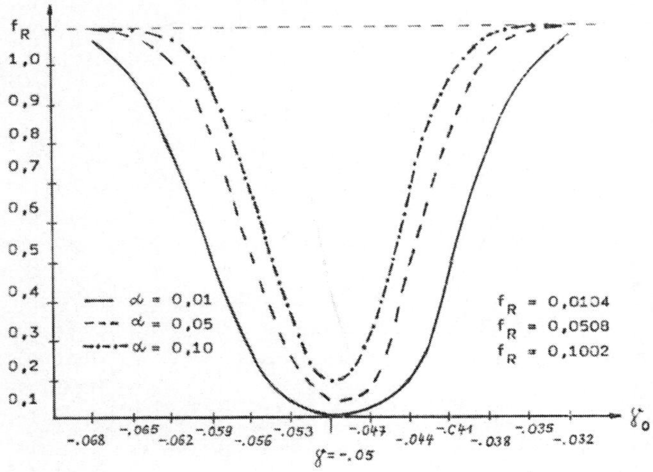

Figure 1

Relative number of rejections for 10 000 tests (n = 10, normal case)

Relative rejections for 1000 tests

Table 1

Means, empirical and asymptotic variances of the estimates of α from N samples of size $n = 4$ ($N = 5000$), $n = 6$ ($N = 5000$) and $n = 14$ ($N = 1000$)

$-\beta$	$-10^2\gamma$	n	$10^3\,\bar{a}$	$10^4\,s_a^2$	$10^4 \frac{\hat{C}\hat{E} - \hat{D}^2}{\hat{\Delta}}$
30	3	4	520	104309	80077
		6	644	98099	61368
		14	263	39213	33850
	5	4	147	19734	18716
		6	125	13575	12998
		14	128	7283	6552
	7	4	55	10502	9899
		6	52	6740	6330
		14	27	3087	2929
	9	4	35	7197	7308
		6	2	4458	4415
		14	—12	1820	1921

Table 1 continued

$-\beta$	$-10^2\gamma$	n	$10^3\bar{a}$	$10^4 s_a^2$	$10^4 \frac{\hat{C}\hat{E}-\hat{D}^2}{\hat{\Delta}}$
50	3	4	310	93815	80077
		6	279	68084	61368
		14	210	37041	33050
	5	4	70	19159	18716
		6	77	13069	12998
		14	25	6371	6552
	7	4	30	9699	9899
		6	45	6442	6330
		14	100	2755	2929
	9	4	23	6829	7308
		6	20	4437	4415
		14	1	1963	1921
70	3	4	301	87822	80077
		6	124	62070	61368
		14	21	33320	33850
	5	4	75	19257	18716
		6	54	12948	12998
		14	22	6227	6552
	7	4	27	10035	9899
		6	20	6330	6330
		14	38	2781	2929
	9	4	20	7254	7308
		6	26	4426	4415
		14	29	1784	1921

Table 2

Means, empirical and asymptotic variances of the estimates of β from N samples of size $n=4$ ($N=5000$), $n=6$ ($N=5000$) and $n=14$ ($N=1000$)

$-\beta$	$-10^2\gamma$	n	$-\bar{b}$	$10^4 s_b^2$	$10^4 \frac{n\hat{E}-\hat{B}^2}{n\hat{\Delta}}$
30	3	4	30.252	99455	78040
		6	30.625	60817	54754
		14	30.287	31097	26131
	5	4	30.135	26630	26069
		6	30.142	18877	18251
		14	30.166	10384	8984
	7	4	30.059	19764	19102
		6	30.048	14715	14455
		14	30.035	8241	8541
	9	4	30.035	17000	17031
		6	30.019	13501	13557
		14	29.987	9167	9079
50	3	4	50.323	89294	78040
		6	50.307	61103	54754
		14	50.190	29000	26131
	5	4	50.041	26746	26069
		6	50.090	17873	18251
		14	50.042	9095	8984
	7	4	50.048	18550	19102
		6	50.033	14599	14455
		14	50.032	7676	8541
	9	4	50.039	16981	17031
		6	50.023	13757	13557
		14	50.021	9207	9079

Table 2 continued

$-\beta$	$-10^2\gamma$	n	$-\bar{b}$	$10^4 s_b^2$	$10^4 \frac{n\hat{E}-\hat{B}^2}{n\hat{\Delta}}$
70	3	4	70.297	85366	78040
		6	70.125	55038	54754
		14	70.173	25661	26131
	5	4	70.081	26924	26069
		6	70.154	18477	18251
		14	70.061	8424	8984
	7	4	70.054	18955	19102
		6	70.014	14379	14455
		14	70.027	8627	8541
	9	4	70.018	17216	17031
		6	70.020	13641	13557
		14	70.038	8382	9079

Table 3

Means, empirical and asymptotic variances of the estimates of γ from N samples of size $n=4$ ($N=5000$), $n=6$ ($N=5000$) and $n=14$ ($N=1000$)

$-\beta$	$-10^2\gamma$	n	$-10^2\bar{c}$	$10^9 s_c^2$	$10^9 \frac{n\hat{C}-\hat{A}^2}{\beta^2\hat{\Delta}}$
30	3	4	3.0419	42825	40897
		6	3.0131	31315	31024
		14	3.0003	17485	17600
	5	4	5.0441	57989	55525
		6	5.0096	36954	36577
		14	4.9830	19003	19723
	7	4	7.1139	117038	102754
		6	7.0338	57809	55664
		14	7.0120	28176	27958
	9	4	9.2821	281944	214550
		6	9.0685	94522	91272
		14	9.0320	42822	41707
50	3	4	3.0117	15535	14723
		6	3.0048	11273	11169
		14	2.9942	6571	6336
	5	4	5.0236	20416	19989
		6	5.0050	12792	13168
		14	5.0035	6762	7100
	7	4	7.0358	38258	36991
		6	7.0011	20216	20039
		14	7.0122	10343	10065
	9	4	9.0888	80225	77238
		6	9.0182	32718	32858
		14	8.9923	14700	15015
70	3	4	3.0015	7751	7512
		6	3.0054	5519	5698
		14	2.9894	3060	3233
	5	4	5.0079	10475	10198
		6	5.0000	6538	6718
		14	5.0044	3672	3623
	7	4	7.0194	19887	18873
		6	7.0043	10319	10224
		14	6.9992	4777	5135
	9	4	9.0462	41813	39407
		6	9.0071	17001	16764
		14	8.9940	7160	7660

Table 4

Results of (A accepted, R rejected) chi-square tests for comparing the distribution of **c** with a normal one and the distribution of $(n-3)s^2$ with a chi-square distribution with $n-3$ d · f.

			distribution of **c**			distribution of $(n-3)s^2$		
$-\beta$	$-10^2\gamma$	n	k	$\sum_{j=1}^{k} x_j^2$	$H_{0\gamma}$	k	$\sum_{j=1}^{k} x_j^2$	$H_{0\sigma^2}$
30	3	4	17	120.8	R	17	22.9	A
		6	20	20.9	A	19	16.9	A
		14	21	13.4	A	16	16.6	A

−β	$-10^2\gamma$	n	distribution of **c**			distribution of $(n-3)s^2$		
			k	$\sum_{j=1}^{k} x_j^2$	$H_{0\gamma}$	k	$\sum_{j=1}^{k} x_j^2$	$H_{0\sigma^2}$
	5	4	18	110.6	R	17	20.7	A
		6	18	66.2	R	19	15.6	A
		14	20	27.2	A	16	15.3	A
	7	4	24	351.7	R	17	9.7	A
		6	20	82.7	R	19	15.3	A
		14	23	15.0	A	16	10.1	A
	9	4	26	484.0	R	17	18.7	A
		6	22	57.9	R	19	11.6	A
		14	17	13.5	A	16	5.8	A
50	3	4	12	23.9	R	17	8.9	A
		6	12	8.1	A	19	14.6	A
		14	6	2.3	A	16	13.9	A
	5	4	13	78.8	R	17	7.9	A
		6	12	31.9	R	19	13.6	A
		14	14	8.4	A	16	89.2	R
	7	4	17	154.2	R	17	13.4	A
		6	14	35.4	R	19	21.3	A
		14	17	22.8	A	16	12.3	A
	9	4	21	257.2	R	17	13.4	A
		6	16	38.3	R	19	16.7	A
		14	20	10.0	A	16	20.4	A
70	3	4	9	11.0	A	17	16.2	A
		6	11	9.2	A	19	24.4	A
		14	18	15.6	A	16	13.3	A
	5	4	10	27.2	R	17	14.2	A
		6	12	13.9	A	19	24.7	A
		14	20	21.4	A	16	15.9	A
	7	4	12	48.9	R	17	10.5	A
		6	13	23.5	R	19	21.3	A
		14	21	21.0	A	16	12.5	A
	9	4	17	164.4	R	17	7.4	A
		6	14	51.0	R	19	15.5	A
		14	14	17.1	A	16	8.8	A

Table 5

Parameters of the empirical distribution of **c** and $\mathbf{s}^2$ respectively from N samples (N = 5000 for n = 4, 6; N = 1000 for n = 14)

			distribution of **c**		distribution of $\mathbf{s}^2$			
−β	$-10^2\gamma$	n	g_1	g_2	Mean	Variance	g_1	g_2
30	3	4	−0.2954	0.1246	0.9893	1.9026	2.8011	12.0003
		6	−0.0957	0.4051	0.9776	0.6160	1.4768	2.7721
		14	−0.0883	0.0519	0.9932	0.1799	0.8918	1.1473
	5	4	−0.4558	1.0219	1.0272	2.1822	2.8333	11.5235
		6	−0.2493	0.0971	0.9846	0.6570	1.6870	4.8592
		14	−0.0499	−0.3466	1.0196	0.1866	0.6398	0.2735
	7	4	−0.6351	1.0472	1.0086	2.0349	2.7842	11.0562
		6	−0.2426	0.1404	0.9943	0.6490	1.5658	3.2821
		14	−0.0945	0.1060	0.9994	0.1673	0.7754	0.8410
	9	4	−1.4675	5.0798	1.0102	2.0862	2.6872	9.5870
		6	−0.3088	0.1625	0.9906	0.6241	1.6669	4.5617
		14	−0.1564	−0.0600	0.9999	0.1800	0.8390	0.8368
50	3	4	−0.0918	0.0153	0.9643	1.8377	2.8475	12.4419
		6	−0.0085	−0.0699	1.0010	0.6568	1.5592	3.3761
		14	0.0886	0.0451	0.9910	0.1777	0.9095	1.1234
	5	4	−0.2569	0.0905	0.9870	1.9614	2.6927	10.0983
		6	−0.1432	0.1361	1.0091	0.6908	1.7057	4.3269
		14	0.0055	−0.1713	1.0268	0.1925	0.9765	1.8374
	7	4	−0.3518	0.2509	1.0232	2.1501	2.8586	11.3022
		6	−0.1815	0.0141	0.9966	0.6859	1.6741	4.1981
		14	−0.1349	−0.0559	0.9863	0.1747	0.7730	0.7504
	9	4	−0.5945	0.6436	0.9654	1.8487	2.8936	12.9046
		6	−0.1884	−0.0734	0.9894	0.6624	1.6542	3.8212
		14	−0.0703	−0.0732	0.9713	0.1751	0.8741	1.0290

$-\beta$	$-10^2\gamma$	n	distribution of **c** g_1	g_2	distribution of s^2 Mean	Variance	g_1	g_2
70	3	4	−0.0576	0.0072	0.9904	1.9733	2.6857	9.6423
		6	−0.0130	0.0675	1.0120	0.6926	1.5665	3.2508
		14	−0.0012	−0.1426	0.9950	0.1713	0.7858	1.0694
	5	4	−0.1764	0.1338	0.9785	1.8452	2.7524	11.7852
		6	−0.0746	−0.0458	1.0004	0.6653	1.6502	3.9828
		14	0.0845	0.1715	0.9886	0.1848	0.9558	1.9057
	7	4	−0.2854	0.3534	0.9825	1.9450	3.0306	14.8935
		6	−0.1119	0.0279	1.0045	0.6342	1.5309	3.5318
		14	−0.1994	0.1685	1.0024	0.1976	0.8870	0.8054
	9	4	−0.3523	0.1894	1.0074	2.0231	2.6581	9.6403
		6	−0.1895	0.0479	1.0100	0.6845	1.6362	4.0455
		14	−0.0976	−0.1617	1.0055	0.1772	0.6417	0.2281

Table 6

Percentages of acceptance in 10,000 simulations of testing $H_0: \gamma = \gamma_0$ against $H_A: \gamma \neq \gamma_0$ for $\beta = -30$ (n = 10)

$-10^2\gamma_0$	$-10^2\gamma$	$\alpha^* = 0.01$ N_A	$\alpha^* = 0.05$ N_A	$\alpha^* = 0.10$ N_A
3	1.2	36.54	8.52	3.00
	1.5	55.62	20.98	10.07
	1.8	74.02	40.74	24.80
	2.1	87.29	62.54	46.84
	2.4	95.04	81.13	69.06
	2.7	98.30	92.19	84.84
	3.0	99.07	95.01	90.28
	3.3	97.84	91.02	84.18
	3.6	93.69	79.54	68.48
	3.9	85.31	62.23	47.53
	4.2	72.45	41.89	28.43
	4.5	55.51	24.92	14.04
	4.8	38.65	12.67	5.78
5	3.2	42.76	11.09	4.37
	3.5	61.66	25.56	13.06
	3.8	78.03	45.50	29.84
	4.1	89.21	67.06	51.61
	4.4	95.71	83.28	71.84
	4.7	98.61	92.82	86.10
	5.0	98.92	95.15	90.01
	5.3	97.78	90.87	83.52
	5.6	93.83	79.58	68.65
	5.9	85.98	64.06	49.99
	6.2	73.59	45.19	30.91
	6.5	59.07	28.15	16.82
	6.8	42.43	15.39	7.70
7	5.2	62.01	24.47	12.01
	5.5	76.18	41.63	24.84
	5.8	87.18	60.01	43.28
	6.1	93.57	76.77	62.26
	6.4	97.12	88.04	78.39
	6.7	98.81	93.77	87.80
	7.0	99.07	95.20	90.11
	7.3	98.09	92.03	86.00
	7.6	95.41	84.75	75.67
	7.9	90.47	73.33	61.39
	8.2	82.97	59.63	44.85
	8.5	72.48	43.80	29.47
	8.8	60.28	29.67	18.09
9	7.2	78.24	43.75	26.82
	7.5	87.06	59.68	41.81
	7.8	92.75	73.26	57.57
	8.1	96.09	84.07	72.31
	8.4	97.84	91.32	83.07
	8.7	98.80	94.44	89.11
	9.0	99.06	94.81	90.21
	9.3	98.16	92.56	86.58
	9.6	96.26	87.78	79.74
	9.9	93.22	79.89	69.23
	10.2	88.30	69.59	56.70
	10.5	81.81	57.61	44.03
	10.8	73.53	45.51	31.43

Table 7

Percentages of acceptance in 10,000 simulations of testing $H_0: \gamma = \gamma_0$ against $H_A: \gamma \neq \gamma_0$ for $\beta = -50$ (n = 10)

$-10^2\gamma_0$	$-10^2\gamma$	$\alpha^* = 0.01$ N_A	$\alpha^* = 0.05$ N_A	$\alpha^* = 0.10$ N_A
3	1.2	1.24	0.06	0
	1.5	7.27	0.43	0.13
	1.8	25.39	4.16	1.21
	2.1	55.59	21.51	10.52
	2.4	82.74	55.61	69.76
	2.7	95.86	85.14	75.26
	3.0	98.92	95.14	89.78
	3.3	95.49	84.19	74.27
	3.6	81.18	54.52	39.43
	3.9	54.72	22.57	11.47
	4.2	26.12	5.50	2.08
	4.5	8.28	0.80	0.22
	4.8	1.98	0.09	0.01
5	3.2	2.21	0.01	0
	3.5	10.62	0.67	0.10
	3.8	31.40	6.12	2.03
	4.1	61.29	26.25	13.52
	4.4	85.34	59.72	43.95
	4.7	96.71	86.72	77.10
	5.0	98.96	94.92	89.98
	5.3	95.83	84.79	75.26
	5.6	83.07	59.07	44.08
	5.9	60.17	27.86	16.24
	6.2	32.77	8.58	3.45
	6.5	13.23	1.63	0.36
	6.8	3.76	0.17	0.06
7	5.2	10.39	0.51	0.08
	5.5	26.34	4.42	1.13
	5.8	51.03	17.37	7.87
	6.1	75.20	42.13	26.43
	6.4	90.99	70.49	55.55
	6.7	97.67	89.66	81.08
	7.0	99.08	94.89	89.96
	7.3	96.77	88.11	79.78
	7.6	88.68	68.62	55.26
	7.9	72.61	43.31	28.79
	8.2	50.83	20.25	10.90
	8.5	29.68	7.54	3.13
	8.8	13.90	2.13	0.53
9	7.2	28.31	4.77	1.28
	7.5	48.83	15.01	6.22
	7.8	69.87	34.24	19.47
	8.1	85.23	58.73	41.91
	8.4	94.30	79.60	66.78
	8.7	98.12	91.55	84.50
	9.0	98.93	95.18	89.59
	9.3	97.47	90.09	83.03
	9.6	92.44	77.15	65.52
	9.9	82.16	57.57	43.29
	10.2	67.25	36.71	23.37
	10.5	49.27	19.61	10.24
	10.8	31.93	8.56	3.58

Table 8

Percentages of acceptance in 10,000 simulations of testing $H_0: \gamma = \gamma_0$ against $H_A: \gamma \neq \gamma_0$ for $\beta = -70$ $(n = 10)$

$-10^2\gamma_0$	$-10^2\gamma$	$\alpha^* = 0.01$ N_A	$\alpha^* = 0.05$ N_A	$\alpha^* = 0.10$ N_A
3	1.2	0.01	0	0
	1.5	0.21	0	0
	1.8	2.70	0.10	0.02
	2.1	21.31	2.80	1.00
	2.4	61.91	28.00	15.20
	2.7	92.50	75.65	61.94
	3.0	99.06	94.97	89.80
	3.3	91.97	74.86	62.63
	3.6	62.10	29.36	16.39
	3.9	22.75	4.13	1.39
	4.2	3.77	0.19	0.04
	4.5	0.31	0.01	0.01
	4.8	0.02	0	0
5	3.2	0	0	0
	3.5	0.22	0	0
	3.8	4.60	0.10	0
	4.1	26.37	4.52	1.35
	4.4	67.86	32.47	18.45
	4.7	93.41	78.61	66.22
	5.0	99.10	94.99	89.86
	5.3	92.51	76.65	63.90
	5.6	65.44	33.08	20.19
	5.9	27.61	6.49	2.58
	6.2	6.63	0.46	0.08
	6.5	0.83	0.02	0
	6.8	0.04	0	0
7	5.2	0.31	0.01	0.01
	5.5	3.09	0.05	0.01
	5.8	16.30	1.69	0.36
	6.1	46.66	14.20	5.92
	6.4	79.10	48.46	32.54
	6.7	95.46	83.71	72.61
	7.0	99.09	94.80	89.75
	7.3	94.34	81.74	71.43
	7.6	76.80	47.56	32.91
	7.9	45.76	16.42	7.83
	8.2	18.21	2.96	1.06
	8.5	4.73	0.35	0.11
	8.8	0.88	0.06	0.01
9	7.2	3.46	0.07	0.01
	7.5	13.84	1.09	0.14
	7.8	36.69	8.27	2.83
	8.1	6.04	30.16	16.43
	8.4	87.80	63.70	47.09
	8.7	96.94	88.26	79.19
	9.0	98.89	94.77	89.64
	9.3	96.06	85.81	77.14
	9.6	84.87	62.41	47.43
	9.9	64.81	33.09	20.34
	10.2	38.70	12.37	5.14
	10.5	18.33	2.89	0.89
	10.8	6.23	0.50	0.09

Table 9

Percentages of acceptance of $H_0: \gamma_0 = -0.05$ for different values γ under the alternative hypothesis for $n = 4$

$-10^2\gamma$	$\alpha^* = 0.01$ N_A	$\alpha^* = 0.05$ N_A	$\alpha^* = 0.10$ N_A
3.2	95.07	75.36	53.25
3.8	96.75	84.28	67.79
4.4	98.42	91.27	82.89
5.0	99.12	95.08	89.79
5.6	98.23	90.81	82.26
6.2	96.52	83.26	66.64
6.8	94.46	74.33	52.32

Table 10

Percentages of acceptance of $H_0: \gamma_0 = -0.05$ for different values γ under the alternative hypothesis for $n = 9$

$-10^2\gamma$	$\alpha^* = 0.01$ N_A	$\alpha^* = 0.05$ N_A	$\alpha^* = 0.10$ N_A
3.2	3.52	0.11	0
3.5	13.77	1.28	0.20
3.8	35.77	8.07	2.82
4.1	62.99	29.56	15.99
4.4	85.46	61.64	46.13
4.7	96.31	86.64	77.73
5.0	98.77	94.20	89.25
5.3	95.47	84.70	75.94
5.6	83.31	60.10	45.66
5.9	61.79	31.03	18.94
6.2	39.27	11.02	4.85
6.5	16.89	2.58	0.83
6.8	5.70	0.42	0.12

Table 11

Percentage of rejection of $H_0: \gamma = \gamma_0$ if H_0 is true from 10000 runs for each of 24 $(\gamma_1, \gamma_2, \beta, \gamma)$-combinations $(n = 9)$

γ_1	γ_2	$-\beta$	$-10^2\gamma$	$\alpha^* = 0.01$	$\alpha^* = 0.05$	$\alpha^* = 0.10$
0	1.5	30	3	1.12	5.32	10,34
		30	9	1.19	4.84	10.09
		70	3	1.43	5.72	11.05
		70	9	1.31	5.95	10.83
0	4	30	3	1.22	5.30	10.42
		30	9	1.25	4.95	10.18
		30	3	1.42	5.68	11.17
		70	9	1.30	5.92	11.12
0	7	30	3	1.15	5.41	10.58
		30	9	1.34	5.36	10.55
		70	3	1.49	5.69	11.25
		70	9	1.35	6.01	11.11
1	1.5	30	3	1.17	5.57	11.00
		30	9	1.31	5.26	10.42
		70	3	1.54	5.87	10.92
		70	9	1.41	6.04	11.03
2	7	30	3	1.12	5.55	10.92
		30	9	1.50	5.88	11.25
		70	3	1.66	6.01	11.07
		70	9	1.58	6.24	11.34
1.5	4	30	3	1.16	5.54	10.94
		30	9	1.38	5.52	10.67
		70	3	1.57	5.89	11.04
		70	9	1.43	6.22	11.13

Table 12

Power function $10^2(1-\beta)$ of the t-test in the exponential regression of the hypothesis $\gamma = -0.03$ for different distributions, $\beta = -70$, and $\alpha^* = 0.05$ $(n = 9)$

	γ_1- and γ_2-value of the distribution						
$-10^2\gamma$	0 0	0 1.5	0 4	0 7	1 1.5	1.5 4	2 7
48	100.0	99.9	99.9	99.8	100.0	99.9	99.7
42	99.8	99.0	98.3	98.0	98.8	98.0	97.2
36	70.6	71.1	73.0	74.7	70.0	71.6	73.2
30	5.0	5.7	5.7	5.7	5.9	5.9	6.0
24	72.0	71.3	73.4	74.6	72.1	74.4	76.1
18	99.9	99.4	98.7	97.7	99.4	98.8	98.4
12	100.0	99.9	99.9	99.8	100.0	99.9	99.9

Table 13

Power function $10^2(1-\beta)$ of the t-test in the exponential regression of the hypothesis $\gamma = -0.09$ for different distributions, $\beta = -30$, and $\alpha^* = 0.05$ $(n = 9)$

	γ_1- and γ_2-value of the distribution						
$-10^2\gamma$	0 0	0 1.5	0 4	0 7	1 1.5	1.5 4	2 7
108	54.5	54.7	58.4	62.2	55.1	58.6	62.2
102	30.4	31.6	35.0	37.3	32.9	36.0	39.5
96	12.2	13.2	14.4	16.1	13.9	15.0	16.6
90	5.2	4.8	4.9	5.4	5.3	5.5	5.9
84	8.7	10.2	11.3	12.3	10.1	11.2	12.6
78	26.7	28.0	31.6	35.6	27.0	30.7	35.2
72	56.2	55.5	58.9	61.7	55.1	59.3	62.2

Table 14

Empirical first kind risks in percent (10,000 runs) of the t-test $H_{0_\gamma}: \gamma = \gamma_0 = -0.05$ and of analougous tests $H_\alpha: \alpha = \alpha_0 = 0$ and $H_\beta: \beta = \beta_0 = -50$ for n = 4 (1) 10 and half widths d_α, d_β and d_γ of the corresponding confidence intervals on α, β and γ for the three nominal α^*-levels

α^*		n = 4	n = 5	n = 6	n = 7	n = 8	n = 9	n = 10
	$H_{0\alpha}$	1.01	0.84	1.37	0.99	1.04	0.91	1.03
0.01	$H_{0\beta}$	1.01	0.99	1.18	1.04	0.83	0.89	1.05
	$H_{0\gamma}$	0.88	0.95	1.28	0.96	1.03	1.08	1.06
	$H_{0\alpha}$	5.01	5.03	5.03	4.63	4.49	5.23	4.87
0.05	$H_{0\beta}$	4.82	4.92	5.10	4.77	4.61	4.98	4.77
	$H_{0\gamma}$	4.92	5.01	5.02	4.64	5.00	5.30	4.83
	$H_{0\alpha}$	10.09	9.92	9.70	9.72	9.43	10.66	9.59
0.10	$H_{0\beta}$	10.08	9.96	10.24	9.77	9.35	10.07	9.44
	$H_{0\gamma}$	10.21	9.82	9.68	9.76	9.93	10.64	9.58
	d_α	69.90	10.94	5.79	4.70	3.94	3.48	3.16
0.01	d_β	82.18	12.95	6.84	5.55	4.63	4.08	3.70
	d_γ	0.2267	0.0347	0.0183	0.0149	0.0125	0.0111	0.0102
	d_α	13.95	4.74	3.36	2.84	2.51	2.30	2.13
0.05	d_β	16.40	5.61	3.97	3.34	2.95	2.69	2.50
	d_γ	0.0453	0.0151	0.0106	0.0090	0.0080	0.0073	0.0069
	d_α	6.93	3.22	2.49	2.18	1.97	1.82	1.71
0.10	d_β	8.15	3.81	2.94	2.57	2.31	2.14	2.01
	d_γ	0.0225	0.0102	0.0079	0.0069	0.0063	0.0058	0.0055

Table 15

Relative frequencies of rejections (f_l, f_u) and acceptions $H_0: \gamma = \gamma_0$ against $H_A: \gamma_2 \neq \gamma_0$ (values in percent) n = 9 and different σ^2-values

	$\alpha^* = 0.01$			$\alpha^* = 0.05$			$\alpha^* = 0.10$		
σ^2	f_l	f_u	f_A	f_l	f_u	f_A	f_l	f_u	f_A
1	0.5	0.5	99.0	2.4	2.7	94.9	4.7	5.3	90.0
4	0.4	0.8	98.8	2.3	3.0	94.7	4.8	5.8	89.4
9	0.3	0.7	98.9	1.8	3.4	94.9	3.9	6.1	90.0
16	0.4	0.8	98.8	1.8	3.2	95.0	4.1	6.3	89.6

6. References

BATES, M. and WATTS, G.
Parameter transformations for improved approximate confidence regions in nonlinear least squares.
Ann. Statist. 9 (1981) 6 1152–1167.

BEALE, E. M. L.
Confidence regions in non-linear estimation.
J. R. Statist. Soc. B 22 (1960) 41–46.

BISMARCK, M., SCHMERLING, S. and PEIL, J.
Zur Problematik der Konfidenzintervallschätzungen für Parameter bei eigentlich nichtlinearer Regression.
Biom. Zeitschrift 16 (1974) 337–347.

BRAND, R. J., PINNUCK, P. E. and JACKSON, K. L.
Large sample confidence bounds for the logistic response curve and its inverse.
Amer. Statistician 27 (1973) 157–160.

BROEMELING, L. D. and HARTLEY, H. O.
Investigation of the optimality of a confidence region for the parameters of a non-linear regression model.
Ann. Inst. Statist. Math. 20 (1968) 263–269.

DUNCAN, G. T.
An empirical study of Jacknife constructed confidence regions in nonlinear regressions.
Technometrics 20 (1978) 123–129.

FOX, T., HINKLEY, D. and LARNTZ, K.
Jackniking in nonlinear regression.
Technometrics 22 (1980) 29–34.

GALLANT, A. R.
The power of the likelihood rotio test of location in nonlinear regression models.
JASA 70 (1975a) 198–203.

GALLANT, A. R.
Testing the parameters of a nonlinear regression model.
JASA 70 (1975b) 927–932.

GUIARD, V.
Systems of distributions.
in: Rasch and Tiku (ed.): Robustness of statistical methods and nonparametric statistics, VEB Deutscher Verlag der Wissenschaften, Berlin 1984, 43–52.

HALPERIN, M.
Confidence interval estimation in nonlinear regression.
J. R. Statist. Soc. B 25 (1963) 330–333.

HALPERIN, M. and MANTEL, N.
Interval estimation of non-linear parametric functions.
JASA 58 (1963) 611–627.

HAMILTON, D. C., WATTS, D. G. and BATES, D. M.
Accounting for intrinsic nonlinearity in nonlinear regression parameter inference region.
Ann. of Statist. 10 (1982) 386–393.

HARTLEY, H. O.
Exact confidence regions for the parameters in nonlinear regression laws.
Biometrika 51 (1964) 347–353.

JENNRICH, R. J.
Asymptotic properties of nonlinear least squares estimators.
AMS 40 (1969) 633–643.

KHORASANI, P. and MILLIKEN, G. A.
Simultaneous confidence bands for nonlinear regression models.
Commun. Statist. Theor. Meth. 11 (1982) 1241–1253.

MALINVAUD, E.
The consistency of nonlinear regression.
AMS 41 (1970) 956–969.

MARITZ, J. S.
Confidence regions for regression parameters.
Austr. J. Statist. 4 (1962) 4–10.

MILLIKEN, G. A.
A Procedure to test hypotheses for nonlinear models.
Commun. Statist. Theor. Meth. A 7 (1978) 65–79.

MILLIKEN, G. A. and GRAYBILL, F. A.
Extensions of the general linear hypothesis model.
JASA 65 (1970) 797–807.

PFAFF, Th. and PFLANZAGL, J.
On the numerical accuracy of approximation based on Edgeworth Expansions.
J. Statist. Comput. Simult. 11 (1980) 223–239.

RASCH, D.
Elementare Einführung in die Mathematische Statistik.
VEB Deutscher Verlag der Wissenschaften Berlin, 2. Auflage 1970.

RASCH, D.
Eigentlich nichtlineare Regression.
1. Sitzungsbericht der IGMS, Berlin, 1978, 1–45.

RASCH, D. and HERRENDÖRFER, G.
Robustheit statistischer Methoden.
Rostock. Math. Kolloq. 17 (1981) 87–104.

RASCH, D. and SCHIMKE, E.
Zur Verteilung der Einschätzungen der Parameter der exponentiellen Regression.
7. Sitzungsbericht der IGMS, Berlin 1982, 61–79.

RASCH, D. and SCHIMKE, E.
Distribution of Estimators in Exponential Regression – A Simulation Study.
Skandin. Journ. of Statist. 10 (1983).

RASCH, D., SCHIMKE, E. and RIEMANN, D. G.
Zur Verteilung der Schätzungen der Parameter der exponentiellen Regression und der Robustheit des t-Testes gegenüber Nichtlinearität.
Probleme d. Angew. Statist. 7 (1982) 90–122.

SCHMIDT, W. H.
Normal theory approximations to tests for linear hypotheses.
Math. Operationsf. Statist. 10 (1979) 353–366.

SCHMIDT, W. H.
Asymptotische Methoden in nichtlinearen Modellen.
Diss. B Humboldt Univ. Berlin 1980.

SCHMIDT, W. H.
Testing hypotheses in nonlinear regressions (with Discussions).
Math. Operationsf. Statist., Ser. Statistics Vol. 13 (1982) 1, 3–19.

SCHMIDT, W. H. and ZWANZIG, S.
Testing hypothesis in nonlinear regression for non-normal distributions.
in: Rasch and Tiku (ed.): Robustness of statistical methods and nonparametric statistics, VEB Deutscher Verlag der Wissenschaften, Berlin 1984, 134–138.

SUNDARAJ, N.
A method for confidence regions for nonlinear models.
Austral. J. Statist. 20 (1978) 270–274.

WILLIAMS, E. J.
Exact fiducial limits in nonlinear estimation.
J. R. Statist. Soc. B 24 (1962) 125–139.

Sektion Mathematik, KARL-MARX-UNIVERSITÄT Leipzig, GDR

Comparison of Break Points of Estimators

MANFRED RIEDEL

Abstract
The purpose of this note is the generalization of the concept of the break point, the comparison of four break points and finally the illustration of these results for M- and generalized L-estimators.

1. Introduction

The concept of the break point was introduced by Hampel (1968, 1971). The break point is a simple characteristic of robustness of a sequence $(T_n : n = 1, 2, \ldots)$ of estimators. It is a measure for the smallest fraction in the departure from the parametric model which can carry the estimated value beyond all bounds. Maronna (1976), Maronna at al. (1979) and Huber (1981) used a modified version of the break point which is easier to handle. Recently, Rieder (1982) extends the concept to robust tests. Hampel (1975, 1976, 1980) discussed a series of pratical aspects of the break point. The purpose of this note is the generalization of the concept of the break point, the comparison of four versions of the break point and the illustration of these results for M- and generalized L-estimators.

2. Some Definitions

Let $(\Omega, \mathfrak{Q})$ be a measurable space and $\mathbf{M}$ denotes the family of distributions defined on $(\Omega, \mathfrak{Q})$. We consider a parametric family $\mathfrak{P} = \{P_\vartheta \in \mathbf{M} : \vartheta \in \Theta\}$ on Ω and suppose that the parametric space Θ is a topologic space with σ-algebra $\mathfrak{B}$ generated by the topology. In robustness theory, the parametric model is not supposed to be exactly true. For $Q \in \mathbf{M}$ a departure from it is determined by the „ε-neighbourhood" $\mathbf{H}_\varepsilon(Q) \subset \mathbf{M}$. Then a deviation of Q is defined by the class $\{\mathbf{H}_\varepsilon(Q) : \varepsilon \in [0,1]\}$ provided that (i) $H_0(Q) = \{Q\}$ and (ii) for $\varepsilon_1 < \varepsilon_2$, $\mathbf{H}_{\varepsilon_1}(Q) \subset \mathbf{H}_{\varepsilon_2}(Q)$. If we have a deviation $\{H_\varepsilon(P_\vartheta)\}$ for every $\vartheta \in \Theta$ so we may define a deviation $\{\mathbf{H}_\varepsilon\}$ of $\mathfrak{P}$ by setting

$$\mathbf{H}_\varepsilon = \bigcup_{\vartheta \in \Theta} \mathbf{H}_\varepsilon(P_\vartheta).$$

Obviously, an observation X with values in a space $\mathfrak{X}$ endowed with a σ-algebra $\mathfrak{L}$ generates a parametric family $\mathfrak{F} = \{F_\vartheta : F_\vartheta = (P_\vartheta)_X\}$ on $\mathfrak{X}$ and a deviation $\{\mathbf{G}_\varepsilon : \mathbf{G}_\varepsilon = (H_\varepsilon)_X\}$ of $\mathfrak{F}$. Here we have denoted the distribution for an observation X by $(P)_X$. On the other side, if the observations $X_1, X_2 \ldots$ are independent and identically distributed a prametric family $\mathfrak{F}$ determines a parametric family $\mathfrak{P}$. Moreover, a deviation of $\mathfrak{F}$ may be extended to a deviation of $\mathfrak{P}$ in which observations are independent and identically distributed. Hence, so far as we have dealt with independent und identically distributed observations there is an equivalence between the parametric families $\mathfrak{P}$ and $\mathfrak{F}$ as well as between their deviations, respectively.

Let $T_n \in \Theta$ be an estimator based on the observations $X_1, X_2, \ldots, X_n$. As usually in the robustness theory, we are interested in the behaviour of a sequence (T_n) of estimators in a deviation of a parametric family. A simple robustness characteristic is the break point of a sequence of estimators in a deviation of a parametric family. This characteristic is useful in practice by telling us, loosely speaking, "how far" the robustness of an estimator extends. Now, we introduce the break points b_H, b_{Hg}, b_c and b_T.

The break point b_H, an enlargement of the break point used by Hampel (1968, 1971) is defined as follows:

$$b_H := \sup\{\varepsilon : \forall\, \vartheta \in \Theta\ \exists \text{ a compact set } C_\vartheta \subset \Theta \text{ such that for every } Q \in \mathbf{H}_\varepsilon(P_\vartheta) \lim_{n\to\infty} Q(T_n \in C_\vartheta) = 1\}.$$

Originally, Hampel (1968, 1971) restricted himself to a deviation determined by the Prohorov metric.

Let $g : \Theta \to \mathbf{R}$ be a measurable function, then $(g \circ T_n)$ is a sequence of estimators. Denote its break point by b_{Hg}. Obviously, if g maps compact sets in compact sets then it is true that $b_H \leqq b_{Hg}$. On the other hand side, if for any numbers $a, b \in \mathbf{R}$, $a < b$, the set $\{\vartheta : g(\vartheta) \in [a, b]\}$ is compact then it follows that $b_{Hg} \leqq b_H$. It is convenient to say that (T_n) is consistent for $T_\infty(Q)$ under Q if (T_n) converges in probability to the asymptotic value $T_\infty(Q)$. Then the break point b_c is defined as

$$b_c := \sup\{\varepsilon : \forall\, \vartheta \in \Theta\ \exists \text{ a compact set } C_\vartheta \subset \Theta \text{ such that for every } Q \in \mathbf{H}_\varepsilon(P_\vartheta)\ (T_n) \text{ is consistent under Q and } T_\infty(Q) \in C_\vartheta\}.$$

Huber (1981) considered a special case of this break point and used a contamination deviation.

Before we give the definition of the break point b_T we recall the concept of estimators generated by a functional. First of all, denote the set of all distributions on $\mathfrak{X}$ by $\mathbf{M}_\varkappa$. Let $\mathfrak{M}$ be the smallest σ-algebra of subsets from $\mathbf{M}_\varkappa$ determined by the property that any $A \in \mathfrak{L}$ the map $\mathbf{M}_\varkappa \ni G \to G(A)$ is measurable. For a point $\mathbf{x}_n := (x_1, x_2, \ldots, x_n) \in \mathfrak{X}^n$ the empirical distribution can be written as

$$F_{\mathbf{x}_n} := \frac{1}{n} \sum_{i=1}^{n} \delta_{x_i}$$

where δ_x stands for the distribution concentrated at $x \in \mathfrak{X}$. It is now easy to see that the map $\mathfrak{X}^n \ni \mathbf{x}_n \to F_{\mathbf{x}_n} \in \mathbf{M}_\varkappa$ is measurable with respect to the σ-algebra $\mathfrak{M}$ and the product σ-algebra on $\mathfrak{X}^n$.

Consider a functional T defined on $D(T) \subset \mathbf{M}_\varkappa$ with values in Θ which is measurable. The sequence (T_n) of estimators is induced by the functional T if $T_n = T(F_{\mathbf{x}_n}) = T(F_{x_1, x_2, \ldots, x_n})$ for every natural number n. The assumption about the functional T guarantees that T_n is really an estimator. Many of the most common estimators are determined by functionals.

For the functional T we introduce the break point b_T of a deviation $\{\mathbf{G}_\varepsilon\}$ of a parametric family $\mathfrak{F}$ on $\mathfrak{X}$ as follows:

$$b_T := \sup\{\varepsilon : \forall\, \vartheta \in \Theta\ \exists \text{ a compact set } C_\vartheta \subset \Theta \text{ such that for every } F \in \mathbf{G}_\varepsilon(F_\vartheta),\ T(F) \in C_\vartheta\}.$$

We remark that Maronna at al. (1979) considered this break point for a contamination deviation. This break point is essentially asymptotic and is often easier to calculate than other break points.
The break points behaviour is monotone in respect to the deviations. Let two different deviations $\{\mathbf{H}_\varepsilon\}$ and $\{\widetilde{\mathbf{H}}_\varepsilon\}$ be given for a sequence (T_n) of estimators and a parametric family. Moreover, assume that for every ε there exists a number $\widetilde{\varepsilon} \in [0, \varepsilon]$ such that $\mathbf{H}_\varepsilon \subset \widetilde{\mathbf{H}}_{\widetilde{\varepsilon}}$, i.e. the deviation $\{\mathbf{H}_\varepsilon\}$ is embedded in the deviation $\{\mathbf{H}_\varepsilon\}$. Then we get for an arbitrary break point $b(\{\mathbf{H}_\varepsilon\}) \geqq b(\{\widetilde{\mathbf{H}}_\varepsilon\})$. This inequality enables us to build up bounds for break points.

3. Main Results

As before, we consider a sequence (T_n) of estimators and a deviation $\{\mathbf{H}_\varepsilon\}$ of a parametric family $\mathfrak{P}$. We begin with the comparison of the break points b_H and b_c.
Theorem 1. Let Θ be metric and locally compact. Then the break points b_H and b_c satisfy

$$b_c \leqq b_H. \tag{3.1}$$

Proof. Let $\vartheta \in \Theta$ and $\varepsilon < b_c$. By the definition of b_c there exists a compact set $C_\vartheta \subset \Theta$ such that for all $Q \in \mathbf{H}_\varepsilon(P_\vartheta)$ the sequence $((Q)_{T_n})$ converges weakly to $\delta_{T_\infty}(Q) \in C_\vartheta$. Choose now a neighbourhood U of C_ϑ for which the closure $\overline{U}$ is compact. Furthermore, we choose a neighbourhood $C(Q)$ of $T_\infty(Q)$ so that $C(Q) \subset U$. By a well known theorem (see Huber 1981, lemma 2.2, p. 22) we get

$$\limsup_{n\to\infty} Q\left(T_n \in C(Q)^c\right) \leqq \delta_{T_\infty(Q)}\left(C(Q)^c\right) = 0$$

(where C^c denotes the compliment of C); hence it follows

$$\lim_{n\to\infty} Q\left(T_n \in \overline{C(Q)}\right) = 1. \tag{3.2}$$

Since

$$\bigcup_{Q\in \mathbf{H}_\varepsilon(P_\vartheta)} C(Q) \subset \overline{U}$$

and (2) it follows that

$$\lim_{n\to\infty} Q\left(T_n \in \overline{U}\right) = 1.$$

As $\overline{U}$ is compact we obtain the assertion (1), since $\varepsilon < b_c$ and $\vartheta \in \Theta$ were arbitrary.
We now give a sufficient condition for the equality in (1).

Theorem 2. Let Θ be metric and locally compact. Suppose that for any $\varepsilon < b_H$ and for any $Q \in \mathbf{H}_\varepsilon$ the sequence (T_n) is consistent to $T_\infty(Q)$, Then we have $b_c = b_H$.

Proof. It remain to show that $b_H \leqq b_c$. Let $\vartheta \in \Theta$ and $\varepsilon < b_H$. Then there exists a compact set $C_\vartheta \subset \Theta$ such that for every $Q \in \mathbf{H}_\varepsilon(P_\vartheta)$.

$$\lim_{n\to\infty} Q\left(T_n \in C_\vartheta\right) = 1. \tag{3.3}$$

Proving indirectly, we assume that for some $Q \in \mathbf{H}_\varepsilon(P_\vartheta)$, $T_\infty(Q) \notin C_\vartheta$. Thus there is a neighbaurhood $C(Q)$ of $T_\infty(Q)$ such that $\overline{C(Q)}$ is compact and $\overline{C(Q)} \cap C_\vartheta = \emptyset$ because Θ is a Tickinov space. As in proof of theorem 1 we conclude that (2) is valid; but this relation contradicts (3). In this way we have established that for any $Q \in \mathbf{H}_\varepsilon(P_\vartheta)$ we have $T_\infty(Q) \in C_\vartheta$; i.e. $\varepsilon \leqq b_c$. Consequently, we infer $b_c \geqq b_H$, since ε was arbitrary.
Finally, we investigate relations between the break points b_c and b_T. In this case we suppose that (T_n) is induced by a functional T and the deviation is determined by a deviation $\{\mathbf{G}_\varepsilon\}$ of a parametric family $\mathfrak{F}$. The functional T is called regular if for any $Q \in \mathbf{M}$ for which (T_n) is consistent we have $T_\infty(Q) = T((Q)_X)$. The next result follows directly from the definition of the break points b_c and b_T.

Theorem 3. If the functional T is regular then $b_c = b_T$.

4. Applications

Consider now the location model on $\mathbf{R}$ which is given by the parametric family $\mathfrak{F} = \{F_\vartheta : F_\vartheta(x) = F(x - \vartheta),\ \vartheta \in \mathbf{R}\}$ where F is a fixed distribution function on $\mathbf{R}$. As a deviation $\{\mathbf{G}_\varepsilon\}$ we use the gross error deviation of Hampel (1968, 1974) which is given by

$$\mathbf{G}_\varepsilon(H) = \left\{(1-\alpha)H + \alpha\delta_x : \alpha \in [0,\varepsilon],\ x \in \mathbf{R}\right\}$$

with $H \in \mathbf{M}_1 := \mathbf{M}_\mathbf{R}$.
Let Ψ be the set of all bounded functions $\psi : \mathbf{R} \to \mathbf{R}$ which possess right and left-hand limits everywhere and satisfy

$$\psi(\infty) := \lim_{x\to\infty} \psi(x) > 0, \tag{4.1}$$

$$\psi(-\infty) := \lim_{x\to\infty} \psi(-x) < 0, \tag{4.2}$$

$$\forall x \geqq 0,\ 0 \leqq \psi(x) < \psi(\infty), \tag{4.3}$$

$$\forall x \leqq 0,\ \psi(-\infty) < \psi(x) \leqq 0, \tag{4.4}$$

$$\int_\mathbf{R} \psi(x)\, F(dx) = 0. \tag{4.5}$$

Putting

$$\lambda_\psi(t,H) := \int_\mathbf{R} \psi(x-t)\, H(dx),$$

for $\psi \in \Psi$ and $H \in \mathbf{M}_1$ we may define two functionals

$$\underset{\sim}{T}_\psi(H) := \sup\left\{t : \forall u < t,\ \lambda_\psi(u,H) > 0\right\}$$

and

$$\widetilde{T}_\psi(H) := \inf\left\{t : \forall u > t,\ \lambda_\psi(u,H) < 0\right\}.$$

Obviously, we have $-\infty < \underset{\sim}{T}_\psi(H) \leqq T_\psi(H) < \infty$. Next we give the definition of M-estimators: The sequence (T_n) of estimators is called a sequence of M-estimators if for any natural number n the function T_n is measurable and $T_n \in [\underset{\sim}{T}_\psi(F_{\mathbf{x}_n}), \widetilde{T}_\psi(F_{\mathbf{x}_n})]$. It is not hard to see that $\underset{\sim}{T}_\psi(F_{\mathbf{x}_n})$ and $\widetilde{T}_\psi(F_{\mathbf{x}_n})$ are measurable. Further we want to show

Theorem 4. For $\psi \in \Psi$ we have

$$b_{\underset{\sim}{T}_\psi} = b_{\widetilde{T}_\psi} = \min\left(-\psi(-\infty), \psi(\infty)\right) / \left(\psi(\infty) - \psi(-\infty)\right). \tag{4.6}$$

Proof. Because of analogy we only derive the break point of $\underset{\sim}{T}_\psi$. We may set $\vartheta = 0$ since the functional $\underset{\sim}{T}_\psi$ is translation invariant. Further, without loss of generality we suppose that $\psi(\infty) - \psi(-\infty) = 1$ and $\psi(\infty) \leqq -\psi(-\infty)$. It is convenient to put $t_\varepsilon(x) := \underset{\sim}{T}_\psi((1-\varepsilon)F + \varepsilon\sigma_x)$. Using the definition of the functional $\underset{\sim}{T}_\psi$ we see that for any $t < t_\varepsilon(x)$

$$(1-\varepsilon)\lambda_\psi(t,F) + \varepsilon\psi(x-t) > 0 \tag{4.7}$$

and

$$(1-\varepsilon)\lambda_\psi\left(t_\varepsilon(x)-0,F\right) + \varepsilon\psi\left((x-t_\varepsilon(x))+0\right) \leqq 0. \tag{4.8}$$

It is sufficient to show that neither the relation

$$b_{\underset{\sim}{T}_\psi} < \psi(\infty) \leqq -\psi(-\infty) \tag{4.9}$$

nor

$$b_{\underset{\sim}{T}_\psi} > \psi(\infty) \tag{4.10}$$

is true.
In the case (9) we choose ε, $b_{\underset{\sim}{T}_\psi} < \varepsilon < \psi(\infty)$ and see that there exists a sequence $(t_\varepsilon(x_n))$ which is unbounded. We consider first the case where $t_\varepsilon(x_n) \to \infty$ and $x_n - t_\varepsilon(x_n) \to r$ for some r. Applying now (7) to $x := x_n$, $t := t_\varepsilon(x_n) - 1/n$ we get

$$(1-\varepsilon)\psi(-\infty) + \varepsilon\psi(r \mp 0) \geqq 0.$$

In view of (3) and (4) we conclude that

$$(1-\varepsilon)\psi(-\infty) + \varepsilon\psi(\infty) \geqq 0;$$

hence $\varepsilon \geqq -\psi(-\infty)$. The latter inequality contradicts (9). The case where $t_\varepsilon(x_n) \to -\infty$ is satisfied can be treated in a similar manner.
In the case (10) we choose ε, $\psi(\infty) < \varepsilon < b_{\underset{\sim}{T}_\psi}$ and see again from the definition of $b_{\underset{\sim}{T}_\psi}$ that the sequence $(t_\varepsilon(x_n))$ is bounded for any sequence (x_n). Assuming that for some sequence $x_n \to -\infty$ we have $t_\varepsilon(x_n) \to r \leqq 0$ the inequality (7) implies

$$(1-\varepsilon)\lambda_\psi(r \mp 0, F) + \varepsilon\psi(-\infty) \geqq 0.$$

Analogously as above, taking (3) and (4) into account we get $\varepsilon \leqq \psi(\infty)$; this relation contradicts apparently with (10). The case where $x_n \to \infty$ is treated analogously. In this way the proof of (6) is completed.
The following inequalities are useful to establish the consistence of M-estimators.

Lemma. For nondecreasing $\psi \in \Psi$, $H \in \mathbf{M}_1$ and any sequence (T_n) of M-estimators,

$$\underset{\sim}{T}_\psi(H) \leqq \liminf_{n\to\infty} T_n \leqq \limsup_{n\to\infty} T_n \leqq \widetilde{T}_\psi(H) \qquad \text{a.s.}$$

(with respect to the distribution $\hat{H} \in \mathbf{M}$ for which X_1, $X_2, \ldots$ are independent and $(\hat{H})_{X_i} = H$ for $i = 1, 2, \ldots$).

Proof. Let $t > \widetilde{T}\psi(H)$. The strong law of large numbers applied to $X_1 - t, X_2 - t, \ldots$ gives

$$\lim_{n\to\infty} \lambda_\psi(t, F_{\mathbf{x}_n}) = \lambda_\psi(t, H) < 0 \qquad \text{a.s.}$$

For almost all $\omega \in \Omega$ there exists a natural number $n_0(\omega)$ such that $\lambda_\psi(t, F_{\mathbf{x}_n}) < 0$ provided that $n \geqq n_0(\omega)$. Consequently, we get $\widetilde{T}_\psi(F_{\mathbf{x}_n}) \leqq t$ and tending with $n \to \infty$ we obtain

$$\limsup_{n\to\infty} \widetilde{T}_\psi(F_{\mathbf{x}_n}) \leqq t \qquad \text{a.s.}$$

Since t was arbitrary we infer, finally,

$$\limsup_{n\to\infty} T_n \leqq \widetilde{T}_\psi(H) \qquad \text{a.s.} \tag{4.11}$$

Analogously, we get the inequality

$$\liminf_{n\to\infty} T_n \geqq \underset{\sim}{T}_\psi(H) \qquad \text{a.s.} \tag{4.12}$$

Both inequalities (11) and (12) show the assertion of the lemma.
In view of theorem 4 and the lemma it is easy to see that

$$b_H \geqq \min(-\psi(-\infty), \psi(\infty)) / (\psi(\infty) - \psi(-\infty)) \tag{13}$$

for any sequence of M-estimators. If the functionals $\underset{\sim}{T}_\psi$ and $\widetilde{T}_\psi$ coincide then we get equality in (13). In this latter case, in view of the lemma, the sequence of M-estimators considered is consistent and the functional $T_\psi = \underset{\sim}{T}_\psi = \widetilde{T}_\psi$ is regular. Then theorem 3 and (13) imply $b_H = b_c$.
We remark that Huber (1981) restricted himself to monotone increasing functions ψ and considered a deviation determined by the Lévy metric.
Finally, we turn to L-estimators and start with their definition. We endow Ψ with the supremum metric and introduce the σ-algebra generated by the topology. If μ is a distribution on Ψ then two functionals $\underset{\sim}{T}_\mu$ and $\widetilde{T}_\mu$ are defined as follows:

$$\underset{\sim}{T}_\mu(H) := \int_\Psi \underset{\sim}{T}_\psi(H)\mu(d\psi),$$

$$\underset{\sim}{T}_\mu(H) := \int_\Psi \widetilde{T}_\psi(H)\mu(d\psi),$$

provided that both integrals exist. Therefore, in analogy to M-estimators, a sequence (T_n) is called a sequence of L-estimators if T_n is measurable and $T_n \in [\underset{\sim}{T}_\mu(T_{\mathbf{x}_n}), \widetilde{T}_\mu(F_{\mathbf{x}_n})]$.
Assume that for non-decreasing ψ there exists a unique solution $a(\psi, c)$ of the equation $\lambda_\psi(a(\psi, c), F) = c$ for any $c \in (\psi(-\infty), \psi(\infty))$. It is convenient to put $a(\psi, c) = -\infty$ for $c \leqq \psi(-\infty)$ and $a(\psi, c) = \infty$ for $c \geqq \psi(\infty)$. Under this restriction we can derive the break point of $\underset{\sim}{T}_\mu$ and $\widetilde{T}_\mu$.

Theorem 5. Let μ be a distribution concentrated on non-decreasing.
Then

$$b_{\underset{\sim}{T}_\mu} = b_{\widetilde{T}_\mu} = \sup\left\{\varepsilon : \left|\int_\Psi a(\psi, -\psi(-\infty)\varepsilon/(1-\varepsilon))\mu(d\psi)\right| < \infty, \right.$$
$$\left. \left|\int_\Psi a(\psi, -\psi(\infty)\varepsilon/(1-\varepsilon))\mu(d\psi)\right| < \infty\right\}.$$

Proof. Because of analogy we only derive the break point of $\underset{\sim}{T}_\mu$. For increasing ψ the function t_ε (*) also is increasing. Therefore, the lower and upper bound v and w for $\underset{\sim}{T}_\psi(H)$ in $\mathbf{G}_\varepsilon$ satisfies, for $\varepsilon < b_{\underset{\sim}{T}_\psi}$, the equation

$$(1-\varepsilon)\lambda_\psi(v, F) + \varepsilon\psi(-\infty) = 0$$

and

$$(1-\varepsilon)\lambda_\psi(w, F) + \varepsilon\psi(\infty) = 0,$$

respectively; i. e.

$$v = a(\psi, -\psi(-\infty)\varepsilon/(1-\varepsilon)), \quad w = a(\psi, -\psi(\infty)\varepsilon/(1-\varepsilon)).$$

These relations also are true for all ε because the function $a(\psi, \cdot)$ is monotone decreasing for non-decreasing ψ. Since the lower and upper bound of $\underset{\sim}{T}_\mu(H)$ in $\mathbf{G}_\varepsilon$ is expressed by

$$\int_\Psi a(\psi, -\psi(-\infty)\varepsilon/(1-\varepsilon))\mu(d\psi)$$

and

$$\int_\Psi a(\psi, -\psi(\infty)\varepsilon/(1-\varepsilon))\mu(d\psi),$$

respectively the assertion of theorem 5 is shown.
Note that Huber (1981, p. 58) considered the α-quantile mean functional

$$T(H) = \frac{1}{1-2\alpha}\int_\alpha^{1-\alpha} H^{-1}(s)\,ds \qquad (\alpha \in (0, 1/2))$$

which is a special case of $\underset{\sim}{T}_\mu$. In this case μ is concentrated on $\{\psi_p : p \in [\alpha, 1-\alpha]\}$ where

$$\psi_p(x) = \begin{cases} p & \text{if } x > 0 \\ 0 & \text{if } x = 0 \\ -(1-p) & \text{if } x < 0 \end{cases}$$

and the generated measure on $[\alpha, 1-\alpha]$ is the Lebesgue measure normed by $1 - 2\alpha$. Further, we see that

$$a(\psi_p, -\psi_p(-\infty)\varepsilon/(1-\varepsilon)) = F^{-1}((p-\varepsilon)/(1-\varepsilon)),$$

$$a(\psi_p, -\psi_p(\infty)\varepsilon/(1-\varepsilon)) = F^{-1}(p/(1-\varepsilon))$$

provided that the distribution function F is contiuous and increasing. Applying now theorem 5 we obtain $b_T = \alpha$. In this way theorem 5 generalizes a part of theorem 3.1 of Huber (1981, p. 60).

5. References

HAMPEL, F. R.
Contributions to the theory of robust estimation.
Unpublished Ph. D. Thesis, University of California, Berkeley, 1968.

HAMPEL, F. R.
A general qualitative definition of robustness.
Ann. Math. Statist 42 (1971) 1887–1896.

HAMPEL, F. R.
The influence curve and its role in robust estimation.
J. Amer. Statist. Ass. 62 (1974) 1179–1186.

HAMPEL, F. R.
Beyond location parameters: Robust concepts and methods.
Proc. 40th Session I. S. I. Warsaw 1975, Bull. Int. Statist. Inst. Book 1, 46 (1975) 375–382.

HAMPEL, F. R.
On the break point of some rejection rules with mean.
Res. Rep. no. 11 (1976), Fachgruppe für Statistik, ETH Zürich.

HAMPEL, F. R.
Robuste Schätzungen: Ein Anwendungsorientierter Überblick.
Biometrical Journal 22 (1980) 3–21.

HUBER, P. J.
Robust Statistics.
John Wiley, New York, 1981.

MARONNA, R. A.
Robust M-estimators of multivariate location and scatter.
Ann. Statist. 4 (1976) 51–67.

MARONNA, R. A., BUSTOS, O. H., YOHAI, V. J.
Bias- and efficiency-robustness of general M-estimators for regression with random carriers.
Smoothing Techniques For curve Estimation, eds. Th. Casser and M. Rosenblatt, Lecture notes in Mathematics no. 757, Springer, Berlin, 1979, 91–116.

RIEDER, H.
Qualitative Robustness of rank tests.
Unpublished paper (1982).

Facalta' di Scienze Statistiche Demografiche ed Attuariali Istituto di Statistica e Ricerca Sociale "C. Gini"
Universita' di Roma

The Robustness of some Statistical Test. The Multivariate Case

ALFREDO RIZZI

1. Introduction

In this work some results will be shown, that were obtained by means of Monte Carlo experiments and that concern the robustness of the following statistical parameters:

Quadratic form of the mean;
Determinant of the variance and covariance matrix |S|;
Hotelling T^2;
Correlation coefficient;
Covariance;

for bidimensional statistical variables.
A set of 9 distributions is taken into consideration — besides the normal one —, six of which are symmetric; they are described in paragraph 2.
The sample size varies from 5 to 100; the repetitions of the experiments are 3 000.
Studies about robustness — and we refer also to those of Monte Carlo type — more frequently concern the univariate case; when the bidimensional case is taken into consideration, the analytical problems that arise are such that the work becomes very difficult.
In general, our results show that the analyzed parameters are not much robust. This happens, in particular, when the distributions from which the samples are drawn are not symmetric.
Besides, we noticed that the robustness of the parameters does not change with a significant importance when the sample size changes.
The results concerning the normal distribution, compared with the well-known theoretical ones, show that differences — due to the simulating procedure — are found at the most on the third decimal digit.

2. The Distributions

The bivariates that were taken into consideration in this work are the following:

1) Normal distribution;
2) U-shaped distribution with definition set over an ellipse;
3) U-shaped distribution with definition set over a rectangle;
4) Uniform distribution with definition set over an ellipse;
5) Uniform distribution with definition set over a rectangle;
6) χ^2 distribution;
7) Hat-shaped distribution with definition set over an ellipse;
8) Hat-shaped distribution with definition set over a rectangle;
9) Pareto distribution;
10) Gumbell distribution.

3. The Parameters

3.1. Taking into consideration the set of the 10 previously described distributions, for each one of the parameters described in paragraph 1 we computed the quantiles of the sampling distribution of such parameters at the levels 1 %, 5 %, 10 %, 20 %, 50 %, 80 %, 90 %, 95 %, 99 %, as well as the mean, the standard error, the indexes of asymmetry and kurtosis.

3.2. *Distribution of the quadratic form z'z*

It is well-known that if C is a sample of size n drawn from a population $x \rightarrow N(\mu, \Sigma)$

$$C \equiv (x_1, \ldots, x_n) \equiv \begin{pmatrix} x_{11} & \cdots & x_{1n} \\ \vdots & & \vdots \\ x_{m1} & \cdots & x_{mn} \end{pmatrix}$$

then the distribution of the variable x usually has mean μ and variance Σ/n. Besides, if the original population is not normal, then the distribution of the sample is no longer normal, but is asymptotically normal.
The quadratic form

$$n(\bar{x} - \mu)' \Sigma^{-1} (\bar{x} - \mu) = Fq. \quad (1)$$

is distributed like a χ^2 with m degrees of freedom. In the case we are studying — i.e. in the case of the standardized bivariate — the quadratic form is reduced to:

$$n\, z'\, z = Fq. \quad (2)$$

that is distributed like a χ^2 with two degrees of freedom if the population is binormal; otherwise it is distributed like a χ_2^2 asymptotically.
As far as the mean and the variance of (2) are concerned, we obtain:

$E(Fq.) = 2$
$\mathrm{Var}(Fq.) = 4$
$\sigma_{Fq.} = 2$

The last two formulas hold in the normal case.
In the tables we printed mean, variance, indexes of asymmetry and kurtosis, quantiles of the distributions described in paragraph 2 for the various parameters (the quadratic form is found in tables 2.1—2.3). In table 1. the minimum and the maximum values of the quantiles can be read for some values of n, with reference to 8 out of the 10 distributions taken into consideration — Pareto and Gumbell distributions were excluded, since the behaviour of the parameters is absolutely anomalous in correspondence with such distributions —.
In particular, distributions 6a, 6b, 6c and 6d correspond to $\chi^2_{2,3}$, $\chi^2_{3,4}$, $\chi^2_{4,5}$, $\chi^2_{5,6}$, while in the last row we printed the values of the parameters and of the quantiles of the theoretical normal distribution, so that it is possible to appraise the reliability of our simulating procedure.

As far as the quadratic form is concerned, the analysis of the tables shows that:

- The quantiles of the distributions are not much far from each other. So, e.g., the range of the 95th quantile, for $n = 100$, is equal to the interval 5.70–6.40; for $n = 10$, it shrinks to 5.70–6.31. Among the four parameters taken into consideration, the quadric form can be considered as the most robust one, both when the distributions change and when n changes. This also holds when the original population is strongly asymmetric (e.g. $\chi^2_{2,3}$, Pareto and Gumbell distributions).
- The values corresponding to the normal distribution are found on the centre of the range.
- In correspondence with all the distributions, the distribution of the quadratic form is highly skewed to the right; this is also verified when $n = 100$ and does not changes perceptibly when the starting distribution changes.
- The shape of the sampling distribution of the quadric form is far from normal. The indexes of kurtosis, for $n = 100$, are all greater than four; when n increases, they do not tend to decrease. The highest values of such index are found in correspondence with Pareto and Gumbell distributions.
- The width of the range within which the quantiles vary (when the distribution of the population from which they are drawn changes) is rather small. In fact the relative width [1]), e.g., of the 95th quantile is approximately equal to 11 % when $n = 100$; such a value does not change perceptibly when n changes, so that the robustness of the quadratic form is confirmed.

3.3. *Distribution of the determinant of* S

The matrix of variances and covariances corresponds, in the multivariate case, to the univariate case. In the multivariate case we notice that the scalar $|\Sigma|$ – which is called generalized variance — can be associated to an index of multiple variability.
By the same token, the generalized variance of a sample C is given by:

$$|\mathbf{S}| = \left| \frac{1}{n-1} \sum_{i=1}^{n} (\mathbf{x}_i - \bar{\mathbf{x}})(\mathbf{x}_i - \bar{\mathbf{x}})' \right|$$

If the population x is $N(\mu, \Sigma)$, when the sample C changes $|S|$ is distributed like the product of χ^2 with decreasing degrees of freedom, i.e.:

$$|\mathbf{B}| = \frac{(n-1)^{m} |\mathbf{S}|}{|\Sigma|}$$

where the density function of $|\mathbf{B}|$ is:

$$f(|\mathbf{B}|) = \chi^2_{\nu} \cdot \chi^2_{\nu-1} \cdot \ldots \cdot \chi_{\nu-m-1}$$

Therefore we can compute the moments of $|S|$; in the standardized case we obtain:

$$E(|\mathbf{S}|) = \frac{n-2}{n-1}$$

$$\mathrm{Var}(|\mathbf{S}|) = \frac{(2n-1)(2n-4)}{(n-1)^3}$$

Taking into consideration the tables we note that:

- The sampling distribution of the determinant $|S|$ of the variance and covariance matrix drawn from a population with hat-shaped density function is, among the considered distributions, the one that is more similar to the distribution obtained under the hypothesis that the population from which the sample are drawn is normal. The less similar one is $\chi^2_{2,3}$ (we neglect Pareto and Gumbell distributions).
- The indexes of kurtosis show that, as far as the symmetric distributions are concerned, the distribution of $|S|$ is similar to a normal distribution; they are smaller than the index of kurtosis of the sampling distribution of $|S|$ under the hypothesis that the population is normally distributed.
- The standard error of the distribution of $|S|$, when the distributions are symmetric, is smaller than the one of the distribution of $|S|$ under the hypothesis that the population is normally distributed.
- Important differences are not found in the behaviours of the sampling distribution of $|S|$ in the case of symmetric populations over elliptic or rectangular definition sets.
- The distribution of $|S|$, in all the cases taken into consideration, is slightly skewed to the right.
- The width of the range within which the quantiles of $|S|$ vary is not very large; let us consider, e.g., the 95th quantile: for $n = 100$ it lies in the interval 1.11 – 1.74. Besides, when n increases, we notic that the relative width shrinks sharply: it starts from 70 % when $n = 10$ and reaches 44 % when $n = 100$. We should verify whether or not such a trend also holds when n reaches very high values; the obtained results, anyway, let us confirm that $|S|$ is rather robust as regards the various distributions taken into consideration. This holds for values of n higher than those found for the quadratic form.

3.4. *Hotelling T^2 distribution*

Let us consider the usual sample C drawn from a normal population $\mathbf{x} \rightarrow N(\mu, \Sigma)$ and compute the scalar:

$$t^2 = n(\bar{\mathbf{x}} - \mu)' \mathbf{S}^{-1} (\bar{\mathbf{x}} - \mu) \qquad (3)$$

where:

$$\bar{\mathbf{x}} = \frac{1}{n} \sum_{i=1}^{n} (\mathbf{x}_i)$$

and:

$$\mathbf{S}^{-1} = \frac{1}{n-1} \sum_{i=1}^{n} (\mathbf{x}_i - \bar{\mathbf{x}})(\mathbf{x}_i - \bar{\mathbf{x}})'$$

When the sample C changes, the scalar t^2 is distributed according the T^2 variable – studied by Hotelling – with $n-1$ degrees of freedom, whose density function is:

$$f(T^2) = \begin{cases} \dfrac{\Gamma\left(\frac{n}{2}\right)}{(n-1)^{m/2}\,\Gamma\left(\frac{m}{2}\right)\Gamma\left(\frac{n-m}{2}\right)} \left(\dfrac{1+T^2}{n-1}\right)^{n/2} (T^2)^{\frac{m}{2}-1} & \text{for } T^2 > 0 \\ 0 & \text{for } T^2 \leq 0 \end{cases} \qquad (4)$$

It is possible to show that the following T^2 transform:

$$F = \frac{n-m}{(n-1)m} T^2 \qquad (5)$$

is distributed like the Fisher-Snedecor F with m and $n-m$ degrees of freedom.
On the basis of (5) it was possible to compute mean and standard error of T^2 that can be used when resorting to Bienaimé-Tschebyeff theorem.
Recalling the mean value and the variance of Fisher F variable, we obtain on the basis of (5):

[1]) Taking into consideration the interval a_1–a_2 we define absolute width: $a_2 - a_1$, and relative width: $2\,\dfrac{a_2 - a_1}{a_2 + a_1}$.

$$E(T^2)=\frac{2(n-1)}{n-1}E(F_{2,n-2})$$

$$=\frac{2(n-1)}{n-2}\,\frac{n-2}{n-4}=\frac{2(n-2)}{n-4}$$

$$\mathrm{Var}(T^2)=\frac{4(n-1)^2}{(n-2)^2}\mathrm{Var}(F_{2,n-2})$$

$$=\frac{4(n-1)^2}{(n-2)^2}\,\frac{2(n-2)^2(n-2)}{2(n-4)(n-6)}=\frac{4(n-1)^2(n-2)}{(n-4)^2(n-6)}$$

In this case the interval of T^2 is $[0;\mu_{T^2}+K\,\sigma_{T^2}]$ where, for $K=12$, we find at least 99.31 % of the samples.
In the experiment, the remaining 0.69 % was recorded in a vector in increasing order.
In table from 4.1 to 4.3 we printed the values of the parameters of the T^2 distribution for some values of n and for the distributions described in paragraph 2.
By analyzing such tables we remark that:

- The highest differences with respect to the sampling distribution of T^2 for samples drawn from a normal population are found in correspondence with $\chi^2_{2,3}$.
- The width of the range within which the quantiles of T^2 may vary is very small; for $n=100$, for instance, the 95th quantile varies within the interval 6.09–6.79.
- The standard error of the sampling distributions of T^2 for samples drawn from symmetric distributions is very close to the one of the distribution deriving from normal populations. For $n=100$, the maximum difference is 0.15.
- In all the studied cases the sampling distribution shows indexes of kurtosis whose values are very far from zero; for $n=100$ the kurtosis of the normal distribution is 4.8, while the kurtosis of all the other distributions is greater than 6.7.
- The sampling distribution of T^2 is always strongly skewed to the right.
- The sampling distribution of T^2, when the samples are drawn from the Gumbell distribution, shows a behaviour which is similar to the one of $\chi^2_{2,3}$.
- As far as the normal distribution is concerned, the quantiles are found within the left half of the computed range.
- The T^2 test is rather robust when the distributions vary; if we consider the 95th quantile, we see that the relative width of the range within which such a quantile may very is equal to 62 % when $n=10$ and shrinks to 11 % when $n=100$.

3.5. *Correlation coefficient*

Let us consider $\mathbf{x}=(\mathbf{x}_1,\mathbf{x}_2,\dots,\mathbf{x}_n)$ as a size n sample drawn from a bidimensional normal population:

$$\mathbf{x}\Rightarrow(x_1x_2)\Rightarrow N\left\{\begin{pmatrix}\mu_1\\ \mu_2\end{pmatrix},\begin{pmatrix}\sigma_1^2 & \sigma_1\sigma_2\varrho\\ \sigma_1\sigma_2\varrho & \sigma_2^2\end{pmatrix}\right\}$$

The maximum likelihood estimate of ϱ is the following:

$$\hat{\varrho}=r=\frac{\sum_{i=1}^{n}(x_{1i}-\bar{x}_1)(x_{2i}-\bar{x}_2)}{\sqrt{\sum_{i=1}^{n}(x_{1i}-\bar{x}_1)^2\cdot\sum_{i=1}^{n}(x_{2i}-\bar{x}_2)^2}}=\frac{\bar{S}_{12}}{\bar{S}_1\cdot\bar{S}_2}$$

Besides, it is well-known that if $\mathbf{x}$ is normal and $\varrho=0$ then the variable

$$w=\sqrt{n-2}\cdot r/\sqrt{1-r^2}$$

is distributed like a Student t with $n-2$ degrees of freedom.

Since $dw/dr=(1-r)^{-3/2}$, the density function of r is the following:

$$f(r)=\frac{\Gamma\left(\frac{n-1}{2}\right)}{\Gamma\left(\frac{n-1}{2}\right)\sqrt{\pi}}\left(1-r^2\right)^{\frac{n-4}{2}} \qquad (7)$$

This holds under the hypothesis that $\mathbf{x}$ is normal, with independently drawn samples and $\varrho=0$.
From (7) we see that the function is symmetric with respect to the origin, while for $n>4$ its mode is found in correspondence with $r=0$ and the contact order with the r axis in ±1 is of $(n-5)/2$ when n is an odd digit and of $(n-3)/2$ when n is an even digit.
Besides, as the density function is symmetric, the odd order moments are equal to zero – in particular the mean is equal to zero –. The even order moments obtained through the integrating procedure are equal to:

$$E\left[r^{2m}\right]=\frac{\Gamma\left(\frac{n-1}{2}\right)\cdot\Gamma\left(m+\frac{1}{2}\right)}{\sqrt{\pi}\cdot\Gamma\left(\frac{n-1}{2}+m\right)}$$

By using this formula we find the mean and the variance: $E(r^2)=0$; $\mathrm{Var}(r^2)=(n-1)^{-1}=\nu^{-1}$
As far as our experiment is concerned, since $r\in[-1,1]$, it is not necessary to resort to the Bienaimé-Tchebycheff theorem; our procedure was to divide the range of $r\in[-1,1]$ into 1000 parts and to find, on such a basis, the sampling distribution of r^2 under the various hypotheses.
In the tables from 5.1 to 5.3 we printed the parameters and the quantiles of the variable R in correspondence with the ten distributions and with some values of n. We remark that:

- If we take into consideration the symmetric distributions, the computed quantiles are not very far from those of the normal distribution; the differences are smaller than 0.09 for $n=40$, 50, 75, 100 and slightly greater for the other values of n.
- The distribution of r deriving from symmetric distributions over a rectangular definition set (i.e. the third, the fifth and the eight distributions) are more similar to those deriving from a normal distribution than the distributions over an elliptic definition set. So, e.g., for $n=75$ and $n=100$ and at the three levels of 90 %, 95 % and 99 %, the differences between the distributions over a rectangular definition set and the normal distribution are not greater than 0.01.
- The highest differences with respect to the r distribution deriving from a normal distribution are found in correspondence with $\chi^2_{2,3}$; such differences tend to decrease when the degrees of freedom of χ^2 increase.
- Taking into consideration the symmetric distributions, the smallest range of the distribution of the correlation coefficient is found when the samples are drawn from a population with U-shaped density function over an elliptic definition set (distribution 2).
- The standard error of the distributions over an elliptic definition set (distributions 2, 4, 7) is smaller than the one of the distributions over a rectangular definition set.
- The kurtosis indexes of the symmetric distributions reach values very close to zero.

- When the samples are drawn from the Gumbell distribution (distribution 10), the distribution of the correlation coefficient is similar to the one of $\chi^2_{2,3}$.
- The width of the interval within which the quantiles of r may vary is not negligible even when the sample size is large ($n = 75$, $n = 100$). Besides when n increases, the absolute width of the interval decreases, while the relative width increases, at least for n up to 100. As a matter of fact, taking into consideration the 95th quantile, we see that the absolute width shrinks from 0.29 (0.46–0.75) for $n = 10$ to 0.16 (0.13–0.29) for $n = 100$, while the relative one increases from 48 % for $n = 10$ up to 76 % for $n = 100$.
- As far as the normal distribution is concerned, the quantiles are found within the left half of the computed range.

In tables from 6.1 to 6.3, at last, we printed the parameters and the quantiles of the covariance variable

$$s_{x_1x_2} = \sum_{i=1}^{n} (x_{1i} - \bar{x}_1)(x_{2i} - \bar{x}_2)$$

in correspondence with the ten distributions and with some values of n.

As far as the covariance variable is converned, the same remarks hold, that were made about the correlation coefficient. Besides, when n increases, the distributions of r and of $s_{x_1x_2}$ tend to coincide; as it is logical.

4. **Distributions**

1) Normal distribution (Monte Carlo)
2) U-shaped distribution with definition set over an ellipse
3) U-shaped distribution with definition set over a rectangle
4) Uniform distribution with definition set over an ellipse
5) Uniform distribution with definition set over a rectangle

6a) χ^2 distribution with $\nu_1 = 2$ $\nu_2 = 3$
6b) χ^2 distribution with $\nu_1 = 3$ $\nu_2 = 4$
6c) χ^2 distribution with $\nu_1 = 4$ $\nu_2 = 5$
6d) χ^2 distribution with $\nu_1 = 5$ $\nu_2 = 6$

7) Hat-shaped distribution over an elliptic definition set
8) Hat-shaped distribution over a rectangular definition set
9) Pareto distribution
10) Bidimensional Gumbell exponential distribution

N*) Normal distribution

Table 1
Minimum and maximum values with reference to distributions 2 ÷ 8 pag. 23 Between brackets the simulated values of normal bivariate

									Hotelling T^2									
n	Min	1 %	Max	Min	5 %	Max	Min	10 %	Max	Min	90 %	Max	Min	95 %	Max	Min	99 %	Max
10	0.02	(0.02)	0.03	0.08	(0.14)	0.14	0.19	(0.24)	0.25	7.11	(7.23)	11.61	10.71	(10.82)	20.46	19.91	(19.91)	53.89
15	0.02	(0.02)	0.02	0.09	(0.09)	0.14	0.21	(0.22)	0.26	5.75	(6.02)	8.50	8.07	(8.22)	12.98	13.84	(13.84)	33.45
20	0.02	(0.02)	0.02	0.08	(0.09)	0.13	0.19	(0.22)	0.26	5.39	(5.49)	6.85	7.17	(7.61)	11.08	12.12	(12.61)	27.42
30	0.02	(0.02)	0.02	0.10	(0.11)	0.11	0.20	(0.23)	0.24	4.86	(5.13)	6.21	6.35	(6.80)	8.80	9.95	(11.54)	16.65
50	0.02	(0.02)	0.02	0.10	(0.10)	0.10	0.19	(0.19)	0.24	4.77	(5.09)	5.32	6.30	(6.70)	7.63	9.81	(10.47)	14.40
75	0.01	(0.01)	0.02	0.09	(0.09)	0.10	0.21	(0.21)	0.23	4.60	(4.83)	5.18	6.10	(6.66)	7.39	7.75	(10.22)	12.55
100	0.01	(0.01)	0.02	0.08	(0.09)	0.09	0.20	(0.20)	0.23	4.55	(4.79)	5.16	6.09	(6.35)	6.79	9.17	(9.52)	11.82
									|S|									
10	0.04	(0.11)	0.22	0.07	(0.20)	0.39	0.11	(0.27)	0.50	1.29	(1.74)	1.99	1.41	(2.21)	2.97	1.63	(3.14)	6.16
15	0.06	(0.18)	0.41	0.13	(0.30)	0.56	0.18	(0.37)	0.55	1.23	(1.63)	1.97	1.30	(1.97)	2.58	1.48	(2.59)	4.35
20	0.10	(0.25)	0.52	0.18	(0.36)	0.65	0.30	(0.46)	0.68	1.18	(1.52)	1.93	1.25	(1.78)	2.39	1.39	(2.37)	4.11
30	0.18	(0.34)	0.65	0.32	(0.47)	0.73	0.34	(0.55)	0.77	1.16	(1.45)	1.84	1.21	(1.69)	2.30	1.32	(2.11)	3.68
50	0.25	(0.46)	0.74	0.37	(0.56)	0.81	0.45	(0.64)	0.85	1 12	(1.35)	1.64	1.16	(1.52)	2.00	1.23	(1.82)	2.66
75	0.34	(0.53)	0.79	0.49	(0.64)	0.85	0.52	(0.71)	0.88	1.10	(1.31)	1.56	1.13	(1.43)	1.83	1.20	(1.65)	2.54
100	0.40	(0.59)	0.83	0.52	(0.69)	0.88	0.58	(0.74)	0.90	1.09	(1.26)	1.48	1.11	(1.35)	1.74	1.16	(1.54)	2.22
									Quadratic form									
10	0.01	(0.02)	0.02	0.09	(0.10)	0.12	0.19	(0.20)	0.24	4.33	(4.53)	4.67	5.70	(5.93)	6.31	8.47	(9.57)	10.98
15	0.02	(0.02)	0.02	0.09	(0.10)	0.12	0.19	(0.20)	0.24	4.36	(4.53)	4.70	5.70	(5.94)	6.31	8.50	(9.55)	10.90
20	0.02	(0.02)	0.02	0.09	(0.10)	0.12	0.20	(0.20)	0.24	4.37	(4.60)	4.70	5.70	(5.94)	6.33	8.51	(9.33)	10.27
30	0.02	(0.02)	0.02	0.09	(0.10)	0.12	0.20	(0.20)	0.24	4.44	(4.49)	4.70	5.70	(5.94)	6.34	8.84	(9.33)	10.27
50	0.02	(0.02)	0.02	0.09	(0.10)	0.12	0.20	(0.20)	0.24	4.44	(4.61)	4.70	5.70	(5.95)	6.34	8.84	(9.33)	10.28
75	0.02	(0.02)	0.02	0.09	(0.10)	0.12	0.20	(0.20)	0.24	4.40	(4.62)	4.70	5.70	(6.10)	6.47	8.84	(9.32)	10.28
100	0.02	(0.02)	0.02	0.09	(0.09)	0.10	0.20	(0.20)	0.24	4.40	(4.66)	4.74	5.70	(6.10)	6.40	8.98	(9.23)	10.31
									r									
10	−0.87	(−0.72)	−0.63	−0.73	(−0.55)	−0.44	−0.58	(−0.44)	−0.35	0.36	(0.45)	0.60	0.46	(0.56)	0.75	0.62	(0.72)	0.89
15	−0.78	(−0.43)	−0.47	−0.63	(−0.43)	−0.35	−0.51	(−0.35)	−0.27	0.27	(0.33)	0.50	0.35	(0.43)	0.66	0.48	(0.60)	0.80
20	−0.74	(−0.51)	−0.41	−0.56	(−0.38)	−0.29	−0.47	(−0.30)	−0.24	0.23	(0.30)	0.47	0.29	(0.38)	0.57	0.41	(0.53)	0.77
30	−0.65	(−0.43)	−0.33	−0.49	(−0.31)	−0.23	−0.39	(−0.24)	−0.19	0.18	(0.24)	0.51	0.24	(0.30)	0.51	0.33	(0.43)	0.66
50	−0.49	(−0 32)	−0.26	−0.37	(−0.24)	−0.18	−0.30	(−0.19)	−0.14	0.14	(0.19)	0.33	0.18	(0.24)	0.35	0.25	(0.33)	0.50
75	−0.42	(−0.27)	−0.21	−0.33	(−0.19)	−0.14	−0.25	(−0.15)	−0.11	0.11	(0.15)	0.25	0.14	(0.19)	0.32	0.20	(0.27)	0.44
100	−0.40	(−0.24)	−0.17	−0.29	(−0.16)	−0.13	−0.22	(−0.13)	−0.10	0.10	(0.13)	0.23	0.13	(0.16)	0.29	0.18	(0.23)	0.41
									σ_{xy}									
10	−1.14	(−0.83)	−0.58	−0.97	(−0.56)	−0.42	−0.65	(−0.42)	−0.33	0.33	(0.42)	0.58	0.44	(0.57)	0.84	0.59	(0.85)	1.43
15	−1.28	(−0.64)	−0.47	−0.75	(−0.45)	−0.34	−0.53	(−0.33)	−0.27	0.25	(0.31)	0.53	0.33	(0.42)	0.72	0.46	(0.64)	1.18
20	−0.98	(−0.54)	−0.40	−0.59	(−0.38)	−0.29	−0.48	(−0.29)	−0.23	0.23	(0.30)	0.49	0.29	(0.39)	0.61	0.41	(0.58)	1.05
30	−0.96	(−0.45)	−0.32	−0.56	(−0.31)	−0.23	−0.41	(−0.24)	−0.19	0.18	(0.23)	0.42	0.24	(0.30)	0.54	0.34	(0.47)	0.83
50	−0.60	(−0.31)	−0.26	−0.39	(−0.40)	−0.18	−0.29	(−0.19)	−0.14	0.14	(0.18)	0.31	0.18	(0.23)	0.40	0.24	(0.34)	0.63
75	−0.53	(−0.28)	−0.20	−0.35	(−0.19)	−0.14	−0.26	(−0.15)	−0.11	0.11	(0.15)	0.25	0.14	(0.19)	0.32	0.20	(0.28)	0.49
100	−0.50	(−0.24)	−0.18	−0.32	(−0.16)	−0.13	−0.23	(−0.13)	−0.10	0.10	(0.13)	0.22	0.13	(0.16)	0.29	0.18	(0.23)	0.44

Table 2.1.
Quadratic form N = 10

Dist	M	σ	Asym	Curt	1 %	5 %	10 %	20 %	50 %	80 %	90 %	95 %	99 %
1	1.98	1.90	2.20	3.45	0.02	0.10	0.20	0.43	1.41	3.18	4.53	5.93	9.57
2	1.99	1.90	1.67	3.40	0.02	0.10	0.22	0.45	1.38	3.21	4.61	5.93	8.47
3	2.02	1.91	1.67	3.57	0.02	0.12	0.23	0.48	1.43	3.35	4.56	5.84	8.84
4	2.03	1.93	1.80	3.60	0.03	0.12	0.22	0.45	1.36	3.14	4.49	5.70	8.61
5	2.01	1.92	1.85	3.52	0.01	0.09	0.19	0.44	1.38	3.27	4.65	5.87	9.06
6a	1.98	2.27	3.43	20.78	0.02	0.10	0.22	0.44	1.31	3.03	4.36	6.01	10.87
6b	2.03	2.20	2.50	19.32	0.03	0.10	0.20	0.41	1.38	3.20	4.64	6.04	10.83
6c	2.11	2.35	3.65	28.37	0.02	0.11	0.24	0.47	1.43	3.22	4.67	6.31	10.98
6d	1.96	2.01	2.80	15.22	0.02	0.09	0.20	0.42	1.36	3.07	4.33	5.75	9.50
7	1.95	1.90	1.84	4.52	0.02	0.11	0.22	0.45	1.45	3.14	4.49	5.73	8.84
8	2.02	1.94	1.75	4.22	0.02	0.12	0.23	0.46	1.41	3.31	4.63	5.89	8.78
9	1.91	6.08	11.59	170.37	0.01	0.06	0.13	0.27	0.76	1.77	3.23	6.16	13.99
10	1.99	2.34	4.05	33.8	0.01	0.09	0.21	0.42	1.32	2.95	4.36	6.14	11.08
N*	2.0	2.0			0.02	0.10	0.21		1.39		4.61	5.99	9.21

Table 2.2.
Quadratic form N = 50

Dist	M	σ	Asym	Curt	1 %	5 %	10 %	20 %	50 %	80 %	90 %	95 %	99 %
1	2.00	2.08	2.17	7.04	0.02	0.10	0.20	0.45	1.41	3.17	4.51	5.95	9.33
2	1.97	1.96	1.93	5.30	0.02	0.10	0.20	0.43	1.38	3.20	4.57	5.82	9.06
3	2.00	1.95	1.79	4.24	0.02	0.09	0.22	0.44	1.37	3.27	4.61	5.94	8.94
4	1.99	1.91	1.81	4.48	0.02	0.12	0.22	0.48	1.39	3.19	4.56	5.79	8.98
5	1.96	1.93	1.82	4.64	0.02	0.09	0.20	0.43	1.39	3.25	4.40	5.83	8.84
6a	1.98	1.98	1.99	5.68	0.02	0.10	0.24	0.45	1.36	3.17	4.57	6.34	10.28
6b	2.00	1.99	2.05	6.76	0.02	0.09	0.20	0.43	1.41	3.23	4.58	6.04	9.04
6c	1.99	2.03	2.32	9.65	0.02	0.10	0.22	0.43	1.39	3.16	4.51	5.90	9.29
6d	1.96	1.94	1.87	4.88	0.02	0.10	0.20	0.41	1.39	3.14	4.44	5.90	9.80
7	1.97	1.95	1.89	5.08	0.02	0.09	0.20	0.41	1.39	3.24	4.52	5.70	8.99
8	2.01	1.94	1.75	4.03	0.02	0.12	0.22	0.45	1.42	3.24	4.70	5.91	8.84
9	1.90	4.08	15.31	384.01	0.02	0.08	0.16	0.34	1.05	2.47	3.88	5.97	13.99
10	1.92	1.91	2.07	6.57	0.02	0.10	0.22	0.43	1.38	3.04	4.29	5.79	8.92
N*	2.0	2.0			0.02	0.10	0.21		1.39		4.61	5.99	9.21

Table 2.3.
Quadratic form N = 100

Dist	M	σ	Asym	Curt	1 %	5 %	10 %	20 %	50 %	80 %	90 %	95 %	99 %
1	2.02	2.01	1.86	4.53	0.02	0.09	0.20	0.47	1.38	3.25	4.66	6.10	9.23
2	2.03	2.11	2.25	7.59	0.02	0.10	0.22	0.45	1.38	3.21	4.64	6.14	10.28
3	2.03	2.10	2.19	6.92	0.02	0.10	0.21	0.42	1.39	3.20	4.64	6.17	10.16
4	2.04	2.11	2.20	7.03	0.02	0.09	0.22	0.44	1.42	3.18	4.60	6.17	10.34
5	1.94	1.88	1.97	5.75	0.02	0.10	0.22	0.47	1.39	3.11	4.40	5.70	8.38
6a	2.01	2.05	2.18	7.69	0.02	0.09	0.20	0.44	1.39	3.24	4.58	6.01	9.34
6b	2.06	2.05	1.88	4.55	0.02	0.10	0.24	0.47	1.42	3.27	4.74	6.40	9.15
6c	1.95	1.94	1.89	4.67	0.02	0.10	0.21	0.46	1.34	3.12	4.57	5.93	8.84
6d	2.02	1.96	1.90	4.58	0.02	0.10	0.20	0.45	1.35	3.20	4.60	6.01	8.99
7	2.05	2.11	2.17	6.70	0.02	0.10	0.22	0.44	1.41	3.20	4.67	6.19	10.31
8	1.97	1.95	2.04	6.40	0.02	0.10	0.22	0.44	1.39	3.10	4.46	5.87	8.98
9	1.94	3.18	9.82	18.1	0.02	0.09	0.19	0.38	1.15	2.70	4.00	5.95	13.5
10	1.98	1.98	2.30	10.40	0.02	0.10	0.22	0.44	1.38	3.16	4.53	5.87	9.16
N*	2.0	2.0			0.02	0.10	0.21		1.39		4.61	5.99	9.21

Table 3.1.
Determinant of S N = 10

Dist	M	σ	Asym	Curt	1 %	5 %	10 %	20 %	50 %	80 %	90 %	95 %	99 %
1	0.91	0.67	2.10	0.15	0.11	0.20	0.27	0.39	0.73	1.32	1.74	2.21	3.14
2	0.90	0.30	0.10	− 0.21	0.22	0.39	0.50	0.63	0.89	1.16	1.29	1.41	1.63
3	0.90	0.34	0.44	0.28	0.20	0.37	0.47	0.60	0.87	1.16	1.34	1.51	1.78
4	0.90	0.41	0.73	0.54	0.19	0.33	0.42	0.54	0.83	1.22	1.45	1.66	2.03
5	0.90	0.45	0.94	1.32	0.18	0.29	0.39	0.52	0.83	1.24	1.50	1.72	2.20
6a	0.90	1.33	6.43	89.12	0.04	0.07	0.11	0.19	0.50	1.24	1.99	2.97	6.16
6b	0.88	1.06	3.89	25.06	0.05	0.11	0.14	0.23	0.55	1.27	1.96	2.68	5.28
6c	0.88	0.95	3.40	20.70	0.06	0.11	0.16	0.24	0.58	1.26	1.91	2.69	4.62
6d	0.90	0.95	3.94	28.7	0.06	0.13	0.19	0.27	0.61	1.34	1.86	2.51	4.54
7	0.90	0.48	1.07	1.62	0.16	0.28	0.38	0.49	0.81	1.26	1.53	1.83	2.41
8	0.89	0.47	1.06	1.87	0.15	0.26	0.36	0.49	0.81	1.24	1.53	1.78	2.30
9	0.62	3.12	18.43	501.2	0.00	0.00	0.00	0.01	0.08	0.39	1.03	2.39	8.63
10	0.85	1.28	4.89	38.9	0.02	0.05	0.09	0.16	0.43	1.18	1.97	3.01	6.44
N*	0.89	0.65											

Table 4.1.
Hotelling T^2 N = 10

Dist	M	σ	Asym	Curt	1 %	5 %	10 %	20 %	50 %	80 %	90 %	95 %	99 %
1	3	4.05	3.8	23.9	0.02	0.14	0.24	0.52	1.70	4.35	7.23	10.82	19.91
2	3.23	5.80	6.68	68.7	0.02	0.08	0.23	0.46	1.59	4.50	7.51	10.98	23.36
3	3.30	5.84	6.93	77.90	0.02	0.13	0.24	0.51	1.66	4.55	7.48	11.34	27.80
4	3.09	5.05	7.2	105.5	0.02	0.14	0.24	0.52	1.54	4.35	7.11	11.30	22.81
5	3.07	4.61	5.10	45.3	0.02	0.08	0.19	0.47	1.65	4.51	7.29	10.71	21.49
6a	5.21	12.23	9.19	140.83	0.02	0.13	0.24	0.56	1.91	6.07	11.61	20.46	53.89
6b	4.46	8.98	5.14	35.13	0.02	0.13	0.24	0.52	1.86	5.79	10.38	16.75	41.97
6c	4.23	7.65	6.08	57.3	0.03	0.13	0.24	0.56	1.92	5.48	9.73	16.09	35.41
6d	0.75	7.72	11.09	204.11	0.03	0.13	0.24	0.51	1.76	4.93	8.54	12.70	30.86
7	2.99	4.53	5.79	67.5	0.02	0.13	0.25	0.46	1.58	4.23	6.76	10.86	20.72
8	3.11	4.32	3.71	21.0	0.02	0.13	0.24	0.51	1.65	4.66	7.42	10.96	21.0
9	11.21	29.27	8.85	116.30	0.03	0.19	0.34	0.73	2.83	12.75	27.15	47.95	53.88
10	5.8	14.1	9.6	143.4	0.03	0.13	0.24	0.56	2.02	6.8	12.8	21.9	53.9
N*	3	4.24									7.00	10.04	19.46

Table 4.2.
Hotelling T^2 N = 50

Dist	M	σ	Asym	Curt	1 %	5 %	10 %	20 %	50 %	80 %	90 %	95 %	99 %
1	2.13	2.25	0.80	1.04	0.02	0.10	0.19	0.44	1.37	3.39	5.09	6.70	10.47
2	2.09	2.21	2.34	8.83	0.02	0.10	0.19	0.42	1.40	3.36	4.83	6.30	10.50
3	2.12	2.11	2.17	7.24	0.02	0.10	0.21	0.45	1.43	3.42	4.95	6.54	10.60
4	2.09	2.14	2.24	8.00	0.02	0.10	0.21	0.47	1.42	3.30	4.83	6.32	9.90
5	2.09	2.15	2.20	8.00	0.02	0.10	0.19	0.42	1.42	3.45	4.77	6.32	9.81
6a	2.38	2.87	3.10	14.23	0.02	0.10	0.24	0.48	1.49	3.62	5.32	7.63	14.40
6b	2.33	2.79	3.99	8.75	0.02	0.10	0.21	0.45	1.51	3.61	5.31	7.34	13.97
6c	2.27	2.73	2.97	1.75	0.02	0.10	0.22	0.45	1.46	3.48	5.24	7.17	13.08
6d	2.18	2.39	2.73	13.23	0.02	0.10	0.22	0.45	1.43	3.45	5.15	6.65	11.14
7	1.09	2.17	2.25	8.36	0.02	0.10	0.22	0.42	1.43	3.42	4.77	6.30	10.14
8	2.13	2.18	2.12	6.90	0.02	0.10	0.22	0.45	1.45	3.39	4.97	6.39	10.17
9	3.76	6.22	5.20	43.2	0.02	0.13	0.26	0.56	1.80	5.12	8.75	13.83	28.81
10	2.42	3.25	4.54	37.07	0.02	0.10	0.22	0.45	1.43	3.62	5.49	7.77	15.85
N*	2.13	2.23									4.94	6.53	10.41

Table 4.3.
Hotelling T^2 N = 100

Dist	M	σ	Asym	Curt	1 %	5 %	10 %	20 %	50 %	80 %	90 %	95 %	99 %
1	2.07	2.10	1.93	4.80	0.01	0.09	0.20	0.48	1.41	3.29	4.79	6.35	9.52
2	2.10	2.25	2.43	8.92	0.01	0.09	0.20	0.45	1.41	3.32	4.79	6.38	11.02
3	2.09	2.23	2.38	8.36	0.01	0.09	0.20	0.42	1.41	3.27	4.77	6.51	10.51
4	2.10	2.25	2.39	8.49	0.01	0.09	0.20	0.45	1.44	3.27	4.80	6.38	11.03
5	2.00	1.98	2.10	6.72	0.01	0.09	0.23	0.48	1.41	3.13	4.55	5.86	9.20
6a	2.22	2.57	3.42	21.32	0.02	0.08	0.20	0.44	1.43	3.43	5.16	6.79	11.82
6b	2.22	2.38	2.46	9.58	0.01	0.09	0.23	0.48	1.46	3.45	5.13	6.81	10.83
6c	2.12	2.31	2.61	10.82	0.01	0.09	0.23	0.48	1.38	3.29	4.79	6.68	10.92
6d	2.10	2.20	2.35	13.25	0.02	0.09	0.22	0.48	1.41	3.30	4.80	6.41	10.03
7	2.11	2.25	2.36	8.14	0.01	0.09	0.20	0.45	1.43	3.29	4.85	6.41	11.14
8	2.02	2.05	2.24	8.42	0.02	0.09	0.23	0.45	1.41	3.16	4.66	6.09	9.17
9	3.16	4.26	3.51	18.9	0.03	0.14	0.28	0.56	1.76	4.55	7.78	11.41	20.43
10	2.23	2.58	3.23	18.89	0.02	0.09	0.20	0.45	1.44	3.43	5.01	7.00	12.01
N*	2.06	2.11									4.77	6.26	9.82

Table 5.1.
Correlation Coefficient N = 10

Dist	M	σ	Asym	Curt	1 %	5 %	10 %	20 %	50 %	80 %	90 %	95 %	99 %
1	− 0.01	0.34	0.00	− 0.56	− 0.72	− 0.55	− 0.44	− 0.30	0.00	0.31	0.45	0.56	0.72
2	0.00	0.27	0.02	− 0.20	− 0.63	− 0.44	− 0.35	− 0.23	0.00	0.22	0.36	0.46	0.62
3	0.00	0.32	0.03	− 0.48	− 0.70	− 0.53	− 0.43	− 0.28	0.00	0.28	0.43	0.55	0.70
4	0.00	0.29	0.05	− 0.39	− 0.66	− 0.48	− 0.38	− 0.24	− 0.01	0.24	0.39	0.49	0.63
5	0.00	0.33	0.02	− 0.53	− 0.69	− 0.54	− 0.44	0.30	0.00	0.32	0.44	0.56	0.72
6a	0.02	0.46	− 0.01	− 0.99	− 0.87	− 0.73	− 0.61	− 0.44	0.02	0.47	0.60	0.75	0.89
6b	0.01	0.43	− 0.07	− 0.88	− 0.85	− 0.70	− 0.58	− 0.40	0.00	0.42	0.61	0.71	0.85
6c	0.00	0.41	− 0.05	− 0.80	− 0.83	− 0.65	− 0.52	− 0.36	0.02	0.41	0.56	0.67	0.83
6d	0.00	0.40	− 0.01	− 0.75	− 0.80	− 0.64	− 0.53	− 0.36	− 0.01	0.38	0.55	0.66	0.80
7	0.02	0.30	− 0.05	− 0.44	− 0.66	− 0.49	− 0.40	− 0.27	− 0.02	0.26	0.40	0.52	0.65
8	0.01	0.33	− 0.04	− 0.51	− 0.71	− 0.55	− 0.44	− 0.30	− 0.01	0.29	0.55	0.66	0.80
9	0.02	0.30	0.05	− 0.44	− 0.98	− 0.49	− 0.81	− 0.62	0.03	0.62	0.80	0.90	0.97
10	0.00	0.49	0.02	− 1.06	− 0,98	− 0.77	− 0.67	− 0.48	− 0.01	0.49	0.67	0.77	0.99
N*	0.00	0.33										0.52	0.69

Table 5.2.
Correlation Coefficient N = 50

Dist	M	σ	Asym	Curt	1 %	5 %	10 %	20 %	50 %	80 %	90 %	95 %	99 %
1	0.00	0.14	0.04	− 0.15	− 0.32	− 0.24	− 0.19	− 0.12	− 0.00	0.12	0.19	0.24	0.33
2	0.00	0.11	− 0.02	− 0.02	− 0.26	− 0.18	− 0.14	− 0.09	0.00	0.09	0.14	0.18	0.25
3	0.00	0.14	0.00	− 0.19	− 0.32	− 0.23	− 0.19	− 0.12	0.00	0.12	0.18	0.23	0.31
4	0.00	0.11	− 0.02	+ 0.06	− 0.29	− 0.19	− 0.14	− 0.10	− 0.01	0.10	0.14	0.18	0.27
5	0.00	0.14	− 0.04	− 0.05	− 0.33	− 0.23	− 0.18	− 0.12	0.00	0.12	0.18	0.23	0.32
6a	0.01	0.24	− 0.01	− 0.37	− 0.52	− 0.37	− 0.30	− 0.19	0.01	0.22	0.33	0.40	0.52
6b	0.01	0.22	− 0.03	− 0.19	− 0.49	− 0.35	− 0.28	− 0.18	0.01	0.19	0.28	0.35	0.50
6c	0.01	0.20	0.06	− 0.28	− 0.44	− 0.32	− 0.25	− 0.17	0.01	0.18	0.28	0.35	0.47
6d	0.00	0.19	0.09	− 0.17	− 0.43	− 0.32	− 0.25	− 0.17	0.00	0.16	0.25	0.32	0.45
7	0.00	0.12	− 0.03	− 0.02	− 0.29	− 0.20	− 0.16	− 0.11	0.00	0.10	0.16	0.20	0.28
8	0.00	0.14	− 0.01	− 0.23	− 0.31	− 0.23	− 0.18	− 0.12	0.00	0.12	0.18	0.23	0.31
9	0.01	0.44	0.01	− 0.77	− 0.90	− 0.73	− 0.60	− 0.43	0.00	0.40	0.59	0.71	0.89
10	0.00	0.25	0.02	− 0.34	− 0.56	− 0.42	− 0.34	− 0.22	− 0.01	0.22	0.33	0.42	0.55
N*	0.00	0.14										0.23	0.32

Table 5.3.
Correlation Coefficient N = 100

Dist	M	σ	Asym	Curt	1 %	5 %	10 %	20 %	50 %	80 %	90 %	95 %	99 %
1	0.00	0.10	0.00	− 0.02	− 0.24	− 0.16	− 0.13	− 0.08	0.00	0.08	0.12	0.16	0.23
2	0.00	0.08	0.76	0.02	− 0.17	− 0.13	− 0.10	− 0.07	0.00	0.06	0.10	0.13	0.18
3	0.00	0.10	0.08	0.00	− 0.22	− 0.16	− 0.13	− 0.08	0.00	0.08	0.12	0.16	0.22
4	0.00	0.08	0.09	0.00	− 0.19	− 0.13	− 0.11	− 0.07	0.00	0.07	0.11	0.14	0.19
5	0.00	0.10	0.01	− 0.04	− 0.24	− 0.16	− 0.12	− 0.08	0.00	0.08	0.13	0.16	0.24
6a	0.00	0.18	− 0.01	− 0.29	− 0.40	− 0.29	− 0.22	− 0.15	0.00	0.15	0.23	0.29	0.41
6b	0.00	0.15	− 0.02	− 0.24	− 0.35	− 0.26	− 0.19	− 0.13	0.00	0.14	0.20	0.26	0.35
6c	0.00	0.15	− 0.01	− 0.37	− 0.34	− 0.25	− 0.19	− 0.13	0.00	0.13	0.19	0.25	0.33
6d	0.00	0.13	− 0.03	− 0.06	− 0.32	− 0.23	− 0.18	− 0.11	0.00	0.11	0.18	0.23	0.32
7	0.00	0.09	0.08	− 0.01	− 0.20	− 0.14	− 0.11	− 0.08	0.00	0.07	0.11	0.15	0.20
8	0.00	0.10	0.05	− 0.03	− 0.22	− 0.16	− 0.13	− 0.09	− 0.01	0.08	0.12	0.16	0.22
9	0.00	0.38	0,04	− 0.49	− 0.84	− 0.63	− 0.51	− 0.34	0.00	0.33	0.52	0.65	0.85
10	0.00	0.19	0.05	− 0.29	− 0.42	− 0.30	− 0.23	− 0.16	0.00	0.18	0.26	0.32	0.44
N*	0.00	0.10										0.16	0.23

Table 6.1.
Covariance N = 10

Dist	M	σ	Asym	Curt	1 %	5 %	10 %	20 %	50 %	80 %	90 %	95 %	99 %
1	0.00	0.34	0.05	0.72	− 0.83	− 0.56	− 0.42	− 0.27	0	0.27	0.42	0.57	0.85
2	0.00	0.26	0.05	0.16	− 0.58	− 0.42	− 0.33	− 0.22	0	0.21	0.33	0.44	0.59
3	0.00	0.32	0.02	− 0.21	− 0.72	− 0.54	− 0.42	0.29	0	0.27	0.41	0.53	0.71
4	− 0.01	0.28	0.09	− 0.12	− 0.65	− 0.45	− 0.37	− 0.25	− 0.01	0.23	0.37	0.46	0.65
5	0	0.33	0.02	0.53	− 0.80	− 0.54	− 0.42	− 0.29	0	0.28	0.43	0.54	0.79
6a	− 0.01	0.59	− 1.14	8.73	− 1,74	− 0.97	− 0.65	− 0.36	0.01	0.36	0.58	0.84	1.43
6b	− 0.01	0.53	− 0.41	5.97	− 1.60	− 0.85	− 0.59	− 0.32	0.01	0.33	0.55	0.78	1.50
6c	0.01	0.49	− 0.52	6.19	− 1.39	− 0.76	− 0.49	− 0.30	0.02	0.33	0.54	0.76	1.36
6d	0.01	0.45	0.02	3.39	− 1.19	− 0.72	− 0.52	− 0.31	0.01	0.33	0.52	0.71	1.18
7	− 0.01	0.29	0.07	− 0.01	− 0.68	− 0.48	− 0.38	− 0.25	− 0.02	0.24	0.37	0.49	0.68
8	0.00	0.32	0.03	− 0.03	− 0.78	− 0.53	− 0.41	− 0.28	− 0.01	0.27	0.42	0.53	0.73
9	+ 0.02	2.79	+ 6.27	262.24	− 2.14	− 0.92	− 0.48	− 0.20	0.01	0.26	0.65	1.43	3.33
10	0.00	0.64	0.28	8.05	− 1.77	− 1.00	− 0.65	− 0.36	− 0.01	0.37	0.65	1.02	1.82
N*	0.00	0.33											

Table 6.2.
Covariance N = 50

Dist	M	σ	Asym	Curt	1 %	5 %	10 %	20 %	50 %	80 %	90 %	95 %	99 %
1	0	0.14	0.03	0.12	− 0.31	− 0.24	− 0.19	− 0.12	0	0.12	0.18	0.23	0.34
2	0.00	0.11	− 0.02	− 0.02	− 0.26	− 0.18	− 0.14	− 0.09	0	0.09	0.14	0.18	0.24
3	0.00	0.14	0.00	− 0.14	− 0.31	− 0.23	− 0.19	− 0.12	0	0.12	0.17	0.22	0.31
4	0.00	0.11	− 0.31	0.07	− 0.28	− 0.19	− 0.14	− 0.10	− 0.01	0.09	0.14	0.18	0.27
5	0.00	0.14	− 0.04	0.03	− 0.34	− 0.23	− 0.18	− 0.12	0.00	0.12	0.18	0.23	0.32
6a	0.01	0.25	− 0.22	1.33	− 0.67	− 0.41	− 0.31	− 0.19	0.01	0.20	0.31	0.40	0.63
6b	0.00	0.23	− 0.05	1.30	− 0.60	− 0.39	− 0.29	− 0.18	0.01	0.18	0.27	0.36	0.57
6c	0.01	0.21	0.12	0.81	0.50	− 0.33	− 0.25	− 0.17	0.01	0.17	0.27	0.36	0.55
6d	0.00	0.20	0.02	0.74	− 0.49	− 0.33	− 0.25	− 0.17	0.00	0.15	0.24	0.32	0.48
7	0.00	0.12	0.02	0.03	− 0.20	− 0.16	− 0.10	− 0.10	0.00	0.10	0.15	0.20	0.27
8	0.00	0.14	− 0.00	0.14	− 0.32	− 0.23	− 0.18	− 0.12	0.00	0.12	0.18	0.22	0.32
9	− 0.05	1.53	− 17.10	559.73	− 1.19	− 0.65	− 0.43	− 0.23	0.02	0.27	0.67	1.43	1.43
10	0.00	0.28	− 0.16	2.98	− 0.75	− 0.45	− 0.34	− 0.21	− 0.01	0.20	0.32	0.45	0.69
N*	0.00	0.14											

Table 6.3.
Covariance N = 100

Dist	M	σ	Asym	Curt	1 %	5 %	10 %	20 %	50 %	80 %	90 %	95 %	99 %
1	0.00	0.10	0.01	0.13	− 0.24	− 0.16	− 0.13	− 0.08	0.00	0.08	0.13	0.16	0.23
2	0.00	0.08	0.06	0.02	− 0.18	− 0.13	− 0.10	− 0.07	0.00	0.06	0.10	0.13	0.18
3	0.00	0.10	0.06	0.02	− 0.22	− 0.16	− 0.12	− 0.08	0.00	0.08	0.12	0.16	0.22
4	0.00	0.08	0.07	0.00	− 0.19	− 0.13	− 0.10	− 0.07	0.00	0.07	0.11	0.14	0.19
5	0.00	0.10	0.01	0.02	− 0.23	− 0.16	− 0.12	− 0.08	0.00	0.08	0.13	0.16	0.24
6[a]	0.00	+ 0.19	− 0.21	0.94	− 0.50	− 0.32	− 0.23	− 0.15	0.00	0.14	0.22	0.29	0.44
6[b]	0.00	0.16	− 0.06	0.37	− 0.40	− 0.27	− 0.20	− 0.13	0.00	0.13	0.20	0.26	0.38
6[c]	0.00	0.15	− 0.06	0.18	− 0.37	− 0.26	− 0.19	− 0.13	0.00	0.13	0.19	0.25	0.35
6[d]	0.00	0.14	− 0.07	0.15	− 0.35	− 0.23	− 0.18	− 0.11	0.00	0.11	0.19	0.23	0.34
7	0.00	0.09	0.05	0.00	− 0.20	− 0.14	− 0.11	− 0.07	0.00	0.08	0.11	0.14	0.20
8	0.00	0.10	0.06	0.03	− 0.22	− 0.16	− 0.12	− 0.09	− 0.01	0.08	0.12	0.16	0.22
9	0.00	1.18	− 11.36	358	− 0.88	− 0.52	− 0.34	− 0.19	0.02	0.27	0.65	1.00	1.00
10	0.00	0.20	0.10	0.45	− 0.46	− 0.31	− 0.24	− 0.15	0.00	0.17	0.26	0.33	0.50
N*	0.00	0.10											

Computational Centre, Humboldt-University Berlin

Tests for Independence in the Family of Continuous Bivariate Distributions with Finite Contingency

EGMAR ROEDEL

Abstract

For testing the hypothesis of independence in the family of continuous bivariate distributions with finite contingency it is found under some regularity conditions, that always exists a locally most powerful rank test and that this test is fully efficient in the sense of Pitman. A simple formula is given for the calculation of asymptotic relative efficiencies.

1. Introduction

We consider the family $\mathfrak{P} = \{H_\lambda(x, y): 0 \leq \lambda \leq 1\}$ of bivariate distribution functions with the following properties:

1 The extension P_λ of each $H_\lambda(x, y)$ is absolutely continuous with respect to the Lebesgue measure and possessses the density $h_\lambda(x, y)$.
2. The mean-square-contingency of each $H_\lambda(x, y)$ is finite, that means

$$\Phi^2 = \int_{R^2} \frac{h_\lambda^2(x,y)}{f(x)g(y)}\,dx\,dy - 1 < \infty,$$

where $f(x)$ and $g(y)$ are the marginal densities of h
3. $h_0(x, y) = f(x) \cdot g(y)$.

It is known, that under these conditions the following unique series expansion holds (H. O. Lancaster (1958))

$$h_{\lambda_1}(x,y) := h_{\lambda_1,\lambda_2\ldots}(x,y) = \left\{1 + \sum_{i=1}^{\infty} \lambda_i \varphi_i(x)\psi_i(y)\right\} f(x)g(y), \qquad (1)$$

$$1 \geq \lambda_1 \geq \lambda_2 \geq \ldots \geq 0, \{\varphi_i(x)\}_0^\infty, \{\psi_i(y)\}_0^\infty$$

are complete orthonormal systems,
with respect to the weight functions $f(x)$ resp. $g(y)$.
For testing $H_0: \lambda_1 = 0$ against $H_1: \lambda_1 > 0$ it is assumed, that

(A) $\lambda_i = t_i(\lambda_1), \quad i = 2, 3, \ldots;$

$t_i(\lambda_1)$ is continuously differentiable $(i = 2, 3, \ldots)$,

(B) $t_i(\lambda_1)\big|_{\lambda_1=0} = t_i'(\lambda_1)\big|_{\lambda_1=0} = 0 \quad (i = 2, 3, \ldots)$,

(C) $\sum_{i=2}^{\infty} t_i'(\lambda_1)\varphi_i(x)\psi_i(y)$

converges uniformly over
the interval $0 \leq \lambda_1 \leq \lambda_0 < 1$ for some λ_0 and
for each fixed pair of values $(x, y) \in R^2$,

(D) $$0 < \lim_{\lambda_1 \to 0} \int_{R^2} \left|\frac{\partial \ln h_{\lambda_1}(x,y)}{\partial \lambda_1}\right| h_{\lambda_1}(x,y)\,dx\,dy$$

$$= \int_{R^2} \lim_{\lambda_1 \to 0} \left|\frac{\partial \ln h_{\lambda_1}(x,y)}{\partial \lambda_1}\right| h_{\lambda_1}(x,y)\,dx\,dy$$

$$< \infty$$

Let (ξ, η) be a bivariate random variable, $(\xi_1, \eta_1), \ldots, (\xi_n, \eta_n)$ a sample with

$$P_{\lambda_1}(\xi_i < x, \eta_i < y) = H_{\lambda_1}(x, y) \in \mathfrak{P} \quad (i = 1, 2, \ldots, n).$$

and $(R_{\xi_1}, R_{\eta_1}), \ldots, (R_{\xi_n}, R_{\eta_n})$ the corresponding rank statistic.
Shirahata (1975) has shown that under the assumptions 1., 3. and (D) the rank test

$$\pi_{n0}(R_{\xi_1}, R_{\eta_1}, \ldots, R_{\xi_n}, R_{\eta_n}) = \begin{cases} 1, & \text{if } |Z_n| \geq z_\alpha \\ 0, & \text{if } |Z_n| < z_\alpha, \end{cases}$$

where

$$Z_n = \sum_{i=1}^{n} a_n(R_{\xi_i}, R_{\eta_i}),$$

$$a_n(i,j) = n^2 \binom{n-1}{i-1}\binom{n-1}{j-1} \int_{-\infty}^{+\infty}\!\!\int \left.\frac{\partial \ln h_{\lambda_1}(x,y)}{\partial \lambda_1}\right|_{\lambda_1=0} \cdot (F(x))^{i-1} \times$$

$$\times (1 - F(x))^{n-i} (G(y))^{j-1} (1 - G(y))^{n-j} f(x) g(y)\,dx\,dy,$$

$$0 < P_0(|Z_n| \geq z_\alpha) = \alpha < 1, \; F(x) = P(\xi < x), G(y) = P(\eta < y),$$

is a locally most powerful rank test (lmprt) for testing $H_0: \lambda_1 = 0$ against $H_1: \lambda_1 > 0$ to the level of significance α. Applying this result to the family $\mathfrak{P}$ of two-dimensional distributions defined by the conditions 1.—3. and (A)—(D), we get the lmprt π_{n0} with the test statistic

$$Z_n = \sum_{i=1}^{n} u_n(R_{\xi_i}) v_n(R_{\eta_i}), \qquad (2)$$

where

$$u_n(i) = E(\varphi_1(\xi_1)/R_{\xi_1} = i)$$

and

$$v_n(i) = E(\psi_1(\eta_1)/R_{\eta_1} = i); \quad i = 1, 2, \ldots, n.$$

2. Asymptotic Normality of the LMPRT for Independence

In this chapter we use the known results of Ruymgaart (1972) at al. about the asymptotic normality of rank tests for independence.
We assume that the functions $\varphi_1(x)$ and $\psi_1(y)$ in the series expansion (1) are strictly increasing. Analogous results we would get also in the cases, that both functions are strictly decreasing or that the one is decreasing and the other is increasing. It is easy to see, that the following indentities are true.

$$u_n(i) = E(J(U_1)/R_{U_1} = i),$$

$$v_n(i) = E(L(U_1)/R_{U_1} = i),$$

where $U_1, \ldots, U_n$ is a sample from the uniform distribution over [0,1] and the functions J (u), L (u), $0 \leq u \leq 1$. are defined by the relations

$$J(u) = \varphi_1\left(F^{-1}(u)\right)$$

and $$L(u) = \psi_1\left(G^{-1}(u)\right).$$

Theorem: If $\bar{Z}_n = n^{-1} Z_n$, the functions J (u) and L (u) are twice differentiable on the interval (0,1),

$$\left|J^{(k)}(u)\right| \leq D\left(u(1-u)\right)^{-a-k}, \left|L^{(k)}(u)\right| \leq D\left(u(1-u)\right)^{-b-k}$$
$$k = 0,1,2;$$
$$D > 0,\ a = (1/2 - \delta)\, p_0,\ b = (1/2 - \delta)\, q_0,$$
$$0 < \delta < 1/2,\ p_0 > 1,\ q_0 > 1,\ p_0^{-1} + q_0^{-1} = 1$$

and $$\lim_{n\to\infty} n\, E\left(\bar{Z}_n - \lambda_1\right)^2 = \sigma^2 > 0,$$

then holds

$$\lim_{n\to\infty} P_{\lambda_1}\left(\sqrt{n}\ \frac{\bar{Z}_n - \lambda_1}{\sigma} < z\right) = (2\pi)^{-1/2} \int_{-\infty}^{z} e^{-\frac{x^2}{2}}\, dx$$

and this convergence is uniformly in the family of continuous distributions with finite contingency.

The proof follows from theorem 2.1. an the remark 2.2. of Ruymgaart at al. (1972).

3. Asymptotic Relative Efficiencies of Rank Tests for Independence

The common distribution function H_{λ_1} (x, y) of the random variables ξ and η may satisfy the conditions 1.–3. and (A)–(D). We assume further, that all conditions for the existence of the Maximum-Likelihood-Estimation (MLE) $\hat{\lambda}_{1,n}$ of λ are fulfilled and that $\hat{\lambda}_{1,n}$ is asymptotically unbiased, asymptotically efficient and asymptotically normal. That means

$$\lim_{n\to\infty} E_{\lambda_1} \hat{\lambda}_{1,n} = \lambda_1,$$

$$\lim_{n\to\infty} n\, E_{\lambda_1}\left(\hat{\lambda}_{1,n} - \lambda_1\right)^2 = \frac{1}{E_{\lambda_1}\left(\dfrac{\partial \ln h_{\lambda_1}(\xi,\eta)}{\partial \lambda_1}\right)^2} = \sigma^2(\lambda_1) \tag{3}$$

$$\lim_{n\to\infty} P_{\lambda_1}\left(\sqrt{n}\ \frac{\hat{\lambda}_{1,n} - \lambda_1}{\sigma(\lambda_1)} < z\right) = (2\pi)^{-1/2} \int_{-\infty}^{z} e^{-\frac{x^2}{2}}\, dx.$$

Using the orthonormality properties of the functions φ_i (x) and ψ_i (y) (i = 0, 1, 2, ... $\varphi_0(x) \equiv \psi_0(y) \equiv 1$) in the series expansion (1) we get

$$\sigma^2(0) = \frac{1}{E_0\left(\varphi_1^2(\xi)\, \psi_1^2(\eta)\right)} = 1.$$

To compare the power of tests for independence the asymptotic relative efficiency (ARE) in the sense of Pitman will be calculated with the help of a known theorem of Noether (1955). We give now the general formula of the ARE of two rank tests $\pi_n^{(k)}$ (k = 1, 2) for independence based on the tests statistics

$$T_n^{(k)} = \sum_{i=1}^{n} a_n^{(k)}(R_{\xi_i})\, b_n^{(k)}(R_{\eta_i}),$$

where

$$\sum_{i=1}^{n} a_n^{(k)}(i) = \sum_{i=1}^{n} b_n^{(k)}(i) = 0 \tag{4}$$

and

$$\sum_{i=1}^{n} a_n^{(k)^2}(i) = \sum_{i=1}^{n} b_n^{(k)^2}(i) = 1,\ k = 1.2.$$

It was shown by Roedel (1982), that the relation

$$ARE\left(\pi_n^{(1)}, \pi_n^{(2)}\right) = \left(\frac{C_1}{C_2}\right)^2$$

holds, where

$$C_k = \lim_{n\to\infty}\left(\sum_{i=1}^{n} a_n^{(k)}(i)\, u_n^*(i) \sum_{i=1}^{n} b_n^{(k)}(i)\, v_n^*(i)\right),$$

$$u_n^*(i) = \frac{u_n(i)}{\left(\sum_{i=1}^{n} u_n^2(i)\right)^{1/2}},\quad v_n^*(i) = \frac{v_n(i)}{\left(\sum_{i=1}^{n} v_n^2(i)\right)^{1/2}},\quad k = 1,2.$$

The ARE of a rank test π_n with respect to the test $\pi_n^{(0)}$ based on the MLE $\hat{\lambda}_{1,n}$ is given by

$$ARE\left(\pi_n, \pi_n^{(0)}\right) = \lim_{n\to\infty}\left(\sum_{i=1}^{n} a_n(i)\, u_n^*(i) \sum_{i=1}^{n} b_n(i)\, v_n^*(i)\right)^2,$$

where the scores a_n (i) and b_n (i) are assumed to satisfy the standardizations (4). Further it is easy to prove the equations

$$ARE\left(\pi_{n0}, \pi_n^{(0)}\right) = 1,$$

where π_{n0} is the earlier constructed lmprt, and

$$ARE\left(\pi_{n0}, \pi_n(r_{\varphi,\psi})\right) = ARE\left(\pi_n^{(0)}, \pi_n(r_{\varphi,\psi})\right) = 1,$$

where π_n $(r_{\varphi,\psi})$ is the test determined by

$$\pi_n\left(r_{\varphi,\psi}\right) = \begin{cases} 1, & \text{if } \sqrt{n}\ \left|r_{\varphi,\psi}\right| \geq u_{1-\frac{\alpha}{2}} \\ 0 & \text{else} \end{cases}$$

and $r_{\varphi,\psi}$ is the sample correlation of the transformed sample $\{\varphi_1(\xi_i), \psi_1(\eta_i)\}_{i=1}^{n}$. $u_1 - \alpha/2$ denotes the $(1 - \alpha/2$ -quantile of the standardized normal distribution (Roedel (1982)).

4. Examples

Let π_n (r_s), $\pi_n^{(0)}$, π_n (r) and π_{n0} be the tests for independence based on the statistics r_s (Spearman's rank correlation), $\hat{\lambda}_{1,n}$ (MLE of λ_1), r (sample correlation coefficient) resp. Z_n (lmprt).

4.1. The bivariate normal distribution

In this case we have $(\xi, \eta) \sim N(0, \Sigma)$, $\hat{\lambda}_{1,n} = |r| = |r_{\varphi,\psi}|$, $Z_n = \sum_{i=1}^{n} EV_n^{(R_{\xi_i})}\, EV_n^{(R_{\eta_i})}$ where $V_n^{(1)} < V_n^{(2)} < \ldots < V_n^{(n)}$ is a ordered sample of a standardized normal population and

$$\sum = \begin{pmatrix} 1 & \varrho \\ \varrho & 1 \end{pmatrix}.$$

Z_n is the known Fisher-Yates-statistic. It holds

$$\mathrm{ARE}\left(\pi_n(r_s), \pi_n^{(0)}\right) = \mathrm{ARE}\left(\pi_n(r_s), \pi_{n0}\right) = 9\pi^{-2},$$

$$\mathrm{ARE}\left(\pi_{n0}, \pi_n^{(0)}\right) = 1.$$

4.2. The Gumbel-distributions

The following familiy of distributions was introduced by Gumbel (1960)

$$h_{\lambda_1}(x,y) = \left\{1 + \lambda_1 \sqrt{12}\,(F(x) - 1/2)\,\sqrt{12}\,(G(y) - 1/2)\right\} f(x)\,g(y),$$
$$0 \le \lambda_1 \le 1/3.$$

The lmprt π_{n0} bases on the statistic $Z_n = n^{-1} \sum_{i=1}^{n} R_{\xi i} R_{\eta i}$ which is equivalent to Spearman's rank correlation r_s. We have the relations

$$\mathrm{ARE}\left(\pi_{n0}, \pi_n^{(0)}\right) = 1,$$

$$\mathrm{ARE}\left(\pi_n(r), \pi_{n0}\right) = 9\pi^{-2}.$$

5. References

LANCASTER, H. O.
The structure of bivariate distributions.
AMS 37 (1958), 719–736.

SHIRAHATA, S.
Locally most powerful rank tests for independence with censored data.
AS 3 (1975), 241–245.

RUYMGAART, F. H., SHORACK, G. R., van ZWET, W. R.
Asymptotic normality of nonparametric tests for independence.
AMS 43 (1972), 1122–1135.

NOETHER, G.
On a theorem of Pitman.
AMS 26 (1955), 64–68.

ROEDEL, E.
Asymptotic relative efficiency of tests for independence.
Unpublished manuscript 1982.

GUMBEL, E. J.
Bivariate exponential distributions.
JASA 55 (1960), 698–707.

Academy of Agricultural Sciences of the GDR,
Research Centre of Animal Production Dummerstorf-Rostock

Robustness of Many-One Statistics

EBERHARD RUDOLPH

Abstract

Computer simulation techniques were used to investigate the Type I and Type II error rates of one parametric (DUNNETT) and three nonparametric multiple comparison procedures for comparing treatments with a control. It was found that DUNNETT's procedure is robust with respect to violations of the normality assumption. Power comparisons show that for small sample sizes DUNNETT's procedure is superior to the nonparametric procedures also in nonnormal cases, but for larger sample sizes the multiple analogue to WILCOXON and KRUSKAL-WALLIS rank statistics are superior to DUNNETT's procedure in all considered nonnormal cases.

1. Introduction

A problem frequently encountered in applied statistics is the comparison of treatment means with a control mean. Consider $k+1$ sampels of size n_i from populations having unknown means μ_i and variances σ_i^2, $i=0,\ldots,k$. The treatment means are denoted by $\mu_1,\ldots,\mu_k$ and the control mean by μ_0.

The null hypothesis to be tested is

$$H_0 : \mu_0 = \mu_i, \qquad i=1,\ldots,k$$

against the alternatives

$$H_{Ai}: \mu_0 \neq \mu_i, \qquad i=1,\ldots,k.$$

If the data satisfy the usual ANOVA assumptions (independence, homogeneity of variance and normality) DUNNETT's procedure, DUNNETT (1955, 1964), is applicable.

There are also nonparametric procedures for comparing treatments with a control (STEEL (1959), NEMENYI (1963), DUNN (1964), GABRIEL and SEN (1968), PERITZ (1971), PURI and SEN (1971)). The aim of this paper is the investigation of the behaviour of DUNNETT's procedure and some nonparametric procedures based on WILCOXON, KRUSKAL-WALLIS and FRIEDMAN rank statistics by a simulation experiment for $k=2$ if the underlying distribution is not normal.

We limited the investigation to the case of homogeneity of variance $\sigma_i^2 = \sigma^2$ and equal sample sizes $n_i = n$, $i=0,\ldots,k$ and two-sided alternatives. Therefore we get robustness properties for DUNNETT's procedure in the sense of ε-robustness (HERRENDÖRFER (1980), GUIARD (1981)) and because the assumptions of the nonparametric procedures we considered are fullfilled we compare the behaviour of these procedures with that of DUNNETT's procedure in terms of comparison-wise power.

The nonnormality of the underlying distribution is characterized by given values of skewness and kurtosis of a continuous distribution which is generated from a normal one using a power transformation, see FLEISHMAN (1978).

2. Description of the Procedures

We present a short description of the procedures only for the considered case (homogeneity of variance, independence, equal sample sizes, two-sided alternatives). For further explanations see also MILLER (1966).

y_{ij}, $i=0,\ldots,k$; $j=1,\ldots,n$ denote the samples of size n and y_{ij} their realizations.

The test statistics of DUNNETT's procedure (P1) are the usual t-statistics

$$T_{P1}^{(i)} = \frac{\bar{y}_{0.} - \bar{y}_{i.}}{s\sqrt{\frac{2}{n}}}, \qquad i=1,\ldots,k$$

with the variance estimator

$$s^2 = \frac{1}{\nu}\sum_{i,j}(y_{ij} - \bar{y}_{i.})^2, \quad i=0,\ldots,k; \quad \nu=(k+1)(n-1).$$

Then for an experiment-wise error rate α the control and i-th population are declared significantly different if

$$T_{P1}^{(i)} > d(k,\nu,1-\alpha).$$

DUNNETT (1955, 1964) presented tables of the critical values $d(k,\nu,1-\alpha)$ for $\alpha = 0.01$, 0.05 and several values of k and ν.

The first simultaneous technique based on ranking the observations was introduced by STEEL (1959). The test statistics of this procedure (P2) are

$$T_{P2}^{(i)} = \max(\mathbf{R}_{0i}, \mathbf{R}'_{0i}), \quad i=1,\ldots,k.$$

$\mathbf{R}_{0i}$ denotes the WILCOXON two-sample rank statistic for sample pair (0, i) and $\mathbf{R}'_{0i}$ is the conjugate of $\mathbf{R}_{0i}$. The test statistics are compared with the critical values $r(k,n,1-\alpha)$ and the populations are declared significantly different if

$$T_{P2}^{(i)} \geq r(k,n,1-\alpha), \quad i=1,\ldots,k.$$

For approximations of the critical values $r(k,n,1-\alpha)$ see STEEL (1959).

The next investigated nonparametric procedure (P3) based on the KRUSKAL-WALLIS rank statistic was proposed and analyzed by NEMENYI (1963). DUNN (1964) presented a similar procedure using the BONFERRONI inequality.

The test statistics are

$$T_{P3}^{(i)} = |\bar{R}_{i.} - \bar{R}_{0.}|, \quad i=1,\dots,k \quad \text{where}$$

$$\bar{R}_{i.} = \frac{1}{n} \sum_{j=1}^{n} R_{ij}, \quad i=0,\dots,k$$

denote the rank means and R_{ij}, $i=0,\dots,k$; $j=1,\dots,n$, are the ranks if all $k+1$ samples are combined and the individual observations ranked from smallest to largest.

Approximations of the critical values used in procedure P3 are

$$d(k,\infty,1-\alpha)\left[\frac{N(N+1)}{12}\right]^{1/2}\left(\frac{2}{n}\right)^{1/2} \quad \text{with}$$

$N = n(k+1)$ and $d(k,\infty,1-\alpha)$ from DUNNETT's procedure, see MILLER (1966).

The last considered procedure (P4) based on the FRIEDMAN rank statistic was proposed by NEMENYI (1963). This procedure requires that the observations do occur in blocks. Nevertheless we were interested in the behaviour of this procedure with respect to the described multiple test problem.

R'_{ij} denote the ranks of y_{ij} relative to the ordered observations $y_{(1)j} < \dots < y_{(k+1)j}$ in block j, $j=1,\dots,n$. Then procedure P4 uses the test statistics

$$T_{P4}^{(i)} = |\bar{R}'_{i.} - \bar{R}'_{0.}|, \quad i=1,\dots,k$$

with the rank means

$$\bar{R}'_{i.} = \frac{1}{n} \sum_{j=1}^{n} R'_{ij}, \quad i=0,\dots,k.$$

As in procedure P3 we get under using the critical values of DUNNETT's procedure with $\nu=\infty$ approximations of the critical values for procedure P4 in the form

$$d(k,\infty,1-\alpha)\left[\frac{(k+1)(k+2)}{6n}\right]^{1/2} \quad \left(\text{MILLER (1966)}\right).$$

3. Empirical Evaluation of Error Rates

In the null hypothesis case two kinds of error rates are considered: experiment-wise and comparison-wise. The experiment-wise error rate is the proportion of experiments with at least one pair erroneously declared significant, while the comparison-wise error rate is the proportion of pairwise comparisons erroneously declared significant.

A simulation study was performed using an ESER 1040 computer for $k=2$ and $n=6$ or $n=21$ observations per sample.

The simulation made use of the subroutine ZZGD which generates uniformly distributed random numbers in (0,1] and the subroutine NVO1 which transforms uniformly distributed random numbers to normally distributed ones with mean 0 and variance 1. The subroutines ZZGD and NVO1 are described by HERRENDÖRFER, G., (ed.) (1980) For characterizing the nonnormality random numbers with mean 0, variance **1** and given values of skewness γ_1 and kurtosis γ_2 were generated from normal ones under using a power transformation presented by FLEISHMAN (1978). The shapes of distributions generated in such a way are described by NÜRNBERG (1982). We evaluated experiment-wise and comparison-wise error rates in the null hypothesis case and comparison-wise power for some specific nonnull configurations for nominal $\alpha=0.01$ and $\alpha=0.05$ and the following (γ_1,γ_2)-combinations:

γ_1:	0	0	1	0	1.5	0	2
γ_2:	0	1.5	1.5	3.75	3.75	7	7.

The non-null cases were described by the parameters

$$\delta_1 = |\mu_0 - \mu_1| \Big/ \sqrt{\frac{2\sigma^2}{n}} \quad \text{and}$$

$$\delta_2 = |\mu_0 - \mu_2| \Big/ \sqrt{\frac{2\sigma^2}{n}}.$$

10,000 iterated experiments were carried out for each of the experimental conditions and common samples with $\mu=0$ and $\sigma^2=1$ were used across all procedures. The number of 10,000 iterations is a result of planning the simulation experiment for the estimation of a probability in case of ε-robustness with ε equals 20 % of the nominal $\alpha=0.05$, see HERRENDÖRFER, G., (ed.) (1980).

4. Results and Discussion

The simulation results are summarized in tables 1—4. Because for the nominal $\alpha=0.01$ the number of 10,000 iterations is not large enough in the case of ε-robustness with ε equals 20 % these results show only tendencies in the behaviour of the considered procedures.

For $\alpha=0.05$ DUNNETT's procedure is ε-robust with respect to the experiment-wise error rate α (ε equals 20 %). In the case of 6 observations per sample the power of DUNNETT's procedure is greater then the power of the nonparametric procedures for nearly all (γ_1,γ_2)-combinations, but for $n=21$ the nearly same power of the procedures P2 and P3 is greater then that of DUNNETT's procedure in all nonnormal cases.

The differences in comparison-wise power between the procedures P2 and P3 are not great, but the differences in the power between P1 and P2 and between P1 and P3 are quite considerable. Procedure P4 has the smallest power. The figures 1—4 show this behaviour of the four procedures for some representative (γ_1,γ_2)-combinations. From figures 5—6 we get an impression of the behaviour of the power in some non-null cases if the (γ_1,γ_2)-values are varied. The power of DUNNETT's procedure seems to be relatively constant if we increase the degree of nonnormality, but the power of the nonparametric procedures increase rapidly.

Already MILLER (1966) remarked that the comparison of the procedures P1 and P2 paralleled the comparison of the t-test and the WILCOXON-test. The results found in the present study for DUNNETT's procedure and STEEL's procedure agree with those found for the t-test and WILCOXON-test by POSTEN (1982).

Summarizing the results we can say that for small sample sizes DUNNETT's procedure is superior to the nonparametric procedures also in nonnormal cases. For larger sample sizes the procedures P2 and P3 are superior to DUNNETT's procedure.

Finally, for reasonable recommendations to the experimentors we must also know the behaviour of the procedures in case of unequal variances and unequal sample sizes.

Table 1

Simulated significance levels $\times 100$ for the procedures P1, P2, P3, P4 based on samples of 6 observations drawn from distribution with mean $\mu = 0$, variance $\sigma^2 = 1$, skewness γ_1 and kurtosis γ_2

γ_1/γ_2		Nominal $\alpha = 0.01$			Nominal $\alpha = 0.05$		
		α_1	α_2	α	α_1	α_2	α
0/0	P1	0.49	0.58	0.98	2.76	2.68	4.90
	P2	0.13	0.15	0.27	2.61	2.64	4.85
	P3	0.18	0.28	0.46	2.46	2.60	4.72
	P4	0.28	0.27	0.55	2.45	2.46	4.69
0/1.5	P1	0.44	0.50	0.87	2.47	2.13	4.16
	P2	0.23	0.22	0.44	2.36	2.17	4.27
	P3	0.30	0.27	0.57	2.36	2.13	4.31
	P4	0.34	0.34	0.68	2.36	2.43	4.59
1/1.5	P1	0.48	0.64	1.01	2.63	2.64	4.75
	P2	0.22	0.20	0.40	2.70	7.72	5.01
	P3	0.31	0.35	0.66	2.44	2.63	4.75
	P4	0.28	0.41	0.69	2.61	2.72	5.16
0/3.75	P1	0.35	0.37	0.70	2.37	2.37	4.26
	P2	0.21	0.26	0.47	2.53	2.67	4.89
	P3	0.22	0.27	0.49	2.47	2.41	4.61
	P4	0.30	0.24	0.54	2.59	2.42	4.81
1.5/3.75	P1	0.42	0.42	0.77	2.32	2.32	4.23
	P2	0.19	0.25	0.44	2.66	2.36	4.71
	P3	0.27	0.25	0.52	2.31	2.22	4.30
	P4	0.35	0.33	0.68	2.58	2.55	4.98
0/7	P1	0.37	0.38	0.68	1.95	2.42	3.98
	P2	0.24	0.21	0.42	2.42	2.65	4.74
	P3	0.28	0.26	0.52	2.21	2.58	4.55
	P4	0.36	0.35	0.71	2.34	2.66	4.77
2/7	P1	0.38	0.43	0.75	2.27	2.37	4.17
	P2	0.17	0.22	0.39	2.50	2.92	5.08
	P3	0.26	0.33	0.59	2.50	2.48	4.73
	P4	0.31	0.25	0.56	2.68	2.69	4.13

Table 2

Simulated significance levels $\times 100$ for the procedures P1, P2, P3, P4 based on samples of 21 observations drawn from distribution with mean $\mu = 0$, variance $\sigma^2 = 1$, skewness γ_1 and kurtosis γ_2

γ_1/γ_2		Nominal $\alpha = 0.01$			Nominal $\alpha = 0.05$		
		α_1	α_2	α	α_1	α_2	α
0/0	P1	0.54	0.52	0.91	2.76	2.57	4.80
	P2	0.51	0.38	0.79	2.56	2.70	4.82
	P3	0.56	0.46	0.91	2.53	2.53	4.64
	P4	0.39	0.35	0.70	2.55	2.70	4.80
0/1.5	P1	0.32	0.49	0.80	2.60	2.95	5.13
	P2	0.31	0.33	0.63	2.78	2.13	5.34
	P3	0.41	0.39	0.79	2.69	2.96	5.32
	P4	0.37	0.39	0.73	2.49	2.83	4.92
1/1.5	P1	0.50	0.36	0.83	2.70	2.51	4.90
	P2	0.38	0.37	0.74	2.64	2.61	4.95
	P3	0.46	0.42	0.87	2.57	2.40	4.72
	P4	0.33	0.30	0.62	2.22	2.27	4.22
0/3.75	P1	0.42	0.49	0.81	2.65	2.58	4.74
	P2	0.51	0.38	0.79	2.56	2.70	4.82
	P3	0.56	0.46	0.91	2.53	2.53	4.64
	P4	0.39	0.35	0.70	2.55	2.70	4.80
1.5/3.75	P1	0.39	0.52	0.85	2.67	2.70	4.97
	P2	0.37	0.44	0.80	2.71	2.78	5.09
	P3	0.42	0.48	0.88	2.62	2.75	4.98
	P4	0.38	0.34	0.67	2.75	2.47	4.87
0/7	P1	0.28	0.47	0.74	2.47	2.71	4.82
	P2	0.31	0.33	0.63	2.78	2.93	5.34
	P3	0.41	0.39	0.79	2.69	2.96	5.32
	P4	0.37	0.39	0.73	2.49	2.83	4.92
2/7	P1	0.54	0.47	0.94	2.31	2.37	4.30
	P2	0.41	0.30	0.70	2.75	2.65	5.10
	P3	0.52	0.35	0.85	2.65	2.64	5.01
	P4	0.41	0.28	0.67	2.43	2.13	4.33

Table 3

Simulated comparison-wise power $\times 100$ for theoretical $\alpha = 0.05$ for the procedures P1, P2, P3, P4 based on samples of 6 observations drawn from distribution with variance $\sigma^2 = 1$, skewness γ_1, kurtosis γ_2 in the non-null-case with δ_1 and δ_2

γ_1/γ_2	δ_1	1.73		1.73		3.46		3.46		4.85		4.85	
	δ_2	0		1.73		0		3.46		0		4.85	
		$(1-\beta)_1$	$(1-\beta)_2$	$(1-\beta)_1$	$(1-\beta)_2$	$(1-\beta)_1$	$(1-\beta)_2$	$(1-\beta)_1$	$(1-\beta)_2$	$(1-\beta)_1$	$(1-\beta)_2$	$(1-\beta)_1$	$(1-\beta)_2$
0/0	P1	27.61	2.33	27.04	27.40	83.57	2.65	83.64	83.33	98.64	2.85	98.77	98.91
	P2	23.74	2.28	22.73	23.44	76.40	2.62	76.08	75.78	96.42	2.56	96.43	96.69
	P3	23.69	1.16	22.85	22.97	71.99	0.33	71.40	71.57	92.05	0.06	92.04	92.30
	P4	18.53	1.48	17.78	19.07	57.72	0.53	56.80	56.91	78.50	0.04	78.28	78.38
0/1.5	P1	28.92	2.90	29.33	28.96	83.83	2.56	83.34	83.33	98.12	2.41	97.85	97.85
	P2	26.85	2.86	26.49	27.73	76.62	2.70	76.25	76.41	94.08	2.24	94.42	94.14
	P3	26.02	1.64	25.73	25.71	73.37	0.21	73.11	73.00	90.86	0.01	90.98	90.88
	P4	20.63	1.73	20.87	20.45	58.37	0.27	59.14	58.64	76.93	0.05	77.12	76.87
1/1.5	P1	28.59	2.48	28.23	28.65	82.75	2.73	83.35	83.71	97.74	2.82	97.99	97.85
	P2	26.73	2.50	26.11	26.43	75.75	2.86	75.90	76.27	93.21	2.80	93.59	93.30
	P3	25.23	1.90	26.67	26.89	72.83	0.34	72.07	72.07	90.64	0.06	90.55	90.16
	P4	20.44	1.84	21.47	20.66	57.17	0.52	58.25	59.36	76.85	0.12	77.63	76.89
0/3.75	P1	31.38	2.46	30.75	31.12	83.69	2.50	84.06	83.67	97.23	2.55	97.09	97.20
	P2	30.05	2.65	29.86	30.03	77.16	2.79	77.37	77.20	92.83	2.57	91.91	92.40
	P3	29.38	1.17	29.70	29.93	74.00	0.25	75.06	74.90	90.23	0.06	89.80	90.39
	P4	22.85	1.34	22.71	23.18	59.78	0.24	61.39	60.55	76.80	0.08	76.48	77.46
1.5/3.75	P1	31.56	2.33	31.61	31.49	83.80	2.50	83.55	83.47	97.19	2.72	97.27	97.07
	P2	31.72	2.48	31.62	31.39	76.40	2.57	75.86	76.06	91.57	2.69	91.10	91.39
	P3	30.26	1.72	31.85	32.10	74.83	0.50	73.17	73.35	90.00	0.12	88.55	88.67
	P4	23.87	1.83	25.06	25.19	59.38	0.49	59.98	59.85	76.36	0.13	76.67	75.85
0/7	P1	34.04	2.18	33.88	34.74	84.07	2.11	84.19	83.90	96.46	2.14	96.49	96.63
	P2	34.24	2.53	33.07	35.07	76.58	2.55	76.94	77.17	90.86	2.53	90.55	90.85
	P3	34.39	0.89	33.76	34.64	75.71	0.12	76.04	76.27	89.50	0.04	89.38	89.44
	P4	26.36	1.08	26.07	26.52	61.92	0.23	61.80	62.17	76.30	0.07	76.90	77.24
2/7	P1	34.32	1.97	34.45	33.93	84.08	2.28	83.69	83.56	96.22	2.34	96.19	96.52
	P2	36.23	2.46	36.07	36.00	76.29	2.53	76.98	76.61	89.49	2.62	89.66	89.52
	P3	35.24	1.58	36.82	36.82	75.63	0.40	75.66	74.85	89.08	0.10	88.26	88.15
	P4	26.83	1.81	28.67	28.58	61.03	0.40	63.02	61.82	76.09	0.10	76.29	76.58

Table 4

Simulated comparison-wise power $\times 100$ for theoretical $\alpha = 0.05$ for the procedures P1, P2, P3, P4 based on samples of 21 observations drawn from distribution with variance $\sigma^2 = 1$, skewness γ_1, kurtosis γ_2 in the non-null-case with δ_1 and δ_2

γ_1/γ_2	δ_1	1.62		1.62		3.24		3.24		4.86		4.86	
	δ_2	0.0		1.62		0.0		3.24		0.0		4.86	
		$(1-\beta)_1$	$(1-\beta)_2$	$(1-\beta)_1$	$(1-\beta)_2$	$(1-\beta)_1$	$(1-\beta)_2$	$(1-\beta)_1$	$(1-\beta)_2$	$(1-\beta)_1$	$(1-\beta)_2$	$(1-\beta)_1$	$(1-\beta)_2$
0/0	P1	26.54	2.42	26.84	26.55	83.12	2.88	83.53	82.89	99.37	2.47	99.46	99.54
	P2	25.07	2.57	25.24	25.19	80.78	2.92	81.47	80.38	99.19	2.65	99.07	99.30
	P3	24.57	2.20	24.65	24.41	79.78	1.74	80.13	79.45	98.92	0.94	98.87	99.01
	P4	17.81	1.99	17.98	17.83	64.64	1.70	64.72	63.69	94.54	1.05	94.56	94.61
0/1.5	P1	27.09	2.65	26.94	26.66	82.63	2.87	83.27	82.87	99.28	2.55	99.27	99.24
	P2	29.04	2.74	29.01	28.41	85.18	2.88	85.41	85.14	99.58	2.79	99.40	99.49
	P3	28.13	2.38	28.34	28.34	83.94	1.67	84.53	84.02	99.39	0.84	99.26	99.36
	P4	20.25	2.13	20.41	20.14	70.15	1.45	70.28	69.49	95.88	0.93	95.96	96.06
1/1.5	P1	27.68	2.62	27.06	27.28	82.66	2.68	82.98	83.37	99.25	2.52	99.27	99.36
	P2	30.40	2.90	30.09	30.73	85.61	2.88	86.07	86.61	99.48	2.78	99.39	99.51
	P3	29.41	2.82	29.83	30.38	85.70	2.23	84.22	84.90	99.47	1.12	99.10	99.19
	P4	21.48	2.57	21.49	21.49	71.77	2.00	70.72	71.13	96.50	1.10	95.89	95.42
0/3.75	P1	27.76	2.32	28.18	27.68	83.11	2.76	82.41	83.43	99.13	2.51	98.83	98.98
	P2	34.39	2.79	34.39	34.11	89.53	2.92	89.90	89.96	99.67	2.55	99.70	99.72
	P3	33.72	2.12	33.90	33.29	88.83	1.52	88.59	89.10	99.65	0.74	99.57	99.58
	P4	24.31	1.88	24.58	24.19	76.75	1.30	75.25	76.50	97.55	0.71	96.78	97.21
1.5/3.75	P1	27.94	2.44	28.25	27.94	82.85	2.62	83.15	83.02	99.01	2.74	98.99	99.03
	P2	37.17	2.70	37.77	37.31	91.05	2.94	91.11	90.97	99.74	2.85	99.66	99.73
	P3	35.93	2.80	36.97	36.77	91.10	2.15	89.43	89.33	99.76	1.03	99.35	99.37
	P4	26.39	2.55	27.16	26.83	78.45	1.96	77.16	77.51	97.74	1.11	96.78	96.61
0/7	P1	28.58	2.45	28.79	29.07	83.28	2.64	83.34	84.05	98.75	2.57	98.73	98.72
	P2	41.38	2.64	40.66	41.17	93.44	2.92	93.47	93.77	99.81	2.53	99.80	99.83
	P3	40.24	2.04	39.53	40.31	92.79	1.36	92.95	93.23	99.79	0.50	99.83	99.73
	P4	29.77	1.87	29.35	28.99	82.70	1.29	82.32	82.74	97.96	0.59	98.29	98.27
2/7	P1	29.49	2.44	28.91	29.72	82.45	2.42	82.57	82.62	98.45	2.33	98.39	98.51
	P2	46.78	2.64	46.23	46.43	94.12	2.82	94.49	94.41	99.75	2.48	99.74	99.73
	P3	45.80	2.76	45.46	45.51	94.70	2.03	93.05	92.87	99.86	0.98	99.59	99.57
	P4	32.91	2.66	33.96	34.46	84.78	1.91	83.53	83.57	98.45	0.98	97.81	97.78

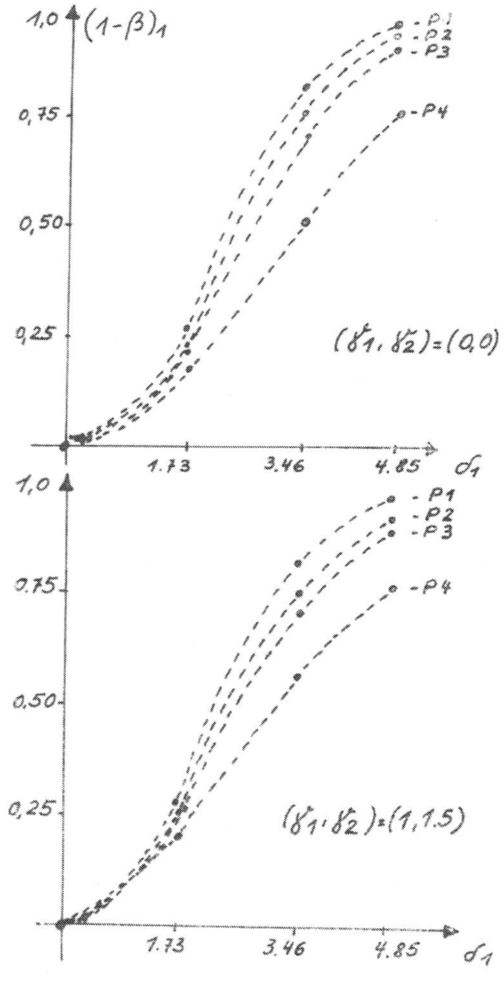

Figure 1

Empirical comparison-wise power $(1-\beta)_1$ in dependence on δ_1 ($\delta_2 = \delta_1$) for nominal $\alpha = 0.05$ and 6 observations per sample in the normal case and nonnormal case with $(\gamma_1, \gamma_2) = (1, 1.5)$

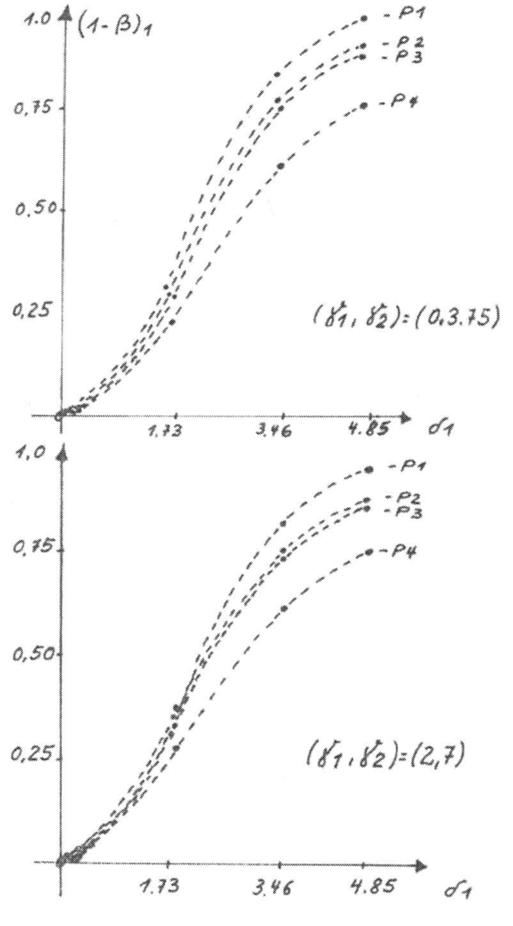

Figure 2

Empirical comparison-wise power $(1-\beta)_1$ in dependence on δ_1 ($\delta_2 = \delta_1$) for nominal $\alpha = 0.05$ and 6 observations per sample in two nonnormal cases

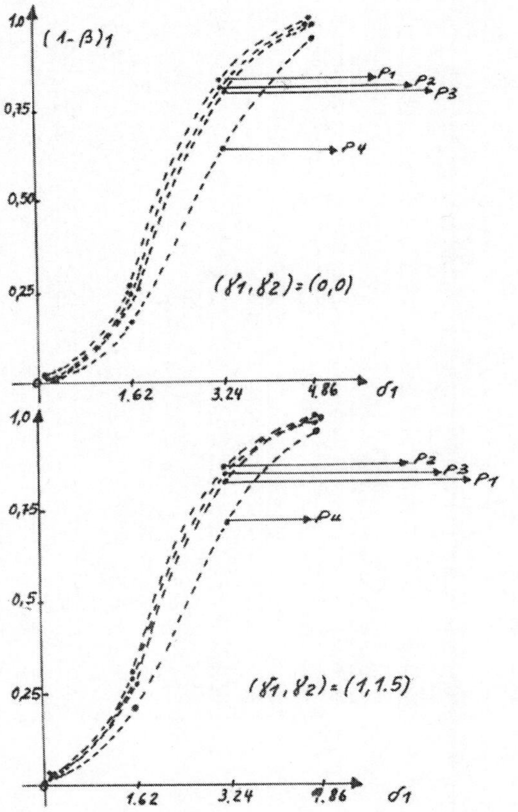

Figure 3
Empirical comparison-wise power $(1-\beta)_1$ in dependence on δ_1 ($\delta_2 = \delta_1$) for nominal $\alpha = 0.05$ and 21 observations per sample in the normal case and nonnormal case with $(\gamma_1, \gamma_2) = (1, 1.5)$

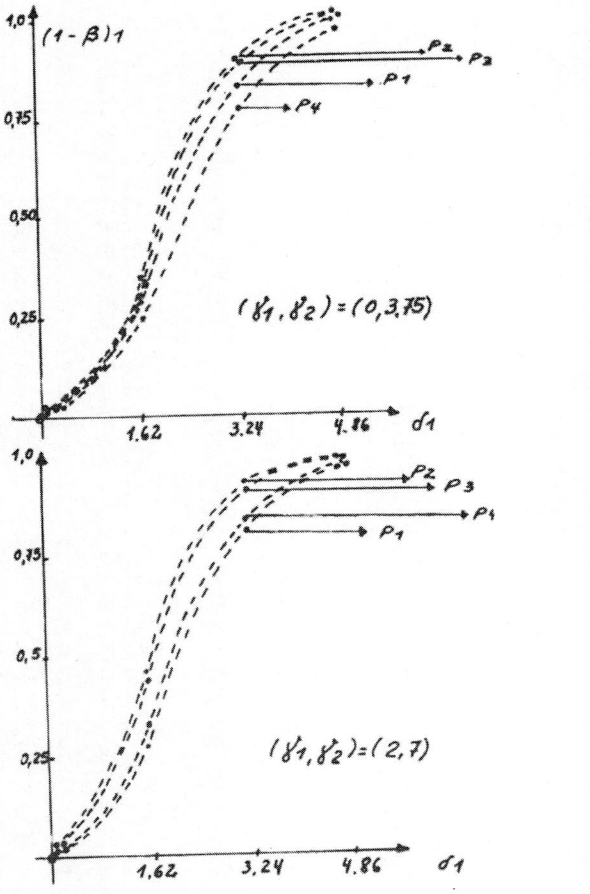

Figure 4
Empirical comparison-wise power $(1-\beta)_1$ in dependence on δ_1 ($\delta_2 = \delta_1$) for nominal $\alpha = 0.05$ and 21 observations per sample in two nonnormal cases

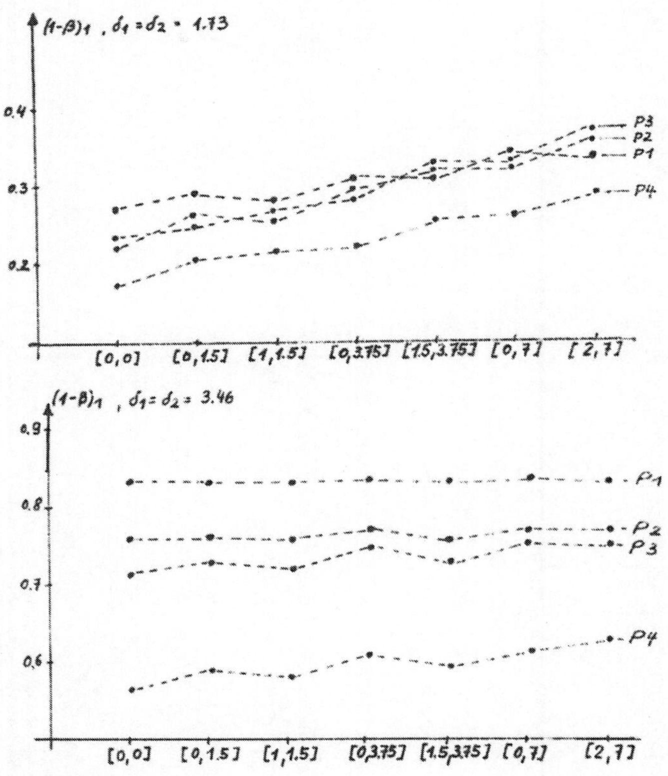

Figure 5
Empirical comparison-wise power $(1-\beta)_1$ in dependence on the degree of nonnormality for nominal $\alpha = 0.05$ and 6 observations per sample in two non-null cases with $\delta_1 = \delta_2$

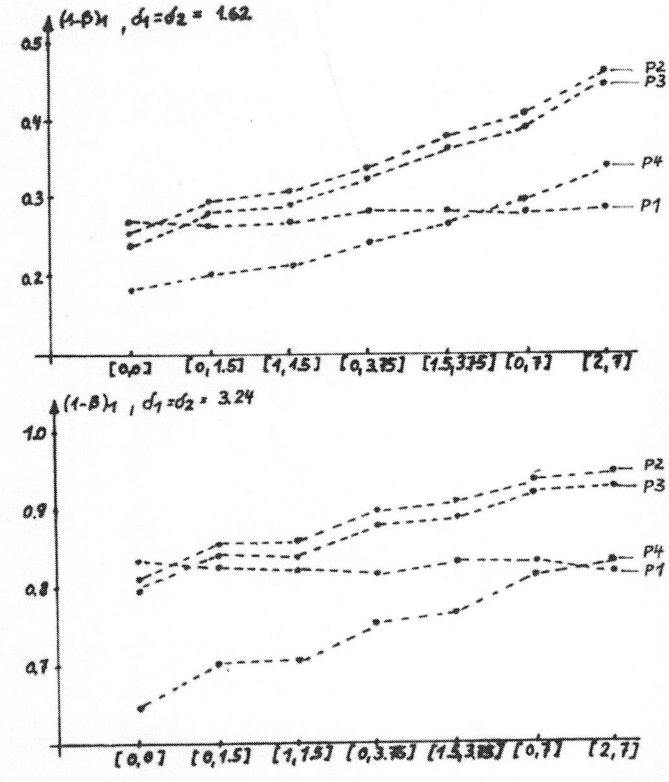

Figure 6
Empirical comparison-wise power $(1-\beta)_1$ in dependence on the degree of nonnormality for nominal $\alpha = 0.05$ and 21 observations per sample in two non-null cases with $\delta_1 = \delta_2$

DUNN, O. J.
Multiple comparisons using rank sums.
Technometrics 6 (1964) 241–52.

DUNNETT, C. W.
A multiple comparisons precedure for comparing several treatments with a control.
J. Amer. Statist. Ass. 50 (1955) 1096–121.

DUNNETT, C. W.
New table for multiple comparisons with a control.
Biometrics 15 (1964) 560–72.

FLEISHMAN, A. I.
A method for simulating non-normal distribution.
Psychometrika 43 (1978) 521–32.

GABRIEL, K. R., SEN, P. K.
Simultaneous test procedures for one-way ANOVA and MANOVA based on rank scores.
Sankhya 30 (1968) 303–12.

GUIARD, V. (ed.)
Robustheit II.
Probleme der angewandten Statistik, Heft 5, (1981) AdL der DDR, Forschungszentrum für Tierproduktion Dummerstorf-Rostock.

HERRENDÖRFER, G. (ed.)
Robustheit I.
Probleme der angewandten Statistik, Heft 4, (1980) AdL der DDR, Forschungszentrum für Tierproduktion Dummerstorf-Rostock.

MILLER, R. G.
Simultaneous Statistical Inference.
McGraw-Hill Co., New York (1966).

NEMENYI, P.
Distribution-free multiple comparisons.
Ph. D. thesis, Princeton University, Princeton, N. J. (1963).

NÜRNBERG, G.
Beiträge zur Versuchsplanung für die Schätzung von Varianzkomponenten und Robustheitsuntersuchungen zum Vergleich zweier Varianzen.
Probleme der angewandten Statistik, Heft 6, (1982) AdL der DDR, Forschungszentrum für Tierproduktion Dummerstorf-Rostock.

PERITZ, E.
On a statistic for rank analysis of variance.
J. of the Royal Statist. Soc., Ser. B. 33 (1971) 137–139.

POSTEN, H. O.
Twa sample Wilcoxon power over the Pearson system and comparison with the t-test.
J. Statist. Comput. Simul. 16 (1982) 1–18.

PURI, M. L., SEN, P. K.
Nonparametric Methods in Multivariate Analysis.
J. Wiley and Sons, New York 1971.

STEEL, R. G. D.
A multiple comparison rank sum test: treatment versus control.
Biometrics 15 (1959) 560–72.

Department of Mathematics, Humboldt University Berlin, GDR
Institute of Mathematics, Academy of Sciences, Berlin, GDR

Testing Hypotheses in Nonlinear Regression for Nonnormal Distributions

WOLFGANG H. SCHMIDT, SILVELYN ZWANZIG

Abstract

The general nonlinear regression model is considered, where the error distribution can be nonnormal. In this paper we present tests, confidence intervals and estimators for both the regression coefficients and the variance of the observations which perform well in the sense of second order asymptotic theory. In particular, robustness results for deviations from the normal distribution are given including distributions with either known or unknown skewness and kurtosis.

1. Introduction

Recently the authors derived in Schmidt, Zwanzig (1983) second order asymptotically efficient tests, confidence regions and estimators for both the regression coefficients and the variance in the normal nonlinear regression model

$$y_t = g(x_t, \xi) + \sigma u_t, \quad t = 1, \ldots, n, \ldots$$

Here u_t constitutes a sequence of independently $N(0,1)$ distributed random variables, g is a known regression function depending on known regressors $x \in R^k$. The unknown parameters to make inference upon are $\xi \in \Xi \subset R^p$ and $0 < \sigma^2 < \infty$.

Popular estimators for ξ and σ^2 are the least-squares estimator (LSE) $\hat{\xi}_n$ and the residual sum of squares $\hat{\sigma}_n^2 = Q(\hat{\xi}_n)$ where $\hat{\xi}$ is a solution of

$$Q(\hat{\xi}_n) = \min_{\xi \in \Xi^c} Q(\xi)$$

with

$$Q(\xi) := \frac{1}{n} \sum_{t=1}^{n} \left(y_t - g(x_t, \xi)\right)^2$$

These estimators are widely used because of its justification from the heuristical point of view even if the underlying distribution is nonnormal. Notice that $\hat{\xi}_n$ and $\hat{\sigma}_n^2$ are maximum-likelihood estimators (MLE) if $u_1 \sim N(0,1)$. Even if there exist some objections to the normal distribution assumption usually the practician cannot specify an alternative fixed distribution. It may happen, however, that he is quite sure that the error distribution is not symmetric or that other global properties fulfilled by the normal distribution are not satisfied.

In such situations natural questions are:

i. Is it possible to construct statistical procedures based on $\hat{\xi}_n$ and $\hat{\sigma}_n^2$ which perform well if there are introduced additional parameters like skewness, kurtosis or other characteristics of the error distribution?
ii. What is the asymptotic deficiency of such procedures?
iii. Do there exist examples with asymptotic deficiency zero, i. e. are the procedures asymptotically robust against deviations from the normal distribution? In particular contaminated families of distributions are of interest.

In the present paper we collect several results giving answers to questions i., ii. and iii. Here we abstain from giving the proofs and all the regularity conditions needed. As in (1983) several smoothness conditions on the regression function and the design of local Lipschitzian type as well as Cramer type conditions on the error distribution are used. For instance, the regularity conditions on the regression function $g(x, \xi)$ are fulfilled, if $g(\cdot\, \xi)$ is a continuous function defined on a compact set $\mathfrak{X}$, and if $g(x, \cdot)$, defined on a compact set Ξ, is identificable and three times continuously partially differentiable. Full proofs and a complete list of the regularity assumptions will be published elsewhere.

2. Some General Definitions and Results

Let us recall some general definitions first.
We consider a statistical framework of independent not necessarily identically distributed random variables

$$y_i \sim P_{i\vartheta} \qquad i = 1, 2, \ldots, n, \ldots$$

$$\text{with } \vartheta \in \Theta \subset R^s$$

It is assumed that $\{P_{i\vartheta} \mid \vartheta \in \Theta\}$ is dominated by some σ-finite measure μ and $p_i(y, \vartheta)$ denotes a μ-density of $P_{i\vartheta}$.

$$\text{Let } \vartheta = \binom{\gamma}{\eta} \quad \text{with } \gamma = \vartheta_1 \in R^1$$

We are interested in testing the composite one-sided hypothesis

$$H: \gamma \leq \gamma_0 \text{ against } K: \gamma > \gamma_0$$

for a given $\gamma_0 \in R^1$. Let $\mathfrak{K} \subset \Theta$ an arbitrary compact subset of Θ.

Definition 1 (Pfanzagl (1973b))

Let $\beta > 0$ be fixed. A sequence $\{\psi_n\}$ of critical functions $\psi_n = \psi_n(Y_1, \ldots, Y_n, \vartheta)$ is called to be asymptotically similar to the level $\alpha + o(n^{-\beta})$ with $0 < \alpha < 1$ if uniformly for ϑ compact subsets $\mathfrak{K} \subset \Theta_0 = \{\vartheta \in \Theta \mid \gamma = \gamma_0\}$.

$$|E_\vartheta \psi_n - \alpha| = o(n^{-\beta})$$

The set of all such sequences will be denoted by the symbol $\Psi_{\alpha,\beta}$

Definition 2 (Pfanzagl (1973)b)

A sequence $\{\psi_n^*\} \in \Psi_{\alpha,\beta}$ is asymptotically efficient of order $2\beta + 1$ if uniformly for ϑ from compact subsets $\mathfrak{K} \subset \Theta_0$

$$E_{\vartheta_n} \psi_n^* \geq E_{\vartheta_n} \psi_n + o(n^{-\beta}) \quad \text{for } h > 0$$

and

$$E_{\vartheta_n} \psi_n^* \leq E_{\vartheta_n} \psi_n + o(n^{-\beta}) \quad \text{for } h < 0$$

with

$$\vartheta_n = \begin{pmatrix} \gamma_0 + n^{-1/2} h \\ \eta \end{pmatrix}$$

for all $\{\psi_n\} \in \Psi_{\alpha,\beta}$ holds. □

Definition 3

Let $\{\psi_n\}$, $\{\psi_n^*\} \in \Psi_{\alpha,\beta}$ be two sequences of critical functions and let $\{\psi_n^*\}$ be asymptotically efficient of order $2\beta+1$. Further it is assumed that

$$E_{\vartheta_n}\psi_n = C(\alpha,h) + n^{-\beta} B(\alpha,h) + o(n^{-\beta})$$

and

$$E_{\vartheta_n}\psi_n^* = C(\alpha,h) + n^{-\beta} B^*(\alpha,h) + o(n^{-\beta})$$

both holding uniformly in $\vartheta \in \mathfrak{K} \subset \Theta_0$

$C(\alpha, h)$, $B(\alpha, h)$ and $B^*(\alpha, h)$ are assumed to be bounded in n and $\vartheta \in \mathfrak{K}$

Then the quantity

$$\Delta_{2\beta+1}(\alpha,h) := B^*(\alpha,h) - B(\alpha,h)$$

is the deficiency of order $2\beta+1$ of $\{\psi_n\}$. □

Notice that only functions ψ_n being independent of ϑ are reasonable tests.

We shall mainly discuss the case $\beta = \frac{1}{2}$, i. e. the second order asymptotic efficiency. Very often asymptotically similar tests to the level $\alpha + o(n^{-1/2})$ are of the form

$$\psi_n = \begin{cases} 1, \text{ if } & \dfrac{\sqrt{n}(\bar{\gamma}_n - \gamma_0)}{N(\bar{\vartheta}_n)} \geq u_{1-\alpha} - n^{-1/2} M(u_\alpha, \bar{\vartheta}_n) \\ 0, \text{ else} \end{cases} \tag{1}$$

where $\bar{\vartheta}_n$ is an estimator of ϑ, u_α is the α-fractile of N (0, 1) and $M(u, \vartheta) = m_0(\vartheta) + m_1(\vartheta)u^2$ is a bounded function in $\vartheta \in \mathfrak{K} \subset \Theta_0$

Converting the acceptance region one gets the upper confidence interval

$$\left(F_n(\bar{\vartheta}_n,\alpha),\ +\infty\right) \tag{2}$$

with

$$F_n(\bar{\vartheta}_n,\alpha) := \bar{\gamma}_n - n^{-1/2} N(\bar{\vartheta}_n)\left(u_{1-\alpha} - n^{-1/2} M(u_\alpha, \bar{\vartheta}_n)\right)$$

which possesses the asymptotic confidence level $1-\alpha+o(n^{-1/2})$ uniformly for $\vartheta \in \mathfrak{K} \subset \Theta_0$.

If, moreover, ψ_n of the form (1) is second order asymptotically efficient then (2) has minimal coverage probability for false parameters

$$\gamma - n^{-1/2} h,\ h > 0$$

up to terms of order $o(n^{-1/2})$.

For $\alpha = \frac{1}{2}$ $(u_\alpha = 0)$ the lower confidence limit results in a second order asymptotically efficient estimator for γ, namely

$$\bar{\gamma}_n^* = F_n\left(\frac{1}{2}, \bar{\vartheta}_n\right) = \bar{\gamma}_n - n^{-1} N(\bar{\vartheta}_n) m_0(\bar{\vartheta}_n) \tag{3}$$

if (1) is second order asymptotically efficient.

It has the properties

$$P_\vartheta\{\bar{\gamma}_n^* \geq \gamma\} \geq \frac{1}{2} + o(n^{-1/2}) \tag{4}$$

and

$$P_\vartheta\{\bar{\gamma}_n^* \leq \gamma\} \geq \frac{1}{2} + o(n^{-1/2}) \tag{5}$$

uniformly for $\vartheta \in \mathfrak{K} \subset \Theta$ (second order asymptotic median unbiasedness).

And for every other sequence of estimators $\tilde{\gamma}_n$ with (4) and (5) it holds

$$P_\vartheta\{\gamma - n^{-1/2}\underline{h} \leq \tilde{\gamma}_n \leq \gamma + n^{-1/2}\bar{h}\}$$
$$\leq P_\vartheta\{\gamma - n^{-1/2}\underline{h} \leq \bar{\gamma}_n^* \leq \gamma + n^{-1/2}\bar{h}\} + o(n^{-1/2})$$

uniformly in $\vartheta \in \mathfrak{K} \subset \Theta$ for every $\underline{h}, \bar{h} > 0$.

Thus it suffices to derive second order asymptotically efficient tests. Second order asymptotically efficient confidence regions and second order asymptotically efficient estimators are then obtained by (2) and (3) automatically.

3. The Asymptotic Envelope Power Function of any Sequence $\{\psi_n\} \in \Psi_{\alpha,1/2}$

A similar argument as in Pfanzagl (1973)a yields for $\beta = \frac{1}{2}$

$$E_{\vartheta_n}\psi_n^* = \Phi\left(u_\alpha + \Lambda_{11}^{-1/2} h\right) + \varphi\left(u_\alpha + \Lambda_{11}^{-1/2} h\right) n^{-1/2} H^*(u_\alpha, h) + o(n^{-1/2}) \tag{6}$$

uniformly in $\vartheta \in \mathfrak{K} \subset \Theta_0$

with

$$H^*(u,h) = \Lambda_{11}^{-5/2} h^2 \left(-\frac{1}{2} L^{(..)(.)} - \frac{1}{3} L^{(.)(.)(.)}\right) + \Lambda_{11}^{-3/2} h^2 \left(L^{(.1)(.)} + \frac{1}{2} L^{(.)(.)(1)}\right) - \frac{1}{6}\Lambda_{11}^{-2} u h L^{(.)(.)(.)} \tag{7}$$

Here and in the sequel we use the following denotations:

$$I(\vartheta) := \left(\frac{1}{n}\sum_{t=1}^{n} E_\vartheta\left(l^{(i)}(y_t,\vartheta)\, l^{(j)}(y_t,\vartheta)\right)\right)_{i,j=1,\ldots,s}$$

is the average Fisher information matrix, where

$$l^{(i)}(y_t,\vartheta) = \frac{\partial}{\partial\vartheta_i} \ln p_t(y_t,\vartheta) \qquad i=1,\ldots,s$$

$$\Lambda(\vartheta) = \left(\Lambda_{ij}(\vartheta)\right)_{i,j=1,\ldots,s} = \left(\Lambda_{ij}\right)_{i,j=1,\ldots,s} = I(\vartheta)^{-1}$$

$$L^{(.)(.)(.)} = \sum_{i_1,i_2,i_3=1}^{s} \Lambda_{1i_1}\Lambda_{1i_2}\Lambda_{1i_3} \frac{1}{n}\sum_{t=1}^{n} E_\vartheta\left(l^{(i_1)}(y_t,\vartheta)\, l^{(i_2)}(y_t,\vartheta)\, l^{(i_3)}(y_t,\vartheta)\right)$$

$$L^{(..)(.)} = \sum_{i_1,i_2,i_3=1}^{s} \Lambda_{1i_1}\Lambda_{1i_2}\Lambda_{1i_3} \frac{1}{n}\sum_{t=1}^{n} E_\vartheta\left(l^{(i_1 i_2)}(y_t,\vartheta)\, l^{(i_3)}(y_t,\vartheta)\right)$$

$$l^{(i_1 i_2)}(y_t,\vartheta) = \frac{\partial}{\partial\vartheta_{i_2}} l^{(i_1)}(y_t,\vartheta)$$

$$L^{(.1)(.)} = \sum_{i_1,i_2=1}^{s} \Lambda_{1i_1}\Lambda_{1i_2} \frac{1}{n}\sum_{t=1}^{n} E_\vartheta\left(l^{(i_1 1)}(y_t,\vartheta)\, l^{(i_2)}(y_t,\vartheta)\right)$$

$$L^{(.)(.)(1)} = \sum_{i_1,i_2=1}^{s} \Lambda_{1i_1}\Lambda_{1i_2} \frac{1}{n}\sum_{t=1}^{n} E_\vartheta\left(l^{(1)}(y_t,\vartheta)\, l^{(i_1)}(y_t,\vartheta)\, l^{(i_2)}(y_t,\vartheta)\right)$$

Further φ denotes the density of N(0, 1) and Φ its distribution function. The formulae (6) and (7) have been given by J. Pfanzagl (1973)a for the independent and identically distributed case.

Usually the asymptotic envelope power function (6) is attained by the sequence of tests (1) based on an MLE $\bar{\vartheta}_n$.

4. **Asymptotically Similar sequences of Tests to the Level α in the Nonlinear Regression Model**

We shall give sequences $\{\psi_n\} \in \Psi_{\alpha,1/2}$ both for $\gamma = \vartheta_1$ where $\vartheta^T = (\xi^T, \sigma^2)$, and $\gamma = \sigma^2$, where $\vartheta^T = (\sigma^2, \xi^T)$, and shall discuss their deficiencies. Two cases are distinguished, namely

i. some moments of $\varepsilon_t = \sigma u_t$ are known, e. g. $\mu_3 = E\varepsilon_1^3$, $\mu_4 = E\varepsilon_1^4$ or $\mu_6 = E\varepsilon_1^6$

ii. the density p(u) of u_t is unknown.

4.1. *Tests on $\gamma = \xi_1$*

4.1.1. $\mu_3 = E\,\varepsilon_1^3$ known

First we introduce some further denotations:
$I_\varphi(\vartheta)$ denotes the average Fisher information matrix under $p(u) = \varphi(u)$ and $I_\varphi^{-1}(\vartheta)$ its inverse:

$$I_\varphi(\vartheta) = \begin{pmatrix} \sigma^{-2}\left(\frac{1}{n}\sum_{t=1}^{n} g^{(i_1)}(x_t,\xi)\, g^{(i_2)}(x_t,\xi)\right)_{i_1,i_2=1,\dots,p} & \begin{matrix}0\\ \vdots \\ 0\end{matrix} \\ 0 \cdots\cdots 0 & \frac{1}{2\sigma^4} \end{pmatrix}$$

$$\Pi^{(i_1)(i_2)(i_3)} := \frac{1}{n}\sum_{t=1}^{n} g^{(i_1)}(x_t,\xi)\, g^{(i_2)}(x_t,\xi)\, g^{(i_3)}(x_t,\xi)$$

$$\Pi^{(i_1 i_2)(i_3)} := \frac{1}{n}\sum_{t=1}^{n} g^{(i_1 i_2)}(x_t,\xi)\, g^{(i_3)}(x_t,\xi)$$

$$\Pi^{(i_1)} := \frac{1}{n}\sum_{t=1}^{n} g^{(i_1)}(x_t,\xi)$$

with $g^{(i_1)}(x_t,\xi) = \frac{\partial}{\partial \xi_{i_1}} g(x_t,\xi)$

$g^{(i_1 i_2)}(x_t,\xi) = \frac{\delta^2}{\partial \xi_{i_1} \xi_{i_2}} g(x_t,\xi)$ and

$$\Pi^{(.)} = \sum_{i_1=1}^{p} \Lambda_{\varphi 1 i_1} \Pi^{(i_1)}$$

$$\Pi^{(.)(.)(.)} = \sum_{i_1,i_2,i_3=1}^{p} \Lambda_{\varphi 1 i_1} \Lambda_{\varphi 1 i_2} \Lambda_{\varphi 1 i_3} \Pi^{(i_1)(i_2)(i_3)}$$

$$\Pi^{(..)(.)} = \sum_{i_1,i_2,i_3=1}^{p} \Lambda_{\varphi 1 i_1} \Lambda_{\varphi 1 i_2} \Lambda_{\varphi 1 i_3} \Pi^{(i_1 i_2)(i_3)}$$

$$\Pi^{(.)(i_1 i_2)} = \sum_{i_3=1}^{p} \Lambda_{\varphi 1 i_3} \Pi^{(i_1 i_2)(i_3)}$$

$$\Pi^{(.)(.1)} = \sum_{i_1,i_2=1}^{p} \Lambda_{\varphi 1 i_1} \Lambda_{\varphi 1 i_2} \Pi^{(i_1)(i_2 1)}$$

$$\Pi^{(.)(.)(1)} = \sum_{i_1 i_2=1}^{p} \Lambda_{\varphi 1 i_1} \Lambda_{\varphi 1 i_2} \Pi^{(i_1)(i_2)(1)}$$

The sequence of tests

$$\psi_n = \begin{cases} 1, & \text{if } \Lambda_{\varphi 11}^{-1/2}(\hat\vartheta_n)\sqrt{n}\,(\hat\gamma_n - \gamma_0) \ge u_{1-\alpha} - n^{-1/2} M_{\mu_3}(u_{1-\alpha}, \hat\vartheta_n) \\ 0, & \text{else} \end{cases} \tag{8}$$

with $\hat\vartheta_n = \begin{pmatrix}\hat\xi_n \\ \hat\sigma_n^2\end{pmatrix}$

$$M_{\mu_3}(u,\vartheta) = m_0(\vartheta,\mu_3) + m_1(\vartheta,\mu_3)\, u^2 \tag{9}$$

$$m_0(\vartheta,\mu_3) = \frac{1}{6\sigma^6}\mu_3 \Lambda_{\varphi 11}^{-3/2} \Pi^{(.)(.)(.)} - \frac{1}{2\sigma^2}\Lambda_{\varphi 11}^{-3/2}\Pi^{(.)(..)} + \frac{1}{2\sigma^2}\Lambda_{\varphi 11}^{-1/2}\sum_{i_1,i_2=1}^{p}\Pi^{(.)(i_1 i_2)}\Lambda_{\varphi i_1 i_2} \tag{10}$$

$$m_1(\vartheta,\mu_3) = -\frac{1}{6\sigma^6}\mu_3 \Lambda_{\varphi 11}^{-3/2}\Pi^{(.)(.)(.)} - \frac{1}{2\sigma^2}\Lambda_{\varphi 11}^{-3/2}\Pi^{(.)(..)} + \frac{1}{2\sigma^4}\mu_3 \Lambda_{\varphi 11}^{-1/2}\Pi^{(.)} \tag{11}$$

belongs to $\Psi_{\alpha,1/2}$
Especially it holds for symmetric error distributions ($\mu_3 = 0$)

$$m_0(\vartheta) = -\frac{1}{2\sigma^2}\Lambda_{\varphi 11}^{-3/2}\Pi^{(.)(..)} + \frac{1}{2\sigma^2}\Lambda_{\varphi 11}^{-1/2}\sum_{i_1,i_2=1}^{p}\Pi^{(.)(i_1 i_2)}\Lambda_{\varphi i_1 i_2}$$

and

$$m_1(\vartheta) = -\frac{1}{2\sigma^2}\Lambda_{\varphi 11}^{-3/2}\Pi^{(.)(..)}$$

Under $p(u) = \varphi(u)$ the sequence of tests (8) is second order asymptotically efficient (see W. Schmidt, S. Zwanzig (1983)). Notice that $\hat\vartheta_n$ is an MLE under $p(u) = \varphi(u)$.

4.1.2. μ_3 unknown

A natural estimator of $\mu_3 = E\,\varepsilon_1^3$ is

$$\hat\mu_{3,n} = \frac{1}{n}\sum_{t=1}^{n}\left(y_t - g(x_t,\hat\xi_n)\right)^3$$

Let us introduce the test

$$\psi_n = \begin{cases} 1, & \text{if } \Lambda_{\varphi 11}^{-1/2}(\hat\vartheta_n)\sqrt{n}\,(\hat\gamma_n - \gamma_0) \ge u_{1-\alpha} - n^{-1/2} M_{\hat\mu_{3,n}}(u_{1-\alpha}, \hat\vartheta_n) \\ 0, & \text{else} \end{cases} \tag{12}$$

with $M_{\mu_3}(u,\vartheta)$ defined by (9), (10) and (11).
Then it holds

$$\{\psi_n\} \in \Psi_{\alpha,1/2}$$

and its power function under local alternatives or local hypothesis ϑ_n equals

$$E_{\vartheta_n}\psi_n = \Phi\left(u_\alpha + \Lambda_{\varphi 11}^{-1/2} h\right) + \varphi\left(u_\alpha + \Lambda_{\varphi 11}^{-1/2} h\right) n^{-1/2} H(u_\alpha, h) + o\left(n^{-1/2}\right) \tag{13}$$

uniformly for $\vartheta \in \mathfrak{K} \subset \Theta_0$

with

$$H(u,h) = -\frac{1}{2\sigma^4}\mu_3 \Lambda_{\varphi 11}^{-1}\Pi^{(.)} hu + \frac{1}{3\sigma^6}\mu_3 \Lambda_{\varphi 11}^{-2}\Pi^{(.)(.)(.)} hu + \frac{1}{6\sigma^6}\mu_3 \Lambda_{\varphi 11}^{-5/2}\Pi^{(.)(.)(.)} h^2 - \frac{1}{2\sigma^2}\Lambda_{\varphi 11}^{-5/2}\Pi^{(..)(.)} h^2 + \frac{1}{\sigma^2}\Lambda_{\varphi 11}^{-3/2}\Pi^{(.1)(.)} h^2 . \tag{14}$$

Thus one might compute the second order deficiency

$$\Delta_2(\alpha,h) = B^*(\alpha,h) - B(\alpha,h)$$

using (14) and (7).
It may happen that $\Delta_2(\alpha, h) = 0$ even if the underlying distribution is nonnormal. This is demonstrated by the following example:
Let $\{Z_t\}$ be a sequence of independent and identically distributed random variables with Lebesgue-density

$$q(z)=\begin{cases} e^{-2} & z>0 \\ 0 & \text{else} \end{cases}$$

and let

$$u_t = Z_t - 1$$

Then all the model assumption are fulfilled.
We shall consider the case $\sigma^2 > 0$ known only.
It turns out

$$\Delta_1(\alpha, h) \equiv 0$$

i. e. the sequence of tests $\{\psi_n\}$ defined by (12) is first order asymptotically efficient.
The second order asymptotic deficiency equals

$$\Delta_2(\alpha,h)=\left(\left(\frac{1}{\sigma}\Lambda_{11}^{-1}\Pi^{(.)}-\frac{5}{6\sigma^3}\Lambda_{11}^{-3}\Pi^{(.)(.)(.)}\right)hu_\alpha\right.$$

$$\left.+\left(\frac{1}{2\sigma^3}\Lambda_{11}^{-3/2}\Pi^{(.)(.)(1)}-\frac{2}{3\sigma^3}\Lambda_{11}^{-5/2}\Pi^{(.)(.)(.)}\right)h^2\right)\varphi\left(u_\alpha+\Lambda_{11}^{-1/2}h\right)$$

For the linear model

$$y_t = x_t^T \xi + \sigma(Z_t - 1), \quad t = 1, 2, \ldots, n, \ldots$$

with Z_t as discribed above the second order asymptotic deficiency $\Delta_2(\alpha, h)$ would vanish if

$$\frac{1}{n}\sum_{t=1}^{n} x_{ti_1} = \frac{1}{n}\sum_{t=1}^{n} x_{ti_1} x_{ti_2} x_{ti_3} = 0 \tag{15}$$

for all triples $i_1, i_2, i_3 = 1, \ldots, p$.
Condition (15) refers to the design only. (15) is fulfilled e. g. for a special D-optimal design over the cube $\{x \in R^p \mid |x_i| \leq 1, \ i = 1, \ldots, p\}$ if the asymptotics are considered for the sequence $n = m2^p$, $m = 1, 2, \ldots$
Let us introduce the commonly used 2^p xp matrix

$$X_0 = \begin{pmatrix} +1 +1 & \begin{matrix}+1\\-1\end{matrix} \\ +1 & \begin{matrix}+1\\-1\end{matrix} \\ \cdots\cdots & \\ -1 & \begin{matrix}+1\\-1\end{matrix} \\ & \\ +1 -1 & \begin{matrix}+1\\-1\end{matrix} \\ -1 +1 & \\ +1 & \vdots \\ -1 & \\ \vdots & \\ -1 -1 & -1 \end{pmatrix}$$

Then the design matrix

$$X = \begin{pmatrix} x_0 \\ \vdots \\ x_0 \end{pmatrix} \qquad x_0 \text{ occurs m-times}$$

fulfils (15).
Another commonly used approach to robustness is to analyse the asymptotic behaviour of the test (12) for contaminated families of distributions.
Let us consider a δ-contaminated distribution around the normal distribution for u_1, namely

$$(1-\delta)\varphi(u)+\delta q(u)$$

$$\text{for } \delta = \delta(n) \in (0,1)$$

where q is a fixed density fulfilling certain regularity conditions.
A natural assumption is

$$\delta \xrightarrow[n\to\infty]{} 0$$

what entails first order asymptotic efficiency of the test (12). In what follows we give the second order asymptotic deficiency.
Let

$$r_t(y_t, \vartheta) = \frac{1}{\sigma} q\left(\frac{y_t - g(x_t, \xi)}{\sigma}\right)$$

be the density of y_t under q and

$$A(\vartheta) := \left(A_{ij}(\vartheta)\right)_{i,j=1\ldots p+1}$$

$$:= \left(\frac{1}{n}\sum_{t=1}^{n} E_{r_t}\left(l_\varphi^{(i)}(y_t,\vartheta)\left(l_q^{(j)}(y_t,\vartheta)-l_\varphi^{(j)}(y_t,\vartheta)\right)\right.\right.$$

$$\left.\left.+\, l_\varphi^{(j)}(y_t,\vartheta)\left(l_q^{(i)}(y_t,\vartheta)-l_\varphi^{(i)}(y_t,\vartheta)\right)\right)\right)_{i,j}$$

(E_{r_j} denotes the expectation computed under the density r for y_t; l_φ, l_q denote the log-likelihoods under φ and q respectively.)
Finally we introduce

$$B(\vartheta) = \Lambda_\varphi(\vartheta) A(\vartheta) \Lambda_\varphi(\vartheta)$$

Then for $\delta = 0(n^{-1/2})$ the second order asymptotic deficiency is given by

$$\Delta_2(\alpha,h) = n^{1/2}\delta\frac{1}{2}B_{11}(\vartheta)\Lambda_{\varphi 11}^{-3/2}h_\varphi\left(u_\alpha+\Lambda_{\varphi 11}^{-1/2}h\right)$$

If $\delta = o(n^{-1/2})$ we obtain $\Delta_2(\alpha, h) = o(1)$ such that ψ_n defined by (12) is second order asymptotically efficient. Summerizing, ψ_n from (12) is first order asymptotically robust if $\delta = o(1)$ and is second order asymptotically robust if $\delta = o(n^{-1/2})$.

4.2. *Tests on* $\gamma = \sigma^2$

Obviously, we assume $\vartheta = \begin{pmatrix}\sigma^2\\ \xi\end{pmatrix}$ from now on in order to be consistent with our denotations. The testing problem under consideration is

$$H: \sigma^2 \leq \sigma_0^2$$

against $K: \sigma^2 > \sigma_0^2$

4.2.1. $\mu_i = E_\vartheta \varepsilon_1^i$ known for $i = 3, 4, 6$

The test

$$\psi_n = \begin{cases} 1, & \text{if } \left(\mu_4-\sigma_0^4\right)^{-1/2}\sqrt{n}\left(\hat{\sigma}_n^2-\sigma_0^2\right) \geq u_{1-\alpha} - \frac{1}{\sqrt{n}} M\left(u_{1-\alpha}, \bar{\vartheta}_n\right) \\ 0, & \text{else} \end{cases} \tag{16}$$

with

$$M(u,\vartheta) = m_0(\vartheta) + m_1(\vartheta)u^2,$$

$$m_0(\vartheta) = \left(\mu_4-\sigma^4\right)^{-3/2}\left(\frac{1}{6}\mathfrak{R} - \frac{\mu_3^2}{\sigma^2}\sum_{i_1,i_2=1}^{p}\Lambda_{\varphi\, i_1 i_2}\Pi^{(i_1)}\Pi^{(i_2)} + \left(\mu_4-\sigma^4\right)\sigma^2 p\right),$$

$$m_1(\vartheta) = \left(\mu_4-\sigma^4\right)^{-3/2}\left(-\frac{1}{6}\mathfrak{R} + \frac{\mu_3^2}{\sigma^2}\sum_{i_1,i_2=}^{p}\Lambda_{\varphi\, i_1 i_2}\Pi^{(i_1)}\Pi^{(i_2)}\right),$$

$$\mathfrak{R} = E_\vartheta\left(\varepsilon_1^2-\sigma^2\right)^3,$$

$$\bar{\vartheta}_n = \begin{pmatrix}\sigma_0^2\\ \hat{\xi}_n\end{pmatrix}$$

defines a sequence $\{\psi_n\}$ belonging to $\Psi_{\alpha,1/2}$.
If u_1 possesses the density

$$p_c(u) = \frac{c^c}{\Gamma(c)}|u|^{2c-1}e^{-cu^2} \quad \text{for some } c > 0$$

we have

$$M(u,\vartheta) = \frac{1}{3\sqrt{c}} + p\sqrt{c} - \frac{1}{3\sqrt{c}} u^2$$

and

$\Delta_2(\alpha, h) \equiv 0$ (see W. H. Schmidt, S. Zwanzig (1983)).

Notice that $\widehat{\sigma^2}_n$ is an MLE under $p_{12}(u) = \varphi(u)$ only. Thus $\{\psi_n\}$ is second order asymptotically robust under all densities $p_c(u)$ with $c > 0$.

4.2.2. μ_3, μ_4 or μ_6 unknown

Again useful estimators for μ_i are

$$\hat{\mu}_{in} = \frac{1}{n}\sum_{t=1}^{n}\left(y_t - g\left(x_t, \hat{\xi}_n\right)\right)^i, \quad i = 2, \ldots, 6$$

and a sequence $\{\psi_n\} \in \Psi_{\alpha,1\,2}$ is defined by

$$\psi_n = \begin{cases} 1, & \text{if } \left(\hat{\mu}_{4n} - \sigma_0^4\right)^{-1/2}\sqrt{n}\left(\hat{\sigma}_n^2 - \sigma_0^2\right) \geq u_{1-\alpha} - \frac{1}{\sqrt{n}} M\left(u_{1-\alpha}\right) \\ 0, & \text{else} \end{cases} \tag{17}$$

with

$M(u) = m_0 + m_1 u^2$

$$m_0 = \left(\hat{\mu}_{4,n} - \sigma_0^4\right)^{-3/2}\left(\frac{1}{6}\hat{\mathfrak{R}} + \sigma_0^2 p\left(\hat{\mu}_{4,n} - \sigma_0^4\right) - \hat{\mu}^2{}_n \sum_{i_1, i_2=1}^{p} \Pi^{(i_1)}\Pi^{(i_2)}\sigma_0^{-2}\Lambda_{\varphi i_1 i_2}\right)$$

$$m_1 = \left(\hat{\mu}_4 - \sigma_0^4\right)^{-3/2}\left(-\frac{1}{6}\hat{\mathfrak{R}} + \frac{1}{2}\left(\hat{\mu}_{6n} - \hat{\mu}_{4n}\sigma_0^2\right) - \hat{\mu}^2_{3n} \sum_{i_1, i_2=1}^{p} \Pi^{(i_1)}\Pi^{(i_2)}\sigma_0^{-2}\Lambda_{\varphi i_1 i_2}\right)$$

and $\hat{\mathfrak{R}} = \hat{\mu}_{6n} - 3\hat{\mu}_{4n}\sigma_0^2 + 2\sigma_0^6$

The asymptotic power of this test results in

$$E_{\vartheta_n}\psi_n = \Phi\left(u_\alpha + \left(\mu_4 - \sigma_0^4\right)^{-1/2} h\right) + \frac{1}{\sqrt{n}}\varphi\left(u_\alpha + \left(\mu_4 - \sigma_0^4\right)^{-1/2} h\right) H\left(u_\alpha, h\right) + o\left(n^{-1/2}\right)$$

with

$$H(u,h) = \left(\mu_4 - \sigma_0^4\right)^{-2}\left[\sigma_0^2\left(\mu_4 - \sigma_0^4\right) - \frac{1}{2}\left(\mu_6 - \sigma_0^2\mu_4\right) + \frac{1}{3}\mathfrak{R}\right] h u + \left(\mu_4 - \sigma_0^4\right)^{-5/2}\left[\frac{1}{6}\mathfrak{R} - \sigma_0^{-2}\left(\mu_4 - \sigma_0^4\right)^2 - \frac{\mu_3^2}{\sigma_0^2}\sum_{i_1, i_2=}^{p}\Lambda_{\varphi i_1 i_2}\Pi^{(i_1)}\Pi^{(i_2)}\right] h^2 \tag{18}$$

uniformly in $\vartheta \in \mathfrak{K} \subset \Theta_0 = \left\{\begin{pmatrix}\sigma_0^2 \\ \xi\end{pmatrix} \middle| \xi \in \Xi\right\}$.

From (18) and (7) several conclusions concerning the asymptotic deficiencies could be drawn.

5. References

PFANZAGL, J.
Asymptotic expansions related to minimum contrast estimators.
Ann. of Statist. 1 (1973), 993–1026.

PFANZAGL, J.
Asymptotically optimum estimation and test procedures.
Proc. Prague Conf. on Asymptotic Methods of Statistics 1 (1973), 201–272.

SCHMIDT, W. H., ZWANZIG, S.
Second order asymptotics in nonlinear regression.
Preprint 04/83, Akademie der Wissenschaften der DDR, Institut für Mathematik (1983).

Systems Research Institute, Polish Academy of Sciences, Warsaw, Poland

The Bootstrap in Nonlinear Regression

PIOTR STANIEWSKI

Abstract

Nonlinear regression models are considered. It is shown that under general assumptions the bootstrap approximation to the distribution of the least squares estimates is asymptotically valid. Both cases of finite and infinite error variances are investigated.

1. Introduction

This paper will develop some asymptotic theory for application of the Efron (1979) bootstrap to nonlinear regression. It was shown that under standard conditions the bootstrap approximation to least squares estimates is valid. These results are a generalization of those of Freedman (1981), who considered only linear models.

A question arises as to which conditions imposed on nonlinear models should be regarded as standard ones. In this paper the conditions of Jennrich (1969) are adopted, which are not very restrictive, if we adopt the assumptions of Malinvaud (1970), Bierens (1981) or Wu (1981), the validity of the bootstrap might be proven along the same lines.

In section 2 the assumptions for nonlinear regression asymptotic are briefly presented. Section 3 gives the bootstrap algorithm for nonlinear regression models. Section 4 deals with consistency and Section 5 with asymptotic normality of the bootstrap approximation to least squares estimates for the cases of both finite and infinite error variances.

2. Nonlinear Regression Models

The following models will be considered.

$$y_i = g(x_i, \vartheta_0) + e_i \quad i = 1, 2, 3, \dots \tag{2.1}$$

where (x_i) is a sequence of $k \times 1$ data vectors, the e_i's are random errors, y_i's are real-valued responses, ϑ_0 is a $p \times 1$ parameter vector and $g(x, \vartheta)$ is a given function on $R^k \times R^p$.

Any vector $\hat{\vartheta}_n$ which minimizes

$$Q_n(\vartheta) = n^{-1} \sum_{j=1}^{n} \left(y_j - g(x_j, \vartheta)\right)^2 \tag{2.2}$$

will be called a least squares estimate of ϑ_0 based on the first n values of y_i and x_i.

To obtain the consistency of the sequence $(\hat{\vartheta}_n)$ we impose the following conditions on the model (2.1):

A1: The function $g(x, \vartheta)$ is continuous on $R^k \times H$, where H is a compact subset of R^p.

B1: The errors e_i, $i = 1, 2, \dots$ are independent, identically distributed random variables with zero mean and finite variance $\sigma^2 > 0$.

C1: For any given data sequence (x_n) of $k \times 1$ vectors, the sequence of functions on $H \times H$

$$\left(n^{-1} \sum_{j=1}^{n} g(x_j, \vartheta^1) g(x_j, \vartheta^2)\right)_{n=1,2,3,\dots} \tag{2.3}$$

has a limit and its convergence is uniform in ϑ^1, ϑ^2.

D1: The function

$$Q(\vartheta) = \lim_{n\to\infty} n^{-1} \sum_{j=1}^{n} \left[g(x_j, \vartheta) - g(x_j, \vartheta_0)\right]^2 \tag{2.4}$$

is equal to 0 only at $\vartheta = \vartheta_0$.

Jennrich (1969) has proven that under assumptions A1, B1, C1 and D1, $\hat{\vartheta}_n$ and $\hat{\sigma}_n^2 = Q_n(\hat{\vartheta}_n)$ are strongly consistent estimators of ϑ_0 and σ^2 respectively.

To ensure the asymptotic normality stronger assumptions are required. The assumptions A1 and C1 should be replaced by:

A2: The function $g(x, \vartheta)$ belongs to $C^2(R^p \times H)$, where H is a convex, compact subset of R^p

C2: For any given data sequence (x_n), the sequence of functions on $H \times H$:

$$\left(n^{-1} \sum_{j=1}^{n} f(x_j, \vartheta^1) h(x_j, \vartheta^2)\right)_{n=1,2,\dots} \tag{2.5}$$

has a limit and its convergence is uniform. In the formula (2.5)

$$f, h = g, H(\partial g/\partial\vartheta_k), (\partial/\partial\vartheta_k)\cdot(\partial/\partial\vartheta_l)\, g; \quad k, l = 1, 2, \dots, p$$

Moreover, the following assumptions should be added:

E1: ϑ_0 is an interior point of H.

F1: The matrix $A = \lim_{n\to\infty} A_n$ is nonsingular. An element $a_{nij}(\vartheta)$ of the matrix A_n is given by the formula:

$$a_{nij}(\vartheta) = n^{-1} \sum_{k=1}^{n} \left[\partial g(x_k, \vartheta)/\partial\vartheta_i\right]\left[\partial g(x_k, \vartheta)/\partial\vartheta_j\right]. \tag{2.6}$$

Jennrich (1969) has shown that when the assumptions A2, B1, C2, D1, E1 and F1 are satisfied, the distribution of $n^{1/2}(\hat{\vartheta}_n - \vartheta_0)$ tends to a p-variate normal distribution $N_p(0, \sigma^2 A(\vartheta_0)^{-1})$ and $A_n(\hat{\vartheta}_n)$ is a strongly consistent estimator of $A(\vartheta_0)$.

We will also consider the case when the assumption B1 is weakened. We take

B2: The errors e_i $i = 1, 2, \dots$ are independent, identically distributed random variables, with zero mean.

For proving weak consistency under B2 observe that the least squares estimator can also be obtained from

$$Q_n^0(\hat{\vartheta}_n) = \inf_{\vartheta\in H} Q_n^0(\vartheta) \tag{2.7}$$

where

$$Q_n^0(\vartheta) = Q_n(\vartheta) - n^{-1} \sum_{j=1}^{n} e_j^2 \tag{2.8}$$

For this case we will need some lemma.

Lemma 2.1.

Let (x_n) be a sequence of real numbers. Let $\limsup_{n\to\infty} n^{-1} \sum_{j=1}^{n} x_j^2 < \infty$. Let (y_{jn}) be a doubly indexed sequence of real numbers that the expression $n y_{jn}/x_i$ tends to zero if only x_j/n tends to zero. Then

$$\lim_{n\to\infty}\prod_{j=1}^{n}(1+y_{jn})=1.$$

Proof:

From lemma's assumption and the inequality

$$\sum_{j=1}^{n}|x_j|<\left(n\sum_{j=1}^{n}x_j^2\right)^{1/2}$$

it follows that there exists some $\delta>0$ that $\sum_{j=1}^{n}\delta\,|x_j|<n$. For arbitrary $0<\varepsilon<1$ put $\mu=\varepsilon\,\delta$. Consider the sequence

$$\prod_{j=1}^{n}\left(1+\mu|x_j|/n\right)_{n=1,2,\ldots}$$

Straightforward computations show that the limit of this sequence is majorized by the sum $\sum_{j=0}^{\infty}\varepsilon^j=(1-\varepsilon)^{-1}$.

It is clear that x_j/n tends to 0 as j, n tend to ∞. From lemma's assumption it follows that for large n and j, say $n, j>s$, we have $|y_{jn}|<\mu\,|x_j|/n$. Thus

$$\prod_{j=1}^{n}\left(1-\mu|x_j|/n\right)\le\prod_{j=1}^{n}(1+y_{jn})\le\prod_{j=1}^{n}\left(1+\mu|x_j|/n\right).$$

We will consider only the right inequality. The left inequality may be treated analoguously. Let $n, j>s$

$$\prod_{j=1}^{n}(1+y_{jn})=\prod_{j=1}^{s}(1+y_{jn})\prod_{j=s+1}^{s+n}(1+y_{in})\le$$

$$\le\prod_{j=1}^{s}(1+y_{jn})\prod_{j=s+1}^{s+n}\left(1+\mu|x_j|/n\right)\le(1-\varepsilon)^{-1}\prod_{j=1}^{s}(1+y_{jn})$$

As the ε is arbitrary and s is finite, the last term on the right may be made arbitrarily close to one. The thesis of the lemma follows immediately.

Lemma 2.2.

Let (x_n) be a sequence of real numbers as in Lemma 2.1. Let the assumption B2 be satisfied. Then $n^{-1}\sum_{j=1}^{n}n_j\,x_j$ tends to 0 in probability as n tends to ∞.

Proof.

As the errors e_j have zero mean, the characteristic function of $n^{-1}\sum_{j=1}^{n}e_j\,x_j$ may be represented as $\Phi_n(t)=$ $=\prod_{j=1}^{n}[1+o(x_jt/n)]$. It follows from Lemma 2.1 that for any given t $\Phi_n(t)$ tend to one, i. e. to the characteristic function of the random variable with a one-point distribution concentrated in 0. The lemma follows.

Lemma 2.3.

When the assumptions A1 and C1 are satisfied, then the sequences $\left(n^{-1}\sum_{j=1}^{n}\sup_{\theta\in H}g^2(x_j,\vartheta)\right)_{n=1,2,\ldots}$ and

$\left(n^{-1}\sum_{j=1}^{n}\inf_{\Theta\in H}g^2(x_j,\Theta)\right)_{n=1,2,\ldots}$ are bounded.

Proof.

From A1 it follows that $g(x,\vartheta)$ is uniformly continuous on compact subsets of $R^k\times H$. Thus for every ε and $\vartheta^*\in H$ we may choose a neighbourhood Ω of ϑ^*, that for each j we have $\left|\sup_{\vartheta\in\Omega}g^2(x_j,\vartheta)-g^2(x_j,\vartheta^*)\right|\le\varepsilon$

From this it follows that

$$n^{-1}\sum_{j=1}^{n}\left[\sup_{\vartheta\in\Omega}g^2(x_j,\vartheta)-g^2(x_j,\vartheta^*)\right]\le\varepsilon$$

so, the sequence $\left(n^{-1}\sum_{j=1}^{n}\sup_{\vartheta\in\Omega}g^2(x_j,\vartheta)\right)_{n=1,2,\ldots}$ is bounded.

We may choose for every ε such set of ϑ^*'s that their neighbourhoods cover H. Then we may choose a finite collection of such neighbourhoods, say $\Omega_1,\ldots,\Omega_M$. For each neighbourhood Ω_i we have

have $\limsup_{n\to\infty}n^{-1}\sum_{j=1}^{n}\sup_{\vartheta\in\Omega}g^2(x_j,\vartheta)<\infty$. Thus

$$\limsup_{n\to\infty}n^{-1}\sum_{j=1}^{n}\sup_{\vartheta\in H}g^2(x_j,\vartheta)\le$$

$$\le\limsup_{n\to\infty}\sum_{i=1}^{M}n^{-1}\sum_{j=1}^{n}\sup_{\vartheta\in\Omega_i}g^2(x_j,\vartheta)<\infty$$

From Lemmas 2.3, 2.2 and the three sequences theorem we have the following lemma.

Lemma 2.4.

Let A1, B2 and C1 be satisfied. We have

$$\operatorname{p\,lim}_{n\to\infty}n^{-1}\sum_{j=1}^{n}e_j\,g(x_j,\vartheta)=0$$

and the convergence is uniform in ϑ.

Taking into consideration Lemma 2.4 and applying reasoning like in the proof of Theorem 6 of Jennrich (1969), we obtain the following theorem:

Theorem 2.1.

Let the model (2.1) with A1, B2, C1 and D1 be given. Then

$\operatorname{p\,lim}_{n\to\infty}\hat{\vartheta}_n=\vartheta_0.$

Bierens (1981) investigated symmetric stable distributions, with characteristic functions of the type

$\Phi(t)=\exp(-\sigma|t|^\alpha)$, $\sigma>0$, $\alpha\in(1,2)$.

Similarly to this Theorem 3.1.6, page 68, we may derive:

Theorem 2.2.

Let the model (2.1) with A2, B2, C2, D1, E1 and F1 be given. If in addition the error distribution is symmetric stable with characteristic exponent $\alpha\in(0,1)$ and scaling parameter $\sigma>0$, then $n^{1-1/\alpha}\left(\hat{\vartheta}_n-\vartheta_0\right)$ converges in distribution to certain nonnormal limit distribution.

3. Bootstrapping

The bootstrap algorithm is the following. After computing $\hat{\vartheta}_n$, the observable n-vector of residuals $\tilde{e}$ is given by $\tilde{e}_i=y_i-g\left(x_i,\hat{\vartheta}_n\right)$ $i=1,2,\ldots,n$, Let us denote $s_n=n^{-1}\sum_{j=1}^{n}\tilde{e}_j$. Let $\hat{F}_n$ be the empirical distribution of e, centered at the mean, so $\hat{F}_n$ puts mass 1/n at $\hat{e}_i=\tilde{e}_i-s_n$ and $\int x\,d\hat{F}_n=0$. Given $y_1,\ldots,y_n$, let $w_{1n},\ldots,w_{mn}$ be conditionally independent with common distribution $\hat{F}_n$. Let

$$y_{in}=g\left(x_i,\hat{\vartheta}_n\right)+w_{in}\quad i=1,2,\ldots,m. \tag{3.1}$$

Errors w_{in} are obtained by resampling the centered residuals. New responses $y_{in},\ldots,y_{mn}$ are generated from the data using the regression model with $\hat{\vartheta}_n$ as the vector of parameters and $\hat{F}_n$ as the distribution of the disturbance terms $w_{1n}, \ldots, w_{mn}$. The bootstrap least squares estimate $\hat{\vartheta}_{mn}$ minimizes the function:

$$L_{mn}(\vartheta) = m^{-1}\sum_{j=1}^{m}\left[y_{jn} - g(x_j, \vartheta)\right]^2 =$$

$$= m^{-1}\sum_{j=1}^{m} w_{jn}^2 + m^{-1}\sum_{j=1}^{m}\left[g(x_j, \hat{\vartheta}_n) - g(x_j, \vartheta)\right]^2 +$$

$$- 2m^{-1}\sum_{j=1}^{m} w_{jn}\left[g(x_j, \hat{\vartheta}_n) - g(x_j, \vartheta)\right] \tag{3.2}$$

It follows immediately from Lemma 2 of Jennrich (1969) that $\hat{\vartheta}_{mn}$ can be chosen a random variable.
Now, when the error variance is infinite, we obtain the bootstrap least squares estimate $\hat{\vartheta}_{mn}$ minimizing the function

$$L^{o}_{mn}(\vartheta) = L_{mn}(\vartheta) - n^{-1}\sum_{j=1}^{m} w_{jm}^2 =$$

$$= m^{-1}\sum_{j=1}^{m}\left[g(x_j, \hat{\vartheta}_n) - g(x_j, \vartheta)\right]^2 +$$

$$- 2m^{-1}\sum_{j=1}^{m} w_{jn}\left[g(x_j, \hat{\vartheta}_n) - g(x_j, \vartheta)\right]. \tag{3.3}$$

The bootstrap principle is that the distribution of $n^{1/2}(\hat{\vartheta}_{mn} - \hat{\vartheta}_n)$, which can be computed directly from the data, approximates the distribution of $n^{1/2}(\hat{\vartheta}_n - \vartheta_o)$. This approximation, as is shown in section 5, is likely to be good.

4. Consistency

Using Mallows metric technique (cf. Bickel and Freedman (1981), Freedman (1981)) the following theorems may be proven.

Theorem 4.1.

Let the model (2.1) with A1, B1, C1 and D1 be given. Let $(\hat{\vartheta}_{mn})$ be a sequence of bootstrap least squares estimators. Along almost all sample sequences, given $y_1, \ldots, y_n$, as m and n tend to ∞, $\hat{\vartheta}_{mn}$ and $\sigma^2_{mn} = |L(\hat{\vartheta}_{mn})|$ tend in conditional probability to ϑ_0 and σ^2 respectively.
And for the infinite variance case:

Theorem 4.2.

Let the model 2.1 with A1, B2, C1 and D1 be given. Let $\hat{\vartheta}_{mn}$ be a sequence of bootstrap least squares estimators. Along almost all sample sequences, given $y_1, \ldots, y_n$ as m and n tend to ∞, $\hat{\vartheta}_{mn}$ tend in conditional probability to $\hat{\vartheta}_0$.
The exact proofs of these theorem may be found in Staniewski (1983).

5. Asymptotic Normality

In lemmas of section 5 we will use the concept of "convergence in conditional probability" or "convergence of conditional distribution" of a doubly indexed sequence of certain random vectors. In all these lemmas the "convergence along almost all sample sequences, given $y_1, \ldots y_n$, as m and n tend to ∞" is meant.

We will use the Taylor expansion of the vector of first derivatives of the objective function

$$(\partial/\partial\vartheta)L_{mn}(\hat{\vartheta}_{mn}) = (\partial/\partial\vartheta)L_{mn}(\hat{\vartheta}_n) + (\hat{\vartheta}_{mn} - \hat{\vartheta}_n)^T(\partial/\partial\vartheta)(\partial/\partial\vartheta^T)L_{mn}(\vartheta^*_{mn}) \tag{5.1}$$

where ϑ^*_{mn} is some mean value satisfying

$$\left|\vartheta^*_{mn} - \hat{\vartheta}_n\right| < \left|\hat{\vartheta}_{mn} - \hat{\vartheta}_n\right| \tag{5.2}$$

By $(\partial/\partial\vartheta)f(\vartheta)$ we denote the vector of partial derivates of $f(\vartheta)$ and by $(\partial/\partial\vartheta)(\partial/\partial\vartheta^T)f(\vartheta)$ the matrix $((\partial/\partial\vartheta_i)(\partial/\partial\vartheta_j)f(\vartheta))$.
From Lemma 3 of Jennrich (1969) it follows that under assumption A2, ϑ^*_{mn} is a random vector.
While considering the asymptotic normality of $m^{1/2}(\hat{\vartheta}_{mn} - \hat{\vartheta}_n)$ the following lemmas will be useful.

Lemma 5.1.

Assume A1, B1, C1, D1 and F1 for, the model (2.1). Let (z_n) be a sequence in R^p. Denote $z_n = (z_n^1, \ldots, z_n^p)$.
Assume that for each i, j = 1, 2, ..., p $b_{ij} = \lim_{n\to\infty} n^{-1}\sum_{k=1}^{n} z_k^i z_k^j$ exists. Then the conditional distribution of the random vector $m^{1/2}\sum_{j=1}^{m} w_j\, z_j$ converges weakly to the p-variate normal distribution $N_p(0, \sigma^2 B)$ where $B = (b_{ij})$.

Lemma 5.2.

Assume A2, B1, C1, D1 and F1 for the model (2.1). For all real doubly indexed sequences (b_{mn}), the random vector $b_{mn}(\partial/\partial\vartheta)L_{mn}(\hat{\vartheta}_{mn})$ converges in conditional probability to zero.

Lemma 5.3.

Assume A2, B1, C2 and D1 for the model (2.1). The matrix $(\partial/\partial\vartheta)(\partial/\partial\vartheta^T)L_{mn}(\vartheta^*_{mn})$ converges in conditional probability to the nonrandom matrix $2A(\vartheta_0)$, where the matrix $A(\vartheta)$ is defined as in assumption F1.
The proofs of these lemmas may be found in Staniewski (1983).

Lemma 5.4.

Assume A2, B1, C2, D1 and F1 for the model (2.1). The distribution of the random vector

$$0{,}5m^{1/2}(\partial/\partial\vartheta)L_{mn}(\hat{\vartheta}_n) = m^{-1/2}\sum_{j=1}^{m} w_{jn}(\partial/\partial\vartheta)g(x_j, \hat{\vartheta}_n) \tag{5.4}$$

converges weakly to the p-variate normal distribution $N_p(0, \sigma^2 A(\vartheta_0))$ where matrix A is defined as in assumption F1.

Proof.

From assumptions B1 or B2 and Tchebyshev inequality it follows that for almost every sequence (y_n), the sequence (e_n) is bounded. From Lemma 5.1 it follows that distribution of the vector $m^{1/2}\sum_{j=1}^{m} w_{jn}(\partial/\partial\vartheta)g(x_j, \vartheta_0)$ tends to $N_p(0, \sigma^2 A(\vartheta_0))$ as m and n tend to ∞. Consider arbitrary, say 1-th coordinate of this vector and the vector (5.4) and denote

$$q_{jn} = (\partial/\partial\vartheta_1)q(x_j, \hat{\vartheta}_n) - (\partial/\partial\vartheta_1)q(x_j, \vartheta_0) \tag{5.5}$$

It suffices to prove that

$$m^{-1/2}\sum_{j=1}^{m} w_{jn} q_{jn} \qquad (5.6)$$

converges to a one-point distribution, concentrated in 0. Conditional characteristic function of (5.6) is equal to:

$$\Phi_{mn}(t)=\prod_{j=1}^{m} n^{-1/2}\sum_{k=1}^{n}\exp\left(m^{-1/2} q_{jn}\hat{e}_k it\right) \qquad (5.7)$$

Let t_0 be arbitrary and consider an arbitrary convergent subsequence $\Phi_{m(j)\,n(j)}(t_0)$. It follows from the boundedness of $(\hat{e}_k)$ that for every m(j) we have

$$\lim_{n(j)\to\infty} \Phi_{m(j)n(j)}(t_0)=1. \text{ Thus}$$

$$\lim_{m(j)n(j)\to\infty} \Phi_{m(j)n(j)}(t_0)=$$

$$=\lim_{m(j)\to\infty}\ \lim_{n(j)\to\infty} \Phi_{m(j)n(j)}(t_0)=1$$

Because t_0 and the limit point were arbitrary, it follows that the limit function of $\Phi_{mn}(t)$ exists and is equal to one for every t. Such limit corresponds to the one point distribution concentrated in 0.

From above lemmas, formula 5.1 and Theorem 4.1 follows easily the

Theorem 5.1.

Assume A2, B1, C2, D1, E1 and F1 for the model (2.1). Along almost all sample sequences, given $y_1,\dots,y_n$, as m and n tend to ∞:

a) the conditional distribution of $m^{1/2}\left(\hat{\vartheta}_{mn}-\hat{\vartheta}_n\right)$ converges weakly to a normal one with mean 0 and variance — convergence matrix σ^2A

b) the matrix

$$L_{mn}(\hat{\vartheta}_{mn})\left[(\partial/\partial\vartheta)(\partial/\partial\vartheta^T)L_{mn}(\hat{\vartheta}_{mn})\right]^{-1}$$

converges in conditional probability to $\sigma^2A(\vartheta_0)^{-1}$.

When we deal with the infinite variance case we have the following theorem.

Theorem 5.2.

Let the assumptions of the Theorem 2.1 be given. The limiting conditional distribution of $m^{1-1/\alpha}\left(\hat{\vartheta}_{mn}-\hat{\vartheta}_n\right)$ is the same as the limiting distribution of $n^{1-1/\alpha}\left(\hat{\vartheta}_n-\vartheta_0\right)$. This theorem may be easily proven by the technique used in the proof of Lemma 5.4.

6. References

BICKEL, P. and FREEDMAN, D.
Some asymptotic theory for the bootstrap.
Ann. Statist. 9 (1981) 1196–1217.

BIERENS, H.
Robust Methods and Asymptotic Theory in Nonlinear Econometric.
Springer, Berlin, 1981.

EFRON, B.
Bootstrap methods: another look at the jacknife.
Ann. Statist. 7 (1979) 1–26.

FREEDMAN, D.
Bootstrapping regression models.
Ann. Statist. 9 (1981) 1218–1228.

JENNRICH, R.
Asymptotic properties of non-linear least squares estimators.
Ann. Math. Statist. 40 (1969) 633–643.

MALINVAUD, E.
The consistence of nonlinear regressions.
Ann. Math. Statist. 41 (1970) 956–969.

STANIEWSKI, P.
Nonlinear static models, asymptotics and bootstrap (in Polish).
Research report ZPZC 54/40/83 of the Systems Research Institute of the Polish Acad. of Sci. (1983).

WU, C.
Asymptotic theory of nonlinear least squares estimation.
Ann. Statist. 9 (1981) 501–513.

Computational Centre, Humboldt University, Berlin, GDR

Sharp Inequalities for Error Probabilities of Tests under Moment Conditions

REINHARD STRÜBY

Abstract

We discuss the method of moment spaces to calculate the extreme values of integrals. The maximization (minimization) is done with respect to measures restricted by some moment conditions. Inequalities for error probabilities will give some knowledge on robustness properties of tests.

1. Introduction

We want to consider a simple parameter hypothesis and a test one would prefer in this situation. For example let $X_1, \dots, X_{n+1}$ be independently distributed random variables. Assuming identical distributions

$$P_\vartheta^{X_1} = \dots = P_\vartheta^{X_{n+1}} = N(\vartheta, 1)$$

the u-test is a most powerful α-test for testing $H : \vartheta = \vartheta_0 = 0$ against $K : \vartheta = \vartheta_1,\ \vartheta_1 > \vartheta_0$.

If we want to obtain some knowledge on the power of this test in nonnormal situations we have to ask for sharp inequalities for the error probability. Especially, one is interested to derive such inequalities if the probability distribution of some or all variables has to satisfy some moment conditions. As a consequence of some fundamental theorems due to Markov such inequalities may be obtained with the help of the calculus of extrema in a general moment space.

2. One Nonnormal Distribution

Let $\alpha\left(P_{\vartheta_0}^{X_{n+1}}\right)$ the error probability of the u-test for $P_\vartheta^{X_1} = \dots = P_\vartheta^{X_n} = N(\vartheta, 1)$, but $P_\vartheta^{X_{n+1}}$ nonnormal with expectation ϑ. If H is true, it holds

$$\alpha\left(P_{\vartheta_0}^{X_{n+1}}\right) = \int_{-\infty}^{+\infty} \left(1 - \Phi\left(n^{-1/2}\left((n+1)^{1/2} u_\alpha - x\right)\right)\right) P_{\vartheta_0}^{X_{n+1}}(dx),$$

where Φ is the standard normal distribution function and u_α is the α-fractile of Φ.

Calculating $\sup \alpha\left(P_{\vartheta_0}^{X_{n+1}}\right)$ over the class of all probability measures with a given power moment vector we obtain a problem of general moment spaces. Because we have to consider integrals over infinite intervals the problem is not to handle with the purely classical method of moment spaces.

3. The Method

Let X be a random variable with the induced Borel σ-algebra $\mathfrak{B}^1$. Continuous and linear independent functions $u_i : R^1 \to R^1$, $i = 0, 1, \dots, k$, are given.

We assume that the probability measure P^X of X is an element of the set $\mathfrak{P}_\gamma$ of all probability measures P with generalized moments

$$\gamma = E_P U \in R^{k+1}, \quad U = (u_0, \dots, u_k)^T,$$

$$E_P u_i = \int_{R^1} u_i(x)\, dP(x), \quad i = 0, 1, \dots, k.$$

One is interested in bounds for

$$\int_{R^1} f(x)\, dP(x), \quad f : R^1 \to R^1,$$

under $P \in \mathfrak{P}_\gamma$.

In order to consider integrals on a closed interval we use a continuous 1-1-transformation

$$h : (-1, +1) \to R^1.$$

Further more it is assumed that there exists a positive and continuous function

$$\omega : (-1, +1) \to R^1$$

so that for

$$\tilde{u}_i(t) = \frac{u_i(h(t))}{\omega(t)}, \quad t \in (-1, +1), \quad i = 0, 1, \dots, k$$

it holds

$$\lim_{t \to -1} \tilde{u}_i(t) = \lim_{t \to +1} \tilde{u}_i(t) = 0, \quad i = 0, 1, \dots, k-1,$$

$$\lim_{t \to -1} \tilde{u}_k(t) = \lim_{t \to +1} \tilde{u}_k(t) = 1.$$

Similarly

$$\tilde{f}(t) = \frac{f(h(t))}{\omega(t)}, \quad t \in (-1, +1),$$

and it is assumed that

$$-\infty < \lim_{t \to -1} \tilde{f}(t) = \lim_{t \to +1} \tilde{f}(t) < +\infty.$$

After continuously extending the functions $\tilde{u}_i$, $i = 0, 1, \dots, k$, and $\tilde{f}$ to the boundary of the interval $(-1, +1)$ we get the equivalent problem

$$\sup_{\tilde{P} \in \tilde{\mathfrak{P}}_\gamma} \int_{[-1,+1]} \tilde{f}(t)\, d\tilde{P}(t),$$

$$E_{\tilde{P}} \tilde{u}_i = \int_{[-1,+1]} \tilde{u}_i(t)\, d\tilde{P}(t), \quad i = 0, 1, \dots, k,$$

$$\tilde{U} = (\tilde{u}_0, \dots, \tilde{u}_k)^T,$$

$\tilde{\mathfrak{P}}_\gamma$ — the set of all measures with $E_{\tilde{P}} \tilde{U} = \gamma$.

The extension can be interpreted as a periodic one to R^1. So we can solve the problem with the help of theorems in periodic moment spaces.

Let $\prod$ be the set of generalized polynomials in $\tilde{u}_0, \tilde{u}_1, \ldots, \tilde{u}_k$ and $\prod_+ \subset \prod$ the set of nonnegative polynomials.

We define

$$\underline{\Pi}(\tilde{f}) = \{\pi \in \Pi \mid \pi(t) \leq \tilde{f}(t),\quad t \in [-1, +1]\},$$

$$\overline{\Pi}(\tilde{f}) = \{\pi \in \Pi \mid \pi(t) \geq \tilde{f}(t),\quad t \in [-1, +1]\}.$$

Definition.

For $\gamma \in R^{k+1}$ let c_γ be defined by: If $\pi = a^T \tilde{U}$, $a \in R^{k+1}$, then $c_\gamma(\pi) = a^T \gamma$.

Definition.

γ is strictly positive, if $c_\gamma(\pi) > 0$ for all $\pi \in \prod_+$, $\pi \neq 0$.

Assumptions.

A1. The functions $\tilde{u}_0, \tilde{u}_1, \ldots, \tilde{u}_k$ are continuous and linear independent.

A2. $\tilde{f}$ is continuous and nonnegative.

A3. $\prod$ contains a polynomial π_+ and $\pi_+(t) > 0$, $t \in [-1, +1]$.

A4. γ is strictly positive.

Then the following theorem holds (Krein, Nudel'man (1977), Karlin, Studden (1966)).

Theorem.

Let A1, ..., A4 fulfilled.

1. $$\min_{\tilde{P} \in \tilde{\mathfrak{P}}_\gamma} \int_{[-1,+1]} \tilde{f}(t)\, d\tilde{P}(t) = \max_{\pi \in \underline{\Pi}(\tilde{f})} c_\gamma(\pi),$$

$$\max_{\tilde{P} \in \tilde{\mathfrak{P}}_\gamma} \int_{[-1,+1]} \tilde{f}(t)\, d\tilde{P}(t) = \min_{\pi \in \overline{\Pi}(\tilde{f})} c_\gamma(\pi).$$

2. The integral attains its minimum (maximum) for $\tilde{P}_0 \in \tilde{\mathfrak{P}}_\gamma$ $(\tilde{P}_1 \in \tilde{\mathfrak{P}}_\gamma)$ if and only if there exist a $\pi_0 \in \underline{\Pi}(\tilde{f})$ $(\pi_1 \in \overline{\Pi}(\tilde{f})$ and

$$\tilde{f} = \pi_0 \left[\mathfrak{V}^1_{[-1,+1]}, \tilde{P}_0\right]$$

$$\left(\tilde{f} = \pi_1 \left[\mathfrak{V}^1_{[-1,+1]}, \tilde{P}_1\right]\right).$$

4. Calculation of Extrema

For $k = 4$, γ essentially built by the skewness and kurtosis, we have to calculate the maximum (minimum) of $c_\gamma(\pi)$ for polynomials interpolating $\tilde{f}$ at **three** points of $[-1, +1]$. The choice of these points is made with regard to the fundamental properties of canonical distributions in moment spaces (Krein, Nudel'man (1977)).

The calculation is possible by optimization methods as discribed in Tichatschke (1981).

Kemperman (1968) has considered the method of moment spaces in a very general way and has given a geometrical insight into the method.

Skibinsky (1976) has calculated the bounds for the probability of subintervals of a finite interval when 3 moments are given.

5. References

KARLIN, S. and STUDDEN, W. J.
Tchebycheff Systems.
With Applications in Analysis and Statistics.
Interscience, New York, 1966.

KEMPERMAN, J. H. B.
The general moment problem, a geometric approach.
Ann. Math. Statist. 39 (1968), 93–122.

KREIN, M. G. and NUDEL'MAN, A. A.
The Markov Moment Problem and Extremal Problems.
American Mathematical Society, Providence, Rhode Island, 1977.

SKIBINSKY, M.
Sharp upper bounds for probability on an interval when the first three moments are known.
Ann. Statist. 4 (1976), 187–213.

TICHATSCHKE, R.
Lineare semi-infinite Optimierungsaufgaben und ihre Anwendungen in der Approximationstheorie.
Wissenschaftliche Schriftenreihe der TH Karl-Marx-Stadt, 1981.

Academy of Agricultural Sciences of the GDR,
Research Centre of Animal Production Dummerstorf-Rostock

Simulation Studies on Robustness of the t- and u-Test against Truncation of the Normal Distribution

FRIEDRICH TEUSCHER

1. Introduction

The aim of this simulation-study was to determine the influence of the truncation of a normal-distributed random variable (in practice occuring e. g. by selection) on the power functions of three tests, the t-test, the u-test and a modified Johnson-test. Particularly we were interested in finding sample sizes ensuring a good behaviour of the tests. More precisely, we looked for an n* so that a test is "ε-robust" for $n > n^*$ (Herrendörfer, G., (ed.) (1980)).
The density function of a truncated normal distributed random variable is given by

$$f_s(x) = \begin{cases} 0 & x < u \\ \dfrac{\varphi(x)}{\Phi(v) - \Phi(u)} & u \leq x \leq v \\ 0 & v < x \end{cases} \quad \text{(see fig. 1)}.$$

Seven truncated normal distributions were generated (see table 1). You can see their skewness and kurtosis in fig. 2.

Description of the Tests

The null hypothesis is $H_0 : \mu = \mu_0$ and the alternative hypotheses H_A are

a) $\mu \neq \mu_0$
b) $\mu > \mu_0$
c) $\mu < \mu_0$

In this paper we regard the t-test, the u-test and a modifed Johnson-test (named the t_J-test) (see N. J. Johnson (1978)). Johnson's test was modified because unpublished results have revealed that it's asymptotic behaviour is worse then that of the t-test if we deal with a truncated normal distribution. We can show, that the original Johnson-test as well as the modified Johnson-test are tests, having first kind risks equal 1 in the cases b) and c), but nevertheless they may have a certain usefulness for a reasonable μ_0.
The t-test is formulated as follows:
H_0 is rejected in case a) if

$$\left| \frac{\bar{x} - \mu_0}{s} \sqrt{n} \right| > t\left(n-1; 1-\frac{\alpha}{2}\right),$$

in case b) if

$$\frac{\bar{x} - \mu_0}{s} \sqrt{n} > t(n-1; 1-\alpha)$$

and in case c) if

$$\frac{\bar{x} - \mu_0}{s} \sqrt{n} < -t(n-1; 1-\alpha)$$

The test statistic for the t_J-test is

$$t_J = \left(\bar{x} - \mu_0 + \frac{g_1}{6s^2 n} + \frac{g_1}{3s^4 \sqrt{n}} (\bar{x} - \mu_0)^2 \right) \frac{\sqrt{n}}{s}.$$

γ_1 was estimated by

$$g_1 = \frac{\frac{1}{n} \sum_{i=1}^{n} (x_i - \bar{x})^3}{\left(\frac{1}{n} \sum_{i=1}^{n} (x_i - \bar{x})^2 \right)^{3/2}}.$$

H_0 is rejected in case a)

$$\text{if} \quad |t_J| > t\left(n-1; 1-\frac{\alpha}{2}\right),$$

in case b)

$$\text{if} \quad t_J > t(n-1; 1-\alpha)$$

and in case c)

$$\text{if} \quad t_J < -t(n-1; 1-\alpha).$$

The u-test corresponds to the t-test for a known variance, i. e. H_0 will be rejected, say, in case a) if

$$\left| \frac{\bar{x} - \mu_0}{\sigma} \sqrt{n} \right| > u_{1-\frac{\alpha}{2}} \qquad \left(\text{with } \Phi\left(u_{1-\frac{\alpha}{2}}\right) = 1 - \frac{\alpha}{2} \right).$$

Some Results

The empirical first kind risks are written down in table 2. The marked empirical $\alpha \cdot 100$ show ε-robustness with $\varepsilon = \alpha \cdot 20\,\%$ where α is the nominal first kind risk).
We can find approximate values of n*, so that the tests are ε-robust for $n > n^*$:

test	distr. num.	n* H_A: a)	H_A: b)	H_A: c)
u	1, 2, 3	5	5	5
	4, 5, 6	5	10	10
	7	5	50	50
t	1, 2	5	5	5
	3	10	5	5
	4, 6	10	50	50
	5	20	50	50
	7	50	50	50
t_J	1, 2	10	5	5
	3	50	50	30
	4	30	50	30
	5, 6	50	30	50
	7	50	50	50

We see that the u-test is very robust, the t-test not so much. The t_J-test can compete with the t-test only in the cases 4, 5 and 6 i. e. if the kurtosis is small, which is not generally known in practice, and even then only for the alternative hypotheses b) and c).
The figures 3, 5, 6 and 10 show the most interesting power functions.
In view of the power and the above said it seems advisable to give the t-test preverence, and not the t_J-test, if the variance isn't known. If the variance is known one can use the u-test, but one must keep in mind, that a good accuracy is achievable only by chosing $n > n^*$.

Table 1
The simulated distributions

Number of the distribution	u	v	μ	$\overline{x}$	σ^2	s^2	γ_1	g_1	γ_2	g_2
1	—6	6	0	—0.002	1	0.999	0	—0.0068		—0.06
2	—2.2551	2.2551	0	—0.0014	0.855	0.854	0	—0.0027	—0.5	—0.5221
3	—1.183	1.183	0	—0.001	0.3857	0.3849	0	—0.0002	—1.0013	—1.0683
4	—1	6	0.2876	0.286	0.6297	0.6276	0.5918	0.5903	0.0014	—0.0531
5	—0.29	2.743	0.6108	0.6093	0.4037	0.4027	0.75	0.751	0	—0.1052
6	—0.7041	2.2778	0.3764	0.375	0.4744	0.4732	0.5	0.502	—0.5	—0.575
7	1	6	1.52513	1.5238	0.1991	0.199	1.3159	1.2564	1.998	no result

Fig. 1 The truncated normal distribution

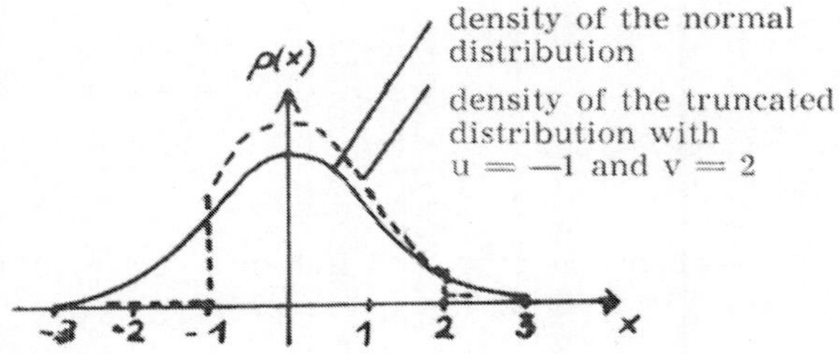

Fig. 2 The simulated distributions in the (γ_1, γ_2)-plane

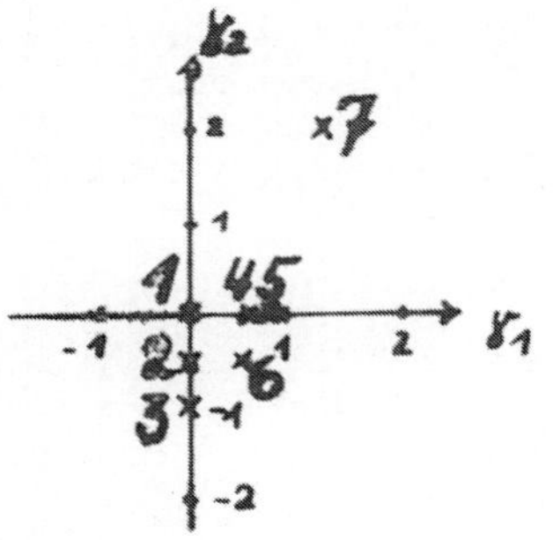

Table 2
The empirical $\hat{\alpha} \cdot 100$ of the simulated tests

num. of the distribution	test→			t-test			t -test			u-test		
	altern. hypotheses			a)	b)	c)	a)	b)	c)	a)	b)	c)
	nominal $\alpha \cdot 100$				5			5		10	2.5	2.5
	γ_1	γ_2	n									
			5	5.3	5.5	5.1	7.3	5.5	5.5	10.1	2.48	2.76
1	0	0	10	5.2	4.8	5.1	4.9	4.8	4.8	10.1	2.46	2.63
			30	4.8	5.0	5.0	4.9	4.6	4.9	9.9	2.38	2.69
			50	5.2	5.0	5.0	5.2	5.0	4.9	10.3	2.78	2.68
			5	5.5	5.6	5.3	7.7	5.5	5.6	10.3	2.52	2.68
2	0	—0.5	10	5.4	4.8	5.1	4.8	4.7	4.7	10.3	2.45	2.54
			30	4.9	4.7	5.0	4.8	4.6	4.9	9.8	2.46	2.71
			50	5.4	5.0	4.9	5.2	5.0	4.9	10.3	2.69	2.72
			5	6.3	5.8	5.5	11.4	7.0	7.3	10.4	2.53	2.47
3	0	—1	10	5.6	4.8	5.1	3.4	3.0	2.9	10.1	2.4	2.66
			30	4.8	4.5	4.9	3.6	3.7	4.2	9.5	2.34	2.66
			50	5.3	4.9	4.9	4.5	4.5	4.5	10.1	2.63	2.71
			5	6.2	3.5	8.0	9.4	6.0	6.4	9.8	3.05	1.64
4	0.6	0	10	5.7	3.3	7.0	3.9	3.8	3.8	9.9	2.85	2.05
			30	5.2	3.6	6.2	4.3	3.8	5.2	9.8	2.53	2.23
			50	5.3	4.3	5.6	4.9	4.5	5.1	9.8	2.9	2.47
			5	7.3	2.7	9.6	13.9	9.4	7.2	9.5	3.28	1.3
5	0.75	0	10	6.2	2.9	8.0	4.7	5.1	2.2	9.9	3.0	1.92
			30	5.6	3.4	6.6	2.8	4.1	2.9	10.0	2.56	2.03
			50	5.2	4.0	6.1	4.2	4.7	4.2	9.6	2.95	2.45
			5	6.6	3.6	8.3	11.1	7.4	6.7	9.9	3.0	1.64
6	0.5	—0.5	10	5.8	3.5	7.0	3.7	4.1	2.6	10.0	2.82	2.09
			30	5.0	3.8	6.1	3.7	4.0	4.3	9.8	2.5	2.33
			50	5.3	4.4	5.6	4.6	4.6	4.6	9.8	2.83	2.5
			5	9.3	1.7	12.9	29.2	23.5	10.2	8.6	3.7	0.45
7	1.3	2.0	10	7.5	1.9	10.6	16.3	18.8	2.1	9.5	3.36	1.26
			30	6.3	2.8	8.0	4.7	8.5	0.1	9.9	2.87	1.71
			50	5.6	3.4	7.0	3.6	6.8	0.02	9.6	3.03	2.05

Fig.3 Power functions of the t-test for n=10 if $H_A: \mu \neq \mu_0$

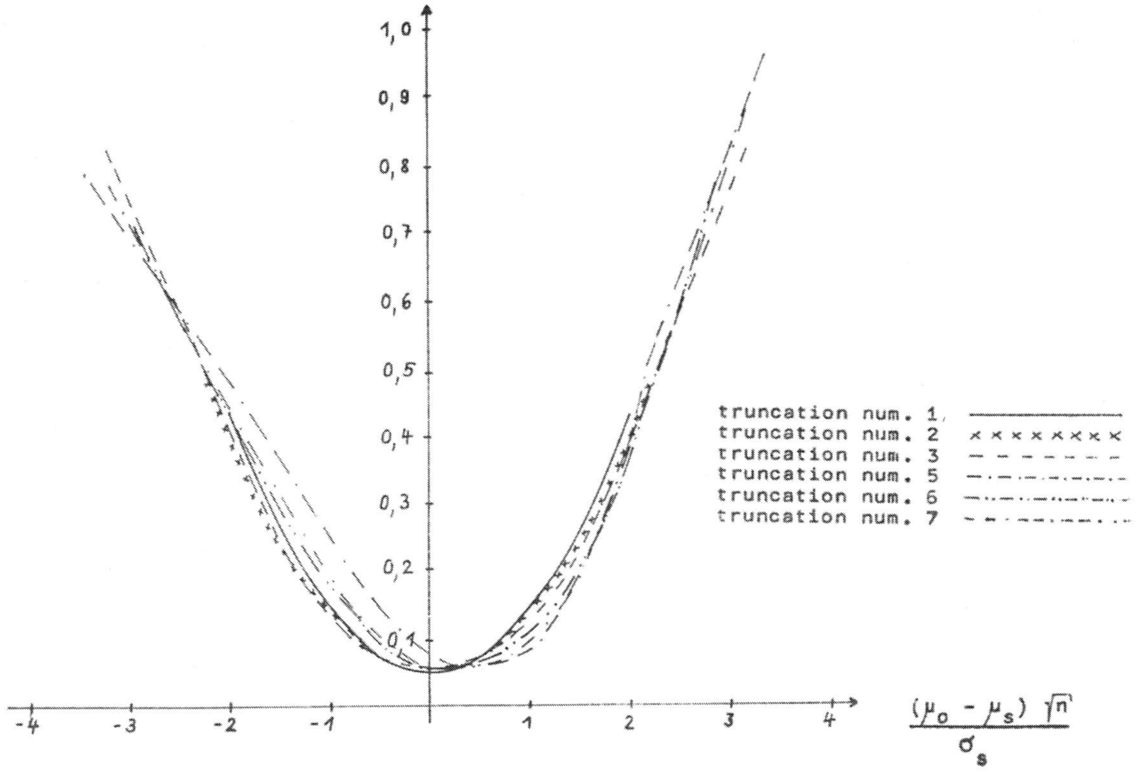

Fig.4 Power functions of the t-test for n=50 if $H_A: \mu \neq \mu_0$

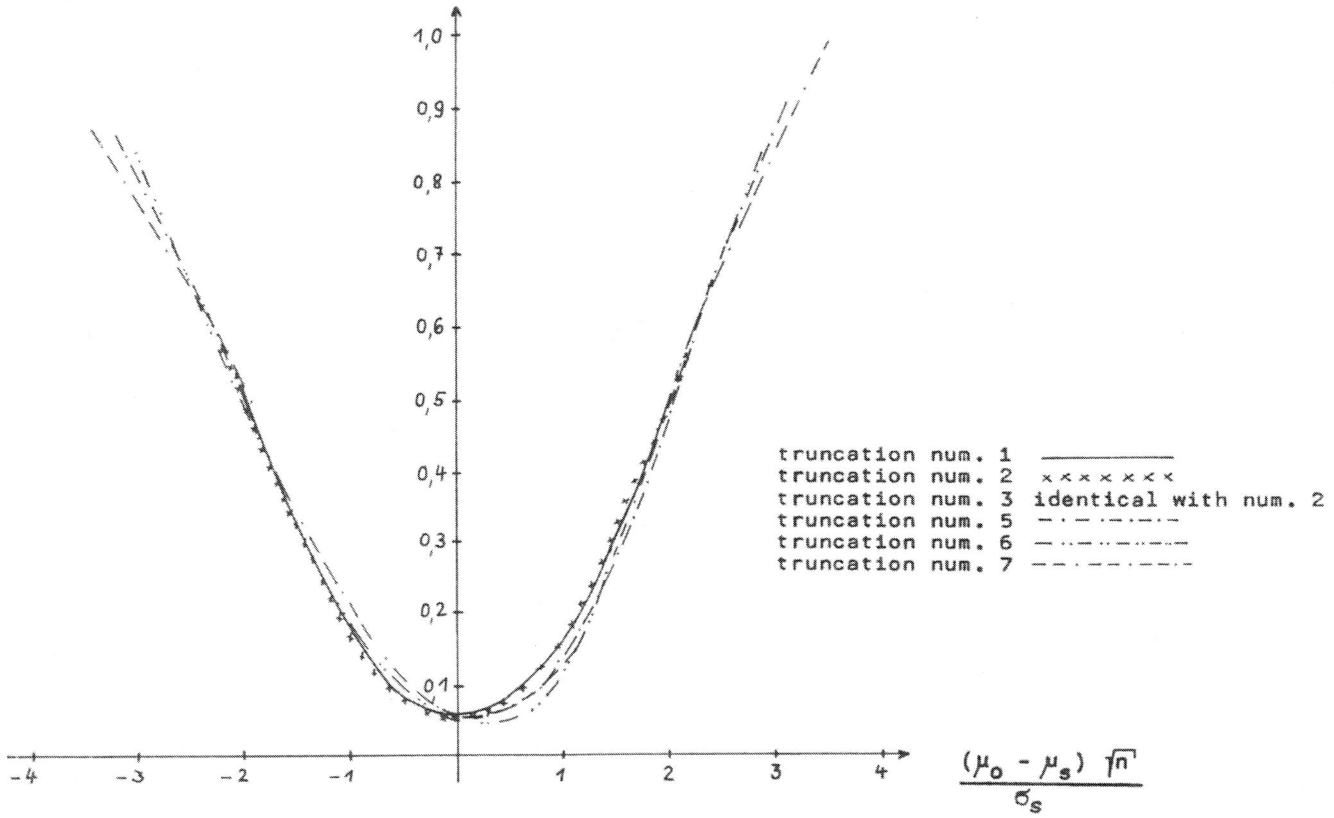

Fig.5 Power functions for n=10 if H_A :b)

t-test and truncation num.1 ———
t-test and truncation num.5 – – – –
t-test and truncation num.6 –·–·–
t_j-test and truncation num. 5 ·······
t_j-test and truncation num. 6 –×–×–×–

$$\frac{(\mu_0 - \mu_s)\sqrt{n}}{\sigma_s}$$

Fig.6 Power functions for n=10 if H_A: c)

t-test and truncation num. 1 ———
(equivalent to the appropriate
t_j-test)

t-test and truncation num. 5 – – – –
t-test and truncation num. 6 –·–·–
t_j-test and truncation num. 5 ·······

t_j-test and truncation num. 6 –×–×–×–×–

$$\frac{(\mu_0 - \mu_s)\sqrt{n}}{\sigma_s}$$

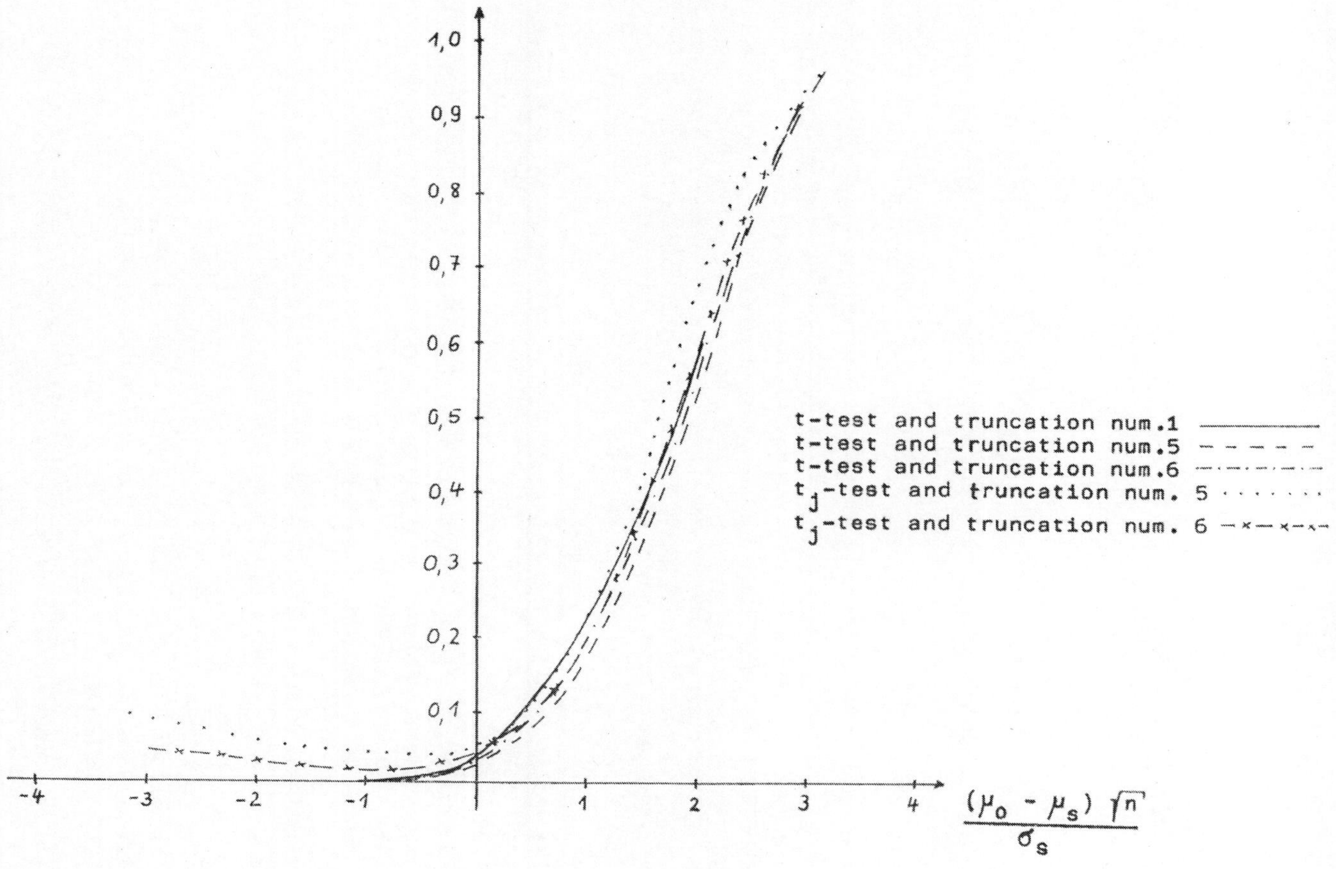

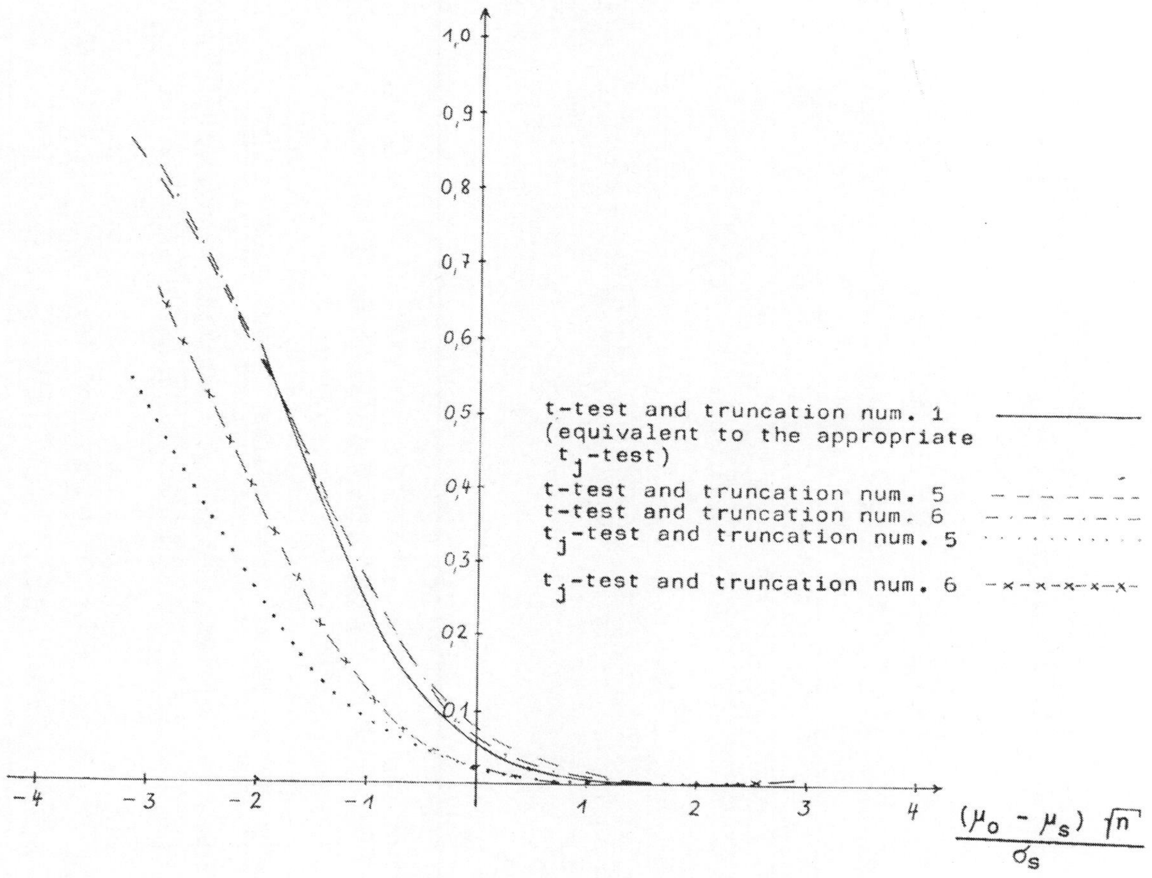

Fig.7 Power functions of the u-test for n=5 if H_A: a)

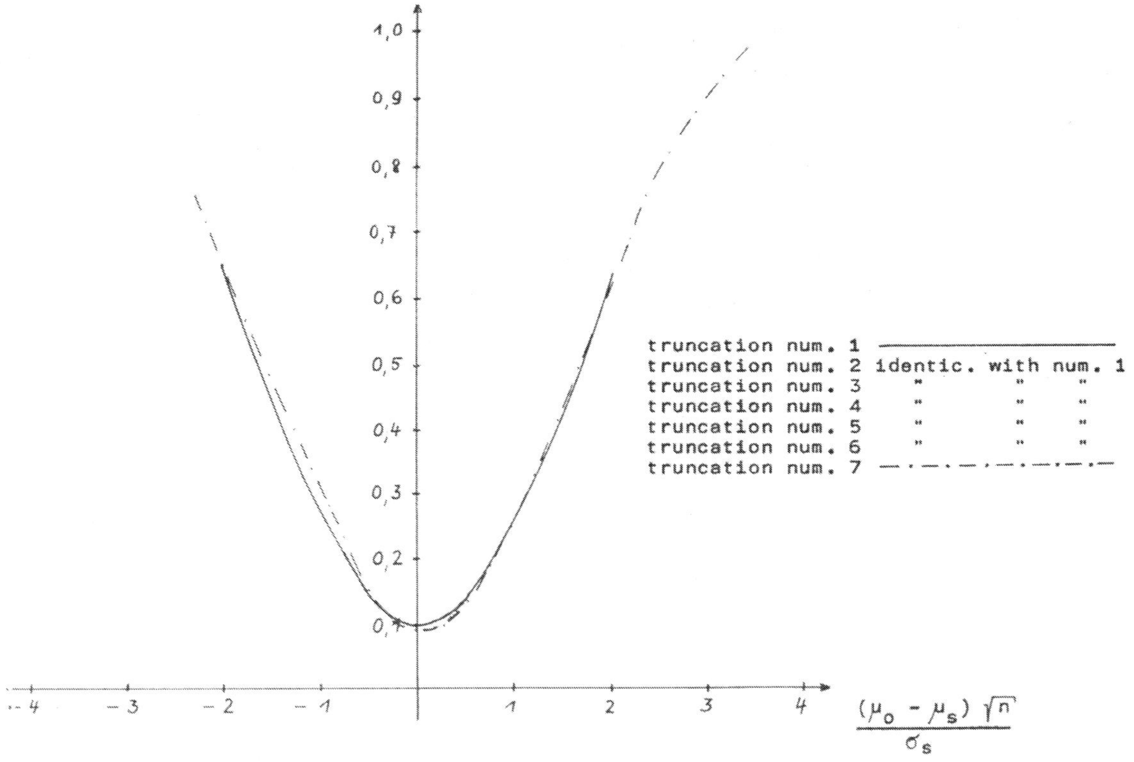

Fig.8 Power functions of the u-test for n=5 if H_A: b)

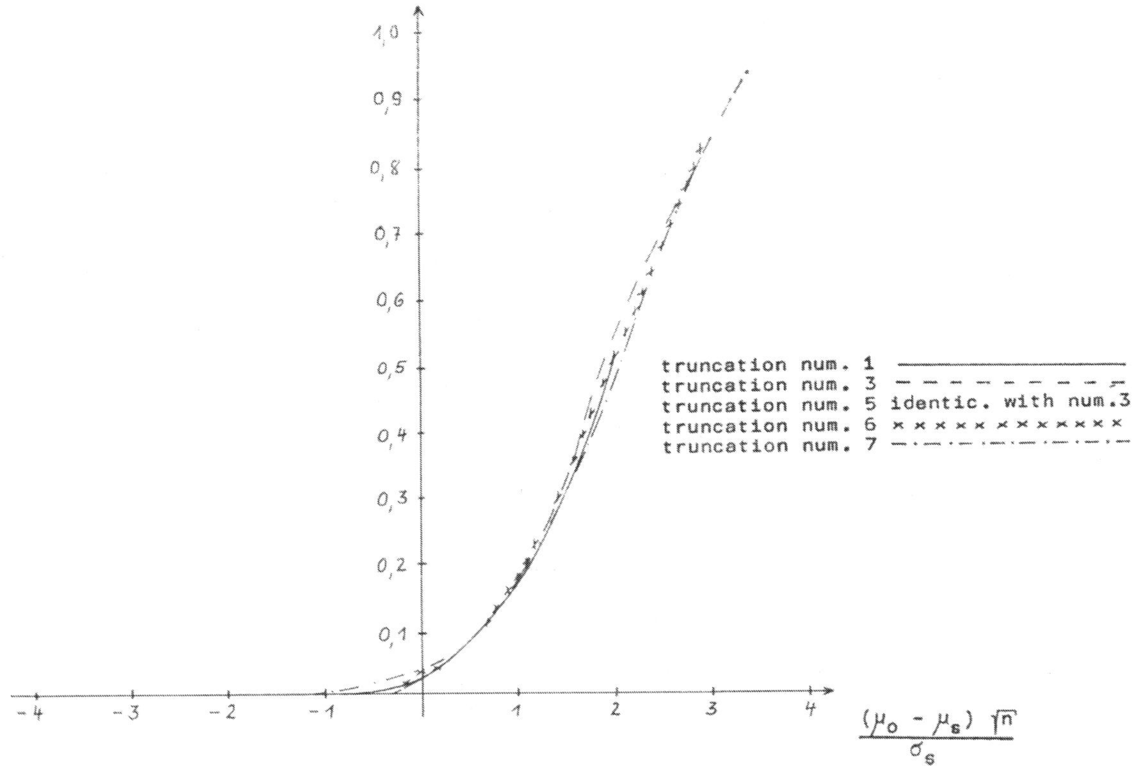

Fig.9 Power functions of the u-test for n=5 if H_A: c)

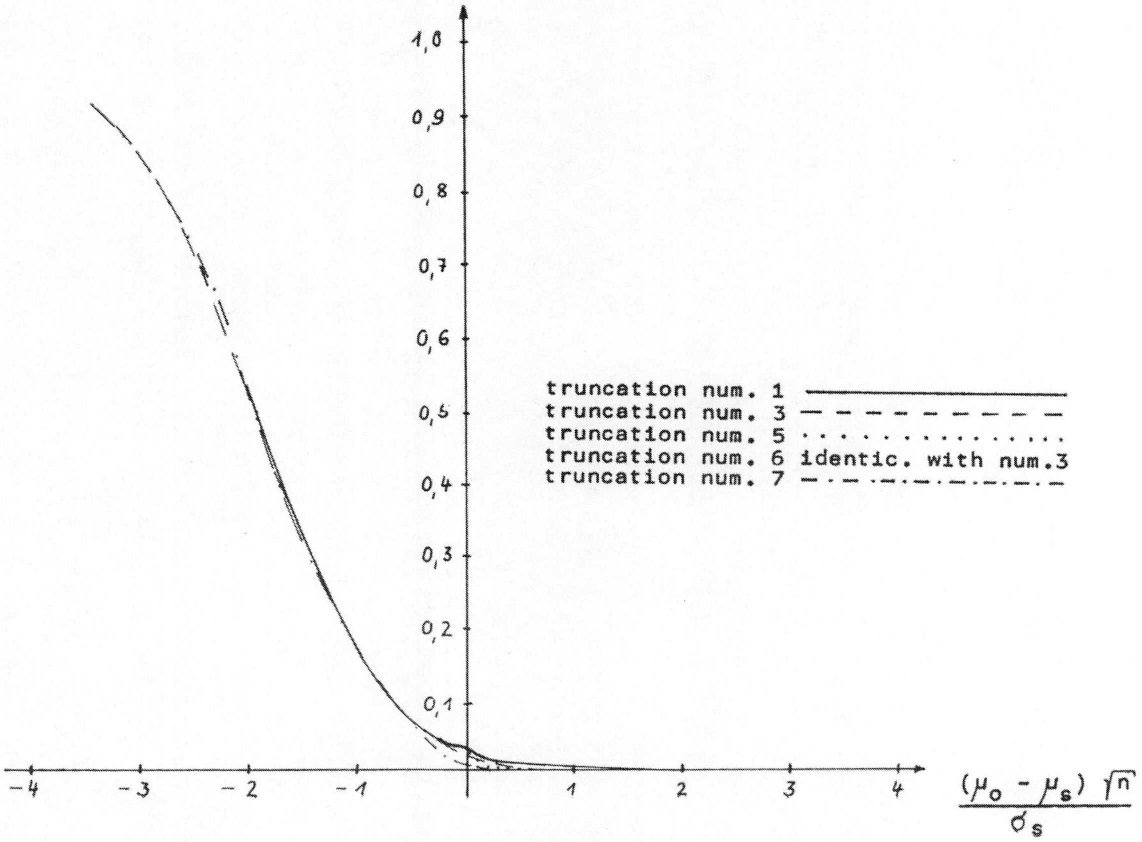

Fig.10 Comparison of the t- and t_j-test for truncation num. 5 if H_A: a)

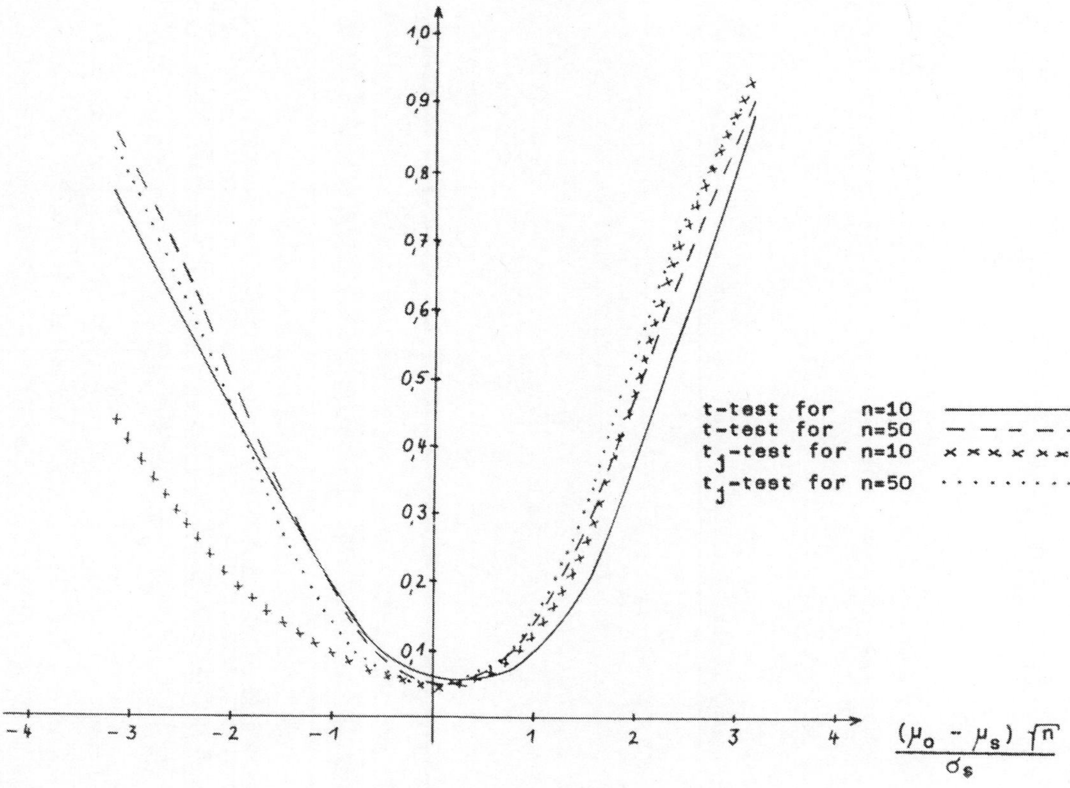

2. **References**

HERRENDÖRFER, G. (ed.)
Probleme der angewandten Statistik. Heft 4.
Robustheit I.
AdL der DDR. Forschungszentrum für Tierproduktion Dummerstorf-Rostock 1980.

JOHNSON, N. J.
Modified t Tests and Confidence Intervalls for Asymmetrical populations.
JASA 73 (1978) 536–544.

RASCH, D., TEUSCHER, F.
Das System der gestutzten Normalverteilungen.
7. Sitzungsbericht der Interessengemeinschaft Mathematische Statistik MG DDR, Berlin 1982, 51–60.

Department of Mathematical Sciences
McMaster University, Hamilton, Ontario, Canada

Robust Location-Tests and Classification Procedures

M. L. TIKU

Abstract

Two sample problems and classifcation procedures are discussed. It is shown that adaptation of the MML estimators leads to tests which have Type I robustness and have remarkably high power.

1. Introduction

To test assumed values of location parameters (means), the classical procedures have been based on the assumption of normality in which case the sample mean $\bar{x} = \sum_{i=1}^{n} x_i/n$ is the minimum variance unbiased estimator of the population mean. In practice, however, the underlying distribution will hardly ever be normal exactly; in fact, the underlying distribution will hardly ever be known exactly. It is therefore imperative to assume that the underlying distribution is one of a reasonably wide class of distributions, which one exactly is not known. Let f(...) denote a family of distributions which consists of the normal distribution and plausible alternatives to the normal. Let μ and σ denote the location parameter (mean) and scale parameter (standard deviation) of a distribution belonging to the family f(...). An unbiased estimator $\hat{\mu}$ of μ will be called more efficient overall (Tiku, 1983) than $\bar{x}$ if

$$\left\{\sum_f \mathrm{Var}(\hat{\mu}) \leq \sum_f \mathrm{Var}(\bar{x})\right\}_{f \in f(\ldots)} \tag{1.1}$$

and Var $(\hat{\mu})$ never greater than Var($\bar{x}$) by more than ε for any f belonging to f(...).

In case of bias, the variance is replaced by the mean square error.

2. MML Estimators of Location and Scale

Let

$$x_{(r+1)}, x_{(r+2)}, \ldots, x_{(n-r)} \tag{2.1}$$

be the censored sample obtained by arranging a random sample $x_1, x_2, \ldots, x_n$ in ascending order of magnitude and censoring (giving zero weights to) the r smallest and largest observations. The MML (modified maximum likelihood) estimators of μ and σ are given by (Tiku, 1967, 1980)

$$\hat{\mu} = \left[\sum_{i=r+1}^{n-r} x_{(i)} + r\beta\{x_{(r+1)} + x_{(n-r)}\}\right]/m \tag{2.2}$$

and

$$\hat{\sigma} = \{B + \sqrt{(B^2 + 4AC)}\}/2\sqrt{\{A(A-1)\}}; \tag{2.3}$$

$A = n - 2r$, $m = n - 2r + 2r\beta$, $B = r\alpha\{x_{(n-r)} - x_{(r+1)}\}$ and

$$C = \sum_{i=r+1}^{n-r} x_{(i)}^2 + r\beta\{x_{(r+1)}^2 + x_{(n-r)}^2\} - m\hat{\mu}^2$$

For $n \geq 10$, the constants α and β are obtained from the equations (Tiku, 1967, 1970)

$$\beta = -f(t)\{t - f(t)/q\}/q \text{ and } \alpha = \{f(t)/q\} - \beta t, \tag{2.4}$$

where $q = r/n$, $F(t) = \int_{-\infty}^{t} f(z)dz = 1 - g$ and $f(z) = \{1/\sqrt{(2\pi)}\}$ $\exp(-\frac{1}{2}z^2)$. For $n < 10$, α and β are obtained from a slightly different set of equations and are given by Tiku (1967, Table 1).

Note that for $r = 0$, $\hat{\mu}$ and $\hat{\sigma}^2$ reduce to the sample mean and variance $\bar{x}$ and s^2, respectively.

For the normal distribution and plausible alternatives to the normal, e. g., long- and short-tailed symmetric distributions with finite mean and variance, (ii) distributions with moderate amount of skewness, and (iii) small proportion (10 % or less) of outliers in a sample, the preferred value of r is the integer value $r = [\frac{1}{2} + 0.1n]$;see Tiku (1980), Spjøtvell and Aastveit (1980) and Stigler (1977).

Under the assumption of normality, $\hat{\mu}$ and $\hat{\sigma}^2$ have for large $A = n(1-2q)$, $q = r/n$ fixed, exactly the same distributional properties as $\bar{x}$ and s^2 (Tiku, 1982a, pp. 626, 627).

3. Two-Sample Problem

Let $x_{i1}, x_{i2}, \ldots, x_{in_i}$, $(i = 1, 2)$, be two independent random samples from populations $(1/\sigma)\, f((x - \mu_i)/\sigma)$. We do not know the functional form f exactly but it belongs to f(...). The classical estimator of $d = \mu_1 - \mu_2$ and σ^2 are given by

$$\bar{d} = \bar{x}_1 - \bar{x}_2 \text{ and } s^2 = \{(n_1-1)s_1^2 + (n_2-1)s_2^2\}/(n_1+n_2-2), \tag{3.1}$$

where $\bar{x}_1$ and $\bar{x}_2$ are the sample means and s_1^2 and s_2^2 are the sample variances.

The corresponding MML estimators are given by

$$\hat{d} = \hat{\mu}_1 - \hat{\mu}_2 \text{ and } \hat{\sigma}^2 = \{(A_1-1)\hat{\sigma}_1^2 + (A_2-1)\hat{\sigma}_2^2\}/(A_1+A_2-2); \tag{3.2}$$

$$A_1 = n_1 - 2r_1 \text{ and } A_2 = n_2 - 2r_2, \left(r_1 = \left[\frac{1}{2} + 0.1n_1\right], r_2 = \left[\frac{1}{2} + 0.1n_2\right]\right)$$

The estimators $\hat{d}$ and $\hat{\sigma}$ are considerably more efficient overall than $\bar{d}$ and s; see Tiku (1980).

A common practical problem is to test the null hypothesis $H_0 : d = 0$. The classical test is based on the statistic

$$t = \bar{d}/s\sqrt{\left(\frac{1}{n_1} + \frac{1}{n_2}\right)}. \tag{3.3}$$

Large values of t lead to the rejection of H_0 in favour of $H_1 : d > 0$.

Under the assumption of normality, the null distribution of t is Student's t with $n_1 + n_2 - 2$ degrees of freedom.

For large samples, the power-function of t is given by ($n_1 = n_2 = n$, say)

$$1-\beta^* = P\{Z \geq Z_\delta - \Delta\}, \quad \Delta = d\sqrt{n}/\sigma\sqrt{2} \tag{3.4}$$

for any underlying distribution f belonging to f(...); Z is a standard normal variate and Z_δ is its $100(1-\delta)$ % point. It is clear that the Type I error (the value of $1-\beta^*$ at $d=0$) and the power $1-\beta^*$ of the above t-test are not sensitive to underlying distributions. In other words, the Type I error and the power of the classical t-test are robust; see also Tan (1982) and Rasch (1983). Similar results hold for small samples (Tan, 1982; Tiku, 1975; Rasch, 1984).

Although robustness of Type I error is a desirable property of a test but not the robustness of its power $1-\beta^*$. One would prefer a test which has robustness of Type I error but has a power-function which is sensitive to underlying distributions and has considerably higher power overall than the above t-test; see Sprott (1982), Tan (1982, p. 2505) and Tiku (1980), for example. One such test is accomplished by using the above MML estimators leading to the statistic (Tiku, 1980)

$$T = \hat{d}/\hat{\sigma}\sqrt{\left(\frac{1}{m_1}+\frac{1}{m_2}\right)}. \tag{3.5}$$

Large values of T lead to the rejection of H_0 in favour of H_1. The null distribution of T is approximately Student's t with $A_1 + A_2 - 2$ degrees of freedom (Tiku, 1980, 1982a).

For large samples, the power-function of T is given by $n_1 = n_2 = n$)

$$1-\beta^* = P\{Z \geq Z_\delta - \Delta^*\}, \quad \Delta^* = 0.983 d\sqrt{n}/E(\hat{\sigma})\sqrt{2}; \tag{3.6}$$

see Tiku (1980, 1982a) for details. Since $E(\hat{\sigma})$ and hence Δ^*, assumes different values for different distributions (Tiku, 1980), the power of the T-test is sensitive to underlying distributions. Moreover, $0.983\sigma/E(\sigma)$ is greater than 1 for most non-normal distributions and, therefore, the T-test is more powerful than the classical t-test for such distributions. For the normal $E(\sigma) = \sigma$ (for large samples) and hence the T-test is only slightly less powerful than the t-test (Tiku, 1982a, p. 621).

The small sample power of the T-test is reported by Tiku (1982a, Table 2) and it is clear from these values that T is more powerful overall than the nonparametric (Wilcoxon, normal-score and Smirnov-Kolmogorov-Massey) tests. For symmetric distributions with finite mean and variance, the T-test is more powerful overall than the analogous tests based on some of the most efficient estimators considered in the Princeton study; see Dunnett (1982) who studies these estimators in the context of multiple comparisons. Realize that most of the estimators considered in the Princeton study (Andrews et al., 1972) are not to be used for skew distributions. One commonly suggested remedy would be a transformation to symmetry, perhaps using the power transformation of Box and Cox (1964). However, the results of Bickel and Doksum (1981) indicate that Box and Cox transformation methods based on normal-theory likelihood have suspect statistical properties; this paper should however be read in conjunction with Box and Cox (1982).

To test $H_0 : d = 0$, when the underlying distributions have unequal variances, the appropriate statistic is the Welch-type statistic (Tiku, 1982a).

$$T^* = \hat{d}\sqrt{\{(\hat{\sigma}_1^2/m_1) + (\hat{\sigma}_2^2/m_2)\}} \tag{3.7}$$

The null distribution of T* is approximately Student's t with degrees of freedom h determined by the equation

$$\frac{1}{h} = \frac{c^2}{(A_1-1)} + \frac{(1-c)^2}{(A_2-1)}, \quad c = \frac{\hat{\sigma}_1^2/m_1}{(\hat{\sigma}_1^2/m_1)+(\hat{\sigma}_2^2/m_2)}. \tag{3.8}$$

The T*-test has robustness of Type I error and is remarkably powerful (Tiku, 1982, pp. 620, 621).

The T*-test can easily be generalized to test a linear contrast $\sum_{i=1}^{k} l_i \mu_i = 0$, $\sum_{i=1}^{k} l_i = 0$, the variances σ_i^2, ($i = 1, 2, \ldots, k$), not necessarily equal. The generalized statistic is given by

$$T^{**} = \left(\sum_{i=1}^{k} l_i\hat{\mu}_i\right)\Big/\sqrt{\left\{\sum_{i=1}^{k} l_i^2(\hat{\sigma}_i^2/m_i)\right\}} \tag{3.9}$$

The null distribution of T** is approximately Student's t with degrees of freedom h* determined by the equation

$$\frac{1}{h^*} = \sum_{i=1}^{k}\left\{\frac{c_i^2}{(A_i-1)}\right\}, \quad c_i = \frac{l_i^2\hat{\sigma}_i^2/m_i}{\sum_{i=1}^{k}(l_i^2\hat{\sigma}_i^2/m_i)}; \tag{3.10}$$

$A_i = n_i - 2r_i$. The statistic T** has remarkable robustness and efficiency properties (Tiku, 1982a).

For an application of the statistic T**, see Tiku (1982a, p. 623).

There are situations (rare though) when one encounters symmetric distributions with infinite mean or variance or extremely skew distributions. Such distributions hardly constitute plausible alternatives to normality, but should be considered on their own right. For distributions of the former type, the MML estimator of d based on samples with $r = [\frac{1}{2} + 0.3n]$ smallest and largest observations censored has remarkably high efficiencies (Tiku, 1980, Table 6). For distributions of the latter type, the MML estimator of d based on samples with $r = [\frac{1}{2} + 0.2n]$ observations censored only on one side (in the direction of the long tail) of the sample has remarkably high efficiencies (Tiku, 1982b, p. 2548). The corresponding T- and T*-statistics have robustness of Type I error and are remarkably powerful as compared to numerous other tests including the Smirnov-Kolmogorov-Massey test which is particularly powerful for extremely skew distributions.

For testing the equality of two population variances, Tiku (1982b) gives a test (based on MML estimators) which has reasonable Type I robustness and has satisfactory power properties. This is motivated by the fact that if x has mean μ and variance σ^2, then $(x-\mu)^2$ has mean σ^2. The problem of testing equality of two variances is thus reduced to testing the equality of means of two extremely skew distributions; the nuisance parameter μ is replaced by the sample mean x; see Tiku (1982b).

4. Robust Classification Procedures

Let x_{1i}, $i = 1, 2, \ldots, n_1$, and x_{2i}, $i = 1, 2, \ldots, n_2$, be independent random samples from populations π_1 and π_2, respectively. Assume that π_1 and π_2 are identical (other than a location shift) and have finite means and a common finite variance. On the basis of these samples, one wants to classify a new independent observation x_0 in π_1 or π_2. Let e_1 and e_2 be the errors of misclassification; e_1 is the probability of classifying x_0 in π_2 when in fact it has come from π_1 and e_2 is the probability of classifying x_0 in π_1 when in fact it has come from π_2. An optimal classification procedure has maximum $1-e_2$ for a fixed e_1; $1-e_2$ is the probability of correctly classifying x_0 in π_2.

Under the assumption of normality, the classical classification procedure is based on the statistic (Anderson, 1951)

$$V=\left\{x_0-\frac{1}{2}(\bar{x}_1+\bar{x}_2)\right\}(\bar{x}_1-\bar{x}_2)/s^2 \tag{4.1}$$

$\bar{x}_1$ and $\bar{x}_2$ are the sample means and s^2 is the pooled sample variance. The observation x_0 is classified in π_1 or π_2 according as (see Balakrishnan et al., 1985, for details).

$$V \gtrless \tilde{C}, \tag{4.2}$$

where ($e_1 \simeq 0.05$)

$$\tilde{C}=\frac{1}{2}\tilde{\delta}-1.64\sqrt{\tilde{\delta}}; \tag{4.3}$$

$$\tilde{\delta}=\{(n_1+n_2-4)/(n_1+n_2-2)\}(\bar{x}_1-\bar{x}_2)^2/s^2. \tag{4.4}$$

However, this classification procedure is not robust to departures from normality.

A robust classification procedure is obtained by replacing the sample means and variances by the corresponding MML estimators. This robust procedure is based on the statistic

$$V_R=\left\{x_0-\frac{1}{2}(\hat{\mu}_1+\hat{\mu}_2)\right\}(\hat{\mu}_1-\hat{\mu}_2)/\hat{\sigma}^2 \tag{4.5}$$

The observation x_0 is classified in π_1 or π_2 according as

$$V_R \gtrless \tilde{C}^* \tag{4.6}$$

where ($e_1 \simeq 0.05$)

$$\tilde{C}^*=\frac{1}{2}\tilde{\delta}^*-1.64\sqrt{\tilde{\delta}^*}; \tag{4.7}$$

$$\tilde{\delta}^*=\{(A_1+A_2-4)/(A_1+A_2-2)\}(\hat{\mu}_1-\hat{\mu}_2)^2/\hat{\sigma}^2,$$

$$(A_1=n_1-2r_1,\ A_2=n_2-2r_2).$$

Under the assumption of normality, the V- and V_R-procedures have exactly the same $1-e_2$ (for a common e_1) for large samples; see Balakrishnan et al., 1985, for a proof.

For most non-normal populations, the V_R-procedure has reasonably stable e_1-values (robustness of Type I error) and has remarkably high values of $1-e_2$ (power). For $\delta=(\mu_1-\mu_2)^2/\sigma^2=3.0$, the standardized distance between the two populations, and $n_1=n_2=n=20$, for example, we have the following simulated (based on 1000 Monte Carlo runs) values of e_1 (supposed to be 0.05) and $1-e_2$:

Normal

e_1		$1-e_2$	
V	V_R	V	V_R
.051	.042	.918	.915

Logistic

e_1		$1-e_2$	
V	V_R	V	V
.046	.052	.916	.924

Outlier model
$(n-1)N(0, 1)$ & $1N(0, 10)$

e_1		$1-e_2$	
V	V_R	V	V_R
.005	.036	.378	.894

Mixture model
$0.90N(0, 1)+0.10N(0, 4)$

e_1		$1-e_2$	
V	V_R	V	V_R
.039	.060	.653	.794

In fact, the V_R-procedure has smaller error rates $\frac{1}{2}(e_1+e_2)$ than the V-procedure almost always (Balakrishnan et al., 1985).

Numerous nonparametric or distribution-free classifcation procedures are available; see Balakrishnan et al., 1985. However, these procedures are very difficult to compute and have also larger error rates than the V_R-procedure (Balakrishnan et al., 1985).

Example:

Consider the following data which represent the values of $(x-2.0)/(0.1)$, x being the pollution levels (measurements of lead in water samples from two lakes);

Sample from π_1: —1.48, 1.25, —0.51, 0.46, 0.60, —4.27, 0.63, —0.14, —0.38, 1.28, 0.93, 0.51, 1.11, —0.17, —0.79, —1.02, —0.91, 0.10, 0.41, 1.11

Sample from π_2: 1.32, 1.81, —0.54, 2.70, 2.27, 2.70, 0.78, —4.62, 1.88, 0.86, 2.86, 0.47, —0.42, 0.16, 0.69, 0.78, 1.72, 1.57, 2.14, 1.62

Observation to be classified is 0.60 (this has in fact come from π_1).

Here, one wants to adopt a classification procedure which does not fix beforehand any of the errors of misclassification e_1 or e_2. The relevant classical procedure is to classify x_0 in π_1 or π_2 according as

$$V \gtrless 0 \tag{4.8}$$

The corresponding robust procedure is to classify x_0 in π_1 or π_2 according as

$$V_R \gtrless 0. \tag{4.9}$$

For the above data,

$\bar{x}_1=-0.063$, $\bar{x}_2=1.036$; $V=(0.60-0.486)(-.063-1.036)=-0.125$;

$\hat{\mu}_1=0.110$, $\hat{\mu}_2=1.247$; $V_R=(0.60-0.678)(.110-1.247)=0.089$.

The V_R-procedure correctly classifies 0.60 in π_1 but not the classical V-procedure. The failure of the V-procedure might be due to a few potential outliers in the two samples (observations —4.27 and —4.62, for example).

5. Bibliography

ANDERSON, T. W.
Classification by multivariate analysis.
Psychometrika 16, (1951), 31—50.

ANDREWS, D. F., BICKEL, P. J., HAMPEL, F. R., HUBER, P. J., ROGERS, W. H. and TUKEY, J. W.
Robust estimates of Location.
Prceeton: Princeton University Press, 1972.

BALAKRISHNAN, N., TIKU, M. L. and EL SHAARAWI, A. H.
Robust univariate two-way classification.
Biometrical J. 27 (1985) to appear.

BICKEL, P. J. and DOKSUM, K.
An analysis of transformations revisited.
J. Amer. Statist. Assoc. 76, (1981) 296—311.

BOX, G. E. P. and COX, D. R.
An analysis of transformations.
J. R. Statist. Soc. B 26, (1964) 211—52.

BOX, G. E. P. and COX, D. R.
An analysis of transformations revisited, rebutted.
J. Amer. Statist. Assoc. 77, (1982) 209—10.

DUNNETT, C. W.
Robust multiple comparisons.
Commun. Statist. Theor. Meth. 11 (22), (1982) 2611–29.

RASCH, D.
Papers by Rasch and his coauthors in these proceedings. (1984).

SPJØTVOLL, E. and AASTVEIT, A. H.
Comparison of robust estimators on data from field experiments.
Scandinavian J. Statist. 7, (1980) 1–13.

SPROTT, D. A.
Robustness and maximum likelihood estimation.
Commun. Statist.-Theor. Meth. 11 (22), (1982) 2513–29.

STIGLER, S. M.
Do robust estimators work with real data?
Ann. Statist. 5, (1977) 1055–98 (with discussion).

TAN, W. Y.
Sampling distributions and robustness of t, F and variance-ratio in two samples and ANOVA models with respect to departure from normality.
Commun. Statist.-Theor. Meth. 11 (22), (1982) 2485–2511.

TIKU, M. L.
Estimating the mean and standard deviation from censored normal samples.
Biometrika 54, (1967) 155–65.

TIKU, M. L.
Monte-Carlo study of some simple estimators in censored normal samples.
Biometrika 57, (1970) 207–11.

TIKU, M. L.
Laguerre series forms of the distributions of classical test-statistics and their robustness in nonnormal situations.
Applied Statistics. (R. P. Gupta, Ed.) American Elsevier Pub. Comp., New York 1975.

TIKU, M. L.
Robustness of MML estimators based on censored samples and robust test statistics.
J. Statistical Planning and Inference 4, (1980) 123–43.

TIKU, M. L.
Testing linear contrasts of means in experimental design without assuming normality and homogeneity of variances.
Biometrical J. 24, (1982a) 613–27.

TIKU, M. L.
Robust statistics for testing equality of means or variances.
Commun. Statist.-Theor. Meth., 11 (22) (1982b) 2543–58.

TIKU, M. L.
MML Estimators and Robust Statistics.
A book manuscript to be published by Marcel Dekker, Inc.: New York 1983.

Academy of Sciences of GDR, Institute of Mathematics, Berlin, GDR

Minimax-Linear Estimation under Incorrect Prior Information

HELGE TOUTENBURG

Abstract

If nonlinear prior information on the parameter β of the familiar linear regression model is available such that β may be assumed to be in a convex set (a concentration ellipsoid), then this information is used optimally by a minimax estimator (MILE). The MILE is of ridge type and is of smaller minimax risk compared with the BLUE so far the prior information is true. The MILE is said to be insensitive against misspecifications of the prior region if its risk stays smaller than the risk of the BLUE. There are given necessary and sufficient conditions for the insensitivity of MILE in typical situations of incorrect prior information.

1. Introduction

Assume the familiar linear model

$$y = X\beta + u, \quad u \sim (0, \sigma^2 I), \quad \text{rank } X = K. \tag{1}$$

Then the BLUE of β is

$$b = (X'X)^{-1}X'y = S^{-1}X'y, \text{ say} \tag{2}$$

with $V(b) = \sigma^2 S^{-1}$.

Assume further to have prior knowledge on β such that β lies in a convex set, especially in an ellipsoid centered in β_0:

$$B = \{\beta: (\beta-\beta_0)'T(\beta-\beta_0) \leq k\} \tag{3}$$

where the matrix T is p.d and $k > 0$ a scalar.

As an example of practical relevance we mention the following special type of prior information on β. If a component-by-component restriction on β of the type

$$a_i \leq \beta_i \leq b_i (i = 1, \dots, K) \tag{4}$$

with known interval limits a_i, b_i is given, then the cuboid defined by the inequalities (4) may be enclosed in an ellipsoid centered in the cuboid's centre $\beta_0 = 1/2\,(a_1 + b_1, \dots, a_K + b_K)$, also. Choosing $T = \text{diag}\,(t_1, \dots, t_K)$ with $t_i = 4K^{-1}(b_i - a_i)^{-2}$ and $k = 1$ gives the ellipsoid which has minimal volume in the set of ellipsoids which contain the corner points of the cuboid and which are centered in β_0.

Another example to construct a prior ellipsoid is as follow. Let an upper bound of the expected value of the response surface be known: $Ey \leqq k$. This leads to

$$(Ey)'(Ey) = \beta'X'X\beta \leq k^2. \tag{5}$$

Such a condition is typical for laws of restricted growth

2. The Presentation of MILE

Using the familiar quadratic risk

$$R(\hat\beta, A) = \text{tr}\, AE(\hat\beta-\beta)(\hat\beta-\beta)' \tag{6}$$

for a linear estimator $\hat\beta$ of β, then b^* is said to be a MILE of β if

$$\min_{\hat\beta} \sup_{\beta\in B} R(\hat\beta, A) = \sup_{\beta \in B} R(b^*, A) \tag{7}$$

holds. If we confine ourselves to a loss matrix A of rank one, i. e. $A = aa'$, we get the MILE with respect to B from (3) when $\beta_0 = 0$ as

$$b^* = (k^{-1}\sigma^2 T + S)^{-1}X'y = D^{-1}X'y, \text{ say}. \tag{8}$$

(For the proof the reader is referred to Kuks and Olman (1971), Läuter (1975), Toutenburg (1976)).
This estimator has

$$\text{bias } b^* = -k^{-1}\sigma^2 D^{-1}, \tag{9}$$

the dispersion matrix

$$V(b^*) = \sigma^2 D^{-1}SD^{-1} \tag{10}$$

and the minimax risk

$$\sup_{\beta'T\beta\leq k} R(b^*, aa') = \sigma^2 a'D^{-1}a. \tag{11}$$

Note: The estimator b^* contains the unknown σ^2. To make b^* practicable, σ^2 has to be estimated or replaced by a prior value. We shall discuss this problem in Section 5.
If the ellipsoid is not centered in the origin but in $\beta_0 \neq 0$ then it holds (Toutenburg (1982)).

Theorem. Under constraint $(\beta - \beta_0)'T(\beta - \beta_0) \leqq k$ the MILE of β is

$$b^*(\beta_0) = \beta_0 + D^{-1}X'(y - X\beta_0). \tag{12}$$

This estimator has

$$\text{bias } b^*(\beta_0) = -k^{-1}\sigma^2 D^{-1}T(\beta-\beta_0), \tag{13}$$

$$V[b^*(\beta_0)] = V(b^*) \quad (\text{as given in (10)}), \tag{14}$$

and

$$\sup_{(\beta-\beta_0)'T(\beta-\beta_0)\leq k} R[b^*(\beta_0), aa'] = \sigma^2 a'D^{-1}a. \tag{15}$$

If follows that the change of the centre of the concentration ellipsoid has influence on the estimator and its bias whereas the dispersion and the minimax risk are not influenced by β_0.

3. Connection of MILE with GLSE

If there is no restriction, or in other words, if $B = E^K$ or equivalently $k = \infty$ holds in (3), we have $\lim_{k\to\infty} D = S$ and therefore

$$\lim_{k\to\infty} b^* = b = S^{-1}X'y. \tag{16}$$

Thus, the BLUE b may be understood as an unrestricted MILE.

Gain in efficiency. To check the gain in efficiency in estimating β by the MILE b^* which uses the prior information $\beta \in B$, we first compare the minimax risks of b^* and the unrestricted GLSE b.

This gives

$$\Delta^k(b,b^*)=\sup_{\beta'T\beta\le k} R(b,aa')-\sup_{\beta'T\beta\le k} R(b^*,aa') \quad (17)$$

$$=\sigma^2 a'\left(S^{-1}-D^{-1}\right)a\ge 0$$

because the matrix in brackets is n. n. d.

If we now compare the MILE b* and b with respect to the unconstrained risk (6) with A = aa' we get (see (9) and (10))

$$R(b,aa')-R(b^*,aa')=\sigma^2 a'D^{-1}CD^{-1}a \quad \text{with}$$

$$C=k^{-2}\sigma^4 T\{S^{-1}+2k\sigma^{-2}T^{-1}-\sigma^{-2}\beta\beta'\}T \quad (18)$$

$$=k^{-2}\sigma^4 T\widetilde{S}^{1/2}\{I-\sigma^{-2}\widetilde{S}^{-1/2}\beta\beta'\widetilde{S}^{-1/2}\}\widetilde{S}^{1/2}T$$

$$\text{where } \widetilde{S}=S^{-1}+2k\sigma^{-2}T^{-1}.$$

Using the familiar theorem (see Yancey et al., 1974)

$I-gg'\geq 0$ iff $g'g\leq 1$

we may conclude that the MILE b* is superior to the unrestricted Aitken estimator b with respect to the unrestricted risk $R(\widehat{\beta}, aa')$ if and only if

$$\sigma^{-2}\beta'(S^{-1}+2k\sigma^{-2}T^{-1})^{-1}\beta\le 1. \quad (19)$$

Note: As a sufficient condition for C n. n. d. we have from (13)

$$T^{-1}-k^{-1}\beta\beta'\ge 0 \quad (20)$$

which gives for T = I, that

$$k^{-1}\le(\beta'\beta)^{-1}. \quad (21)$$

This corresponds of the sufficient condition for superiority of the ridge estimator over ordinary least squares (Theobald (1974)).

4. Minimax Estimation Under Incorrect Prior Information

The aim of the minimax estimator is to decrease the mean square error risk (or its supremum) compared with that of b and therefore to overcome the restrained usefulness of b. This intention is successful so far the used prior information $\beta \in B$ with B from (3) is correctly chosen. In other words, the superiority of b* over b fully depends on the correctness of the prior region B. But it is clear from practice that a misidentification of prior restrictions may happen. Thus the influence of incorrect chosen prior regions has to be investigated. The general situation may be as follows.

The regression model $y = X\beta + u$ is such that β is contained in the ellipsoid

$\beta'T\beta \leq k$ (correct prior information). (22)

Now in practice, the model builder may choose an incorrect ellipsoid

$(\beta - \beta_0)'R(\beta - \beta_0) \leq r$ (incorrect prior information). (23)

This means we have mistakenly used the estimator (see (12))

$$b_R^*(\beta_0)=\beta_0+D_R^{-1}X'(y-X\beta_0) \quad (24)$$

$$\text{where } D_R=(r^{-1}\sigma^2R+S).$$

As (22) was the correct prior information, we have to calculate the minimax risk of $b_R^*(\beta_0)$ with respect to the correct region (22). This gives (see Toutenburg (1982, p. 93))

$$\sup_{\beta'T\beta\le k}\{a' \text{ bias } b_R^*(\beta_0)\}^2=$$

$$=\sup_{\beta'T\beta\le k}\{\tilde{a}'(\tilde{\beta}-\tilde{\beta}_0)\}^2=\{\sqrt{k\tilde{a}'\tilde{a}}+|\tilde{a}'\tilde{\beta}_0|\}^2 \quad (25)$$

$$\text{where } \tilde{a}=a'\sigma^2r^{-1}D_R^{-1}RT^{-1/2},\ \tilde{\beta}_0=T^{1/2}\beta_0,\ \tilde{\beta}=T^{1/2}\beta.$$

Inserting this gives

$$\sup_{\beta'T\beta\le k} R(b_R^*(\beta_0),aa')\le\sigma^2a'D_R^{-1}\{S+\sigma^2r^{-2}R\beta_0\beta_0'R' \quad (26)$$

$$+\sqrt{k}\,\sigma^2r^{-2}RT^{-1}R(\sqrt{k}+2(\beta_0'T\beta_0)^{1/2})\}D_R^{-1}a.$$

As b* (8) is the best estimator with respect to the correct prior region (22), the mistakenly used estimator $b_R^*(\beta_0)$ is less efficient compared with b*. The question arises whether the estimator $b_R^*(\beta_0)$ is better than the GLSE b. This is just the problem of insensitivity of the minimax estimator. That is, we have to investigate whether

$$\sigma^{-2}\Delta^k(b,b_R^*(\beta_0))\ge a'D_R^{-1}UD_R^{-1}a\ge 0 \quad (27)$$

where the matrix U is

$$U=\sigma^2r^{-1}R\{\sigma^2r^{-1}S^{-1}+2R^{-1}-r^{-1}\beta_0\beta_0' \quad (28)$$

$$-r^{-1}(k+2\sqrt{k}(\beta_0'T\beta_0)^{1/2})T^{-1}\}R.$$

That is, U n. n. d. is sufficient to fulfil (27). Now it is very difficult to draw specific conclusions from this general matrix U. In order to give explicity a condition for U n. n. d. we shall confine ourselves to the case R = T, i. e. where incorrectness is caused by translation of the ellipsoid's centre as well as by symmetrical distortion, i. e. changing the length of the axes by the same factor $\sqrt{r/k}$. This case seems to be of practical relevance.

Using R = T, the matrix U (28) becomes

$$U=\sigma^2r^{-2}T^{1/2}\tilde{U}T^{1/2}. \quad (29)$$

Now, U n. n. d. implies that its eigenvalues are nonnegative:

$$\lambda_i(\tilde{U})=\sigma^2\lambda_i(T^{1/2}S^{-1}T^{1/2}-\sigma^{-2}T^{1/2}\beta_0\beta_0'T^{1/2})$$

$$+(2r-k-2\sqrt{k\beta_0'T\beta_0})\ge 0.$$

This is true iff

$$r\ge 1/2\left(k+2\sqrt{k\beta_0'T\beta_0}-\sigma^2\lambda_{min}\{T^{1/2}(S^{-1}-\sigma^2\beta_0\beta_0')T^{1/2}\}\right). \quad (30)$$

Conclusions

(i) If $\beta_0 = 0$ (i. e. the correct and the incorrect prior region have the same centre), then (30) becomes

$$r=1/2\left(k-\sigma^2\lambda_{min}(T^{1/2}S^{-1}T^{1/2})\right). \quad (31)$$

Moreover, this condition (31) is necessary and sufficient to ensure $\Delta^k(b, b_R^*(0)) \geqq 0$ in the case R = T. Clearly, the condition $r \geqq k$ is sufficient to fulfil (31). This means, that in the case

$\beta'T\beta \leq k$ (correct region), $\beta'T\beta \leq r$ (incorrect region)

the 'radius' of the incorrect ellipsoid must be greater than the correct 'radius'.

In other words, if there is some uncertainty on the length of prior intervals as in (4), a 'blowing up' strategy ensures that the resulting MILE stays better than b.

(ii) If $\beta_0 \neq 0$, i. e. if the incorrectness of the prior information is caused by translation of the ellipsoid's centre also, then $r \geqq k$ is not sufficient to guarantee the insensitivity of the mistakenly used MILE $b_R^*(\beta_0)$.

5. **Substitution of σ^2**

The various minimax estimators from above contain the unknown parameter σ^2 and therefore they are not practicable. A way to overcome this is to replace σ^2 by a nonstochastic value $c \geqq 0$ which gives the estimator

$$b_c^* = (k^{-1}cT+S)X'y = D_c^{-1}X'y, \text{ say}, \tag{32}$$

which has

$$\sup_{\beta'T\beta\leq k} R(b_c^*, aa') = \sigma^2 a' D_c^{-1}(S+k^{-1}c^2\sigma^{-2}T)D_c^{-1}a. \tag{33}$$

As the substitution of σ^2 by c may be interpreted as an incorrect chosen prior ellipsoid, namely

$$\beta'T\beta \leq r \quad \text{with} \quad r = k\sigma^2 c^{-1} \tag{34}$$

we may use the results from above to investigate whether $\Delta^k(b, b^*) \geqq 0$. Replacing r in (31) by r from (34) gives

$$\Delta^k(b, b_c^*) \geq 0 \quad \text{iff} \quad c \leq \frac{2\sigma^2}{1-\sigma^2 k^{-1}\lambda_{min}(T^{1/2}S^{-1}T^{1/2})}. \tag{35}$$

Thus

$$c = 2\sigma^2 \tag{36}$$

is sufficient to fulfil (35). To realize (35) or (36), which are just the conditions for the superiority of b^* over b, we need prior knowledge of the type

$$0 < \sigma_1^2 < \sigma^2. \tag{37}$$

As $\sup_{\beta'T\beta \leqq k} R(b^*, aa')$ is monotonically decreasing for $c < \sigma^2$, we may conclude that $c = 2\sigma_1^2$ is the optimal choice of c in the sense

$$\max_{c \leq 2\sigma_1^2} \Delta^k(b, b_c^*) = \Delta^k(b, b_{2\sigma_1^2}^*) \geq 0. \tag{38}$$

Thus we have arrived at a practicable and, in the sense of (38), optimal estimator.

6. **References**

KUKS, J. and OLMAN, W.
Minimaksnaja Linejnaja ozenka koeffizientow regressii. Iswestija Akademija Nauk Estonskoj SSR, 20 (1971), 480–482.

LÄUTER, H.
A minimax linear estimator for linear parameters under restrictions in form of inequalities.
Math. Operationsf. Statistik 6 (1975), 689–696.

THEOBALD, C. M.
Generalizations of mean square error applied to regression.
J. R. Statist. Soc., B, 36 (1974), 103–106.

TOUTENBURG, H.
Minimax-linear and MSE-estimators in generalized regression.
Biometrische Zeitschrift 18 (1976) 91–100.

TOUTENBURG, H.
Prior information in linear models.
John Wiley, Chichester, 1982.

YANCEY, T. A., JUDGE, G. G. and BOCK, M. E.
A mean square error test when stochastic restrictions are used in regression.
Commun. Statist. 3 (1974), 755–768.

Academy of Agricultural Sciences of the GDR Research Centre of Animal Production Dummerstorf-Rostock

The Robustness of Some Procedures for the Two-Sample Location Problem — a Simulation Study (Concept)

ARMIN TUCHSCHERER

Abstract

A comprehensive simulation experiment is to be performed on the most important parametric and nonparametric procedures in order to gain some idea of the robustness of such procedures for comparing two means in respect of violation of the assumptions regarding distribution and (in)homogeneity of variance. The purpose of the study is to find out which of the tests investigated are "robust" and can therefore be recommended for practical application.

1. Introduction

For the purpose of these studies, "robustness" will be understood in the sense of the definition given in Herrendörfer, G. (ed.) (1980), Guiard, V. (ed.) (1981), Rasch, D. and Herrendörfer, G. (ed.) (1982).

Let $\mathfrak{T}_P$ be the set of parametric procedures and $\mathfrak{T}_{NP}$ the set of nonparametric procedures that are to be investigated.

$$R_P := \left\{T_\alpha^P / T_\alpha^P \in \mathfrak{T}_P : \left|\alpha(n_1, n_2, g) - \alpha\right| \le \varepsilon;\ \forall g \in G_2\right\}$$

is the set of ε-robust tests in $\mathfrak{T}_P$, with α being the nominal and $\alpha(n_1, n_2, g)$ the actual first kind risk for a distribution g belonging to the class G_2 of distributions to be investigated. The best ε-robust test $T^*_{P_\alpha}$ from $\mathfrak{T}_P$ then satisfies the condition

$$\pi_{T^*_{P_\alpha}}(n_1, n_2, g, \delta) \ge \pi_{T_{P_\alpha}}(n_1, n_2, g, \delta)$$

$\forall\ T_{P_\alpha} \in R_P$, $\forall\ g \in G_2$ and $\forall\ \sigma > 0$, whereby $\pi_T(.)$ is the power function of the test T. The best nonparametric test T^*_{NP} will then, by analogy, satisfy the condition

$$\pi_{T^*_{NP_\alpha}}(n_1, n_2, g, \delta) \ge \pi_{T_{NP_\alpha}}(n_1, n_2, g, \delta),$$

$$\forall T_{NP} \in \tau_{NP},\ \forall g \in G_2 \text{ and } \forall \delta > 0.$$

The best of the two tests T^*_P and T^*_{NP} will be recommended for application.

Let $G_2 = k(0, \delta^2, \gamma_1, \gamma_2)$ be a class of distributions with the mean 0, variance σ^2, skewness γ_1 and kurtosis γ_2. For the simulation studies, γ_1 and γ_2 values of several characters of importance in agriculture, industry and medicine which where based on estimations (g_1, g_2) of (γ_1, γ_2) and satisfy the inequality $\gamma_2 \geqq \gamma_1^2 - 2$ were chosen for γ_1 and γ_2 (cf. Table 3.3).

2. Description of the tests to be investigated

Let $\mathbf{x}_1$ and $\mathbf{x}_2$ be independent random variables whose first four moments exist, and let

$$\mathbf{x}_i \sim F_i \text{ with } F_i \in k(\mu_i, \sigma_i^2, \gamma_1, \gamma_2);\ i = 1, 2.$$

Two independent random samples $\mathbf{x}_{i1}, \ldots \mathbf{x}_{in_i}$ are to be used to check the hypothesis

$H_0 : \mu_1 = \mu_2$

against the alternatives

$$H_A^{(1)} : \mu_1 > \mu_2$$

$$H_A^{(2)} : \mu_1 < \mu_2$$

$$H_A^{(3)} : \mu_1 \neq \mu_2$$

by means of the parametric procedures described in the following. The subsequent nonparametric tests will be used to test the hypothesis

$H_0 : F_1 = F_2$

against the alternatives

$$H_A^{(1)} : F_1 > F_2$$

$$H_A^{(2)} : F_1 < F_2$$

$$H_A^{(3)} : F_1 \neq F_2\ (\sigma_1^2 = \sigma_2^2).$$

Let $\{x_{i1}, \ldots, x_{in_i}\} =: \mathfrak{X}_i$ $(i = 1, 2)$ denote a realization of the mathematical random sample $\mathbf{x}_{i1} \ldots \mathbf{x}_{in_i}$, and let $x_i^{(j)}$ be the element situated at the j-th place among the elements of the sample $\mathfrak{X}_i$ $(i = 1, 2,\ j = 1, \ldots, n_i)$ which are ordered according to size and $x^{(j)}$ be the j-th element in the common sample $\mathfrak{X}_1 \cup \mathfrak{X}_2$ $(j = 1, \ldots, n_1 + n_2)$, which is also ordered according to size.

Moreover, let

$$Z_j = \begin{cases} 1, & \text{if } x^{(j)} \in X_1 \\ 0, & \text{if } x^{(j)} \in X_2 \end{cases}$$

The critical regions belonging to the alternative hypothesis $H_A^{(e)}$ will be denoted $K^{(e)}$ $(e = 1, 2, 3)$.

2.1. Parametric Procedures

— The t-test:

$$\text{test statistic}: \mathbf{T}_1^P = \frac{\bar{\mathbf{x}}_1 - \bar{\mathbf{x}}_2}{\sqrt{(n_1 - 1)\,\mathbf{s}_1^2 + (n_2 - 1)\,\mathbf{s}_2^2}} \sqrt{\frac{n_1 n_2 (n_1 + n_2 - 2)}{n_1 + n_2}}$$

with

$$\bar{\mathbf{x}}_i = \frac{1}{n_i} \sum_{j=1}^{n_i} \mathbf{x}_{ij} \text{ and}$$

$$\mathbf{s}_i^2 = \frac{1}{n_i - 1} \sum_{j=1}^{n_i} (\mathbf{x}_{ij} - \bar{\mathbf{x}}_i)^2 \qquad (i = 1, 2)$$

critical regions:

$$K^{(1)} = \{t,\ t \in \mathbf{R}:\ t > t(n_1 + n_2 - 2;\ 1 - \alpha)\}$$

$$K^{(2)} = \{t,\ t \in \mathbf{R}:\ t < -t(n_1 + n_2 - 2;\ 1 - \alpha)\}$$

$$K^{(3)} = \{t,\ t \in \mathbf{R}: |t| > t(n_1 + n_2 - 2;\ 1 - \alpha/2)\}$$

– The Welch test
test statistic:

$$\mathbf{T}_2^P = \frac{\bar{\mathbf{x}}_1 - \bar{\mathbf{x}}_2}{\sqrt{\frac{s_1^2}{n_1} + \frac{s_2^2}{n_2}}}$$

critical regions:

$$K^{(1)} = \{t,\ t \in \mathbf{R}:\ t > t(n^*, 1-\alpha)\}$$

$$K^{(2)} = \{t,\ t \in \mathbf{R}:\ t < -t(n^*, 1-\alpha)\}$$

$$K^{(3)} = \{t,\ t \in \mathbf{R}: |t| > t(n^*, 1-\alpha/2)\}$$

with

$$n^* = \left(\frac{s_1^2}{n_1} + \frac{s_2^2}{n_2}\right) \Big/ \left(\frac{s_1^4}{n_1^2(n_1-1)} + \frac{s_1^4}{n_2^2(n_2-1)}\right)$$

– The (1 – 1)-trimmed t-test ($\mathbf{T}_3^P$) and the (1 – 1)-Winsorised t-test ($\mathbf{T}_4^P$)
test statistic:

$$\mathbf{T}_3^P = \frac{\bar{\mathbf{x}}_{1t(1,1)} - \bar{\mathbf{x}}_{2t(1,1)}}{\sqrt{\frac{(n_1-3)s_{1w(1,1)}^2 + (n_2-3)s_{2w(1,1)}^2}{n_1+n_2-2}}\sqrt{\frac{1}{n_1}+\frac{1}{n_2}}} \cdot \frac{n_1+n_2-6}{n_1+n_2-2}$$

$$\mathbf{T}_4^P = \frac{\bar{\mathbf{x}}_{1w(1,1)} - \bar{\mathbf{x}}_{2w(1,1)}}{\sqrt{\frac{(n_1-3)s_{1w(1,1)}^2 + (n_2-3)s_{2w(1,1)}^2}{n_1+n_2-2}}\sqrt{\frac{1}{n_1}+\frac{1}{n_2}}} \cdot \frac{n_1+n_2-6}{n_1+n_2-2}$$

with

$$\bar{\mathbf{x}}_{it(1,1)} = \frac{1}{n_i-2}\sum_{j=1}^{n_i-1} \mathbf{x}_i^{(j)},$$

$$\bar{\mathbf{x}}_{iw(1,1)} = \frac{1}{n_i}\left(2\left(\mathbf{x}_i^{(2)} + \mathbf{x}_i^{(n-1)}\right) + \sum_{j=3}^{n_i-2} \mathbf{x}_i^{(j)}\right),$$

$$s_{iw(1,1)}^2 = \frac{1}{n_i-3}\left[2\left(\left(\mathbf{x}_i^{(2)} - \bar{\mathbf{x}}_{iw(1,1)}\right)^2 + \left(\mathbf{x}_i^{(n_i-1)} - \bar{\mathbf{x}}_{iw(1,1)}\right)^2\right) + \sum_{j=3}^{n_i-2}\left(\mathbf{x}_i^{(j)} - \bar{\mathbf{x}}_{iw(1,1)}\right)^2\right]$$

(i = 1,2).

critical regions: (are the same for the trimmed and Winsorised t-test)

$$K^{(1)} = \{t,\ t \in \mathbf{R}:\ t > t(n_1+n_2-6;\ 1-\alpha)\}$$

$$K^{(2)} = \{t,\ t \in \mathbf{R}:\ t < -t(n_1+n_2-6;\ 1-\alpha)\}$$

$$K^{(3)} = \{t,\ t \in \mathbf{R}: |t| > t(n_1+n_2-6;\ 1-\alpha/2)\}$$

– The (1, 1)-trimmed Welch test ($\mathbf{T}_5^P$) and the (1 – 1)-Winsorised Welch test ($\mathbf{T}_6^P$)
the test statistics:

$$\mathbf{T}_5^P = \frac{\bar{\mathbf{x}}_{1t(1,1)} - \bar{\mathbf{x}}_{2t(1,1)}}{\sqrt{\frac{s_{1w(1,1)}^2}{n_1-2} + \frac{s_{2w(1,1)}^2}{n_2-2}}}$$

$$\mathbf{T}_6^P = \frac{\bar{\mathbf{x}}_{1w(1,1)} - \bar{\mathbf{x}}_{2w(1,1)}}{\sqrt{\frac{s_{1w(1,1)}^2}{n_1-2} + \frac{s_{2w(1,1)}^2}{n_2-2}}}$$

critical regions: (are the same for the trimmed and Winsorised Welch test)

$$K^{(1)} = \{t,\ t \in \mathbf{R}:\ t > t(n^*;\ 1-\alpha)\}$$

$$K^{(2)} = \{t,\ t \in \mathbf{R}:\ t < -t(n^*;\ 1-\alpha)\}$$

$$K^{(3)} = \{t,\ t \in \mathbf{R}: |t| > t(n^*;\ 1-\alpha/2)\}$$

with

$$\frac{1}{n^*} = \frac{c^2}{n_1-3} + \frac{(1-c)^2}{n_2-3}, \qquad c = \frac{s_{1w(1,1)}^2/(n_1-2)}{s_{1w(1,1)}^2/(n_1-2) + s_{2w(1,1)}^2/(n_2-2)}$$

– The onesided trimmed t-test ($\mathbf{T}_7^P$) and the onesided Winsorised Welch test ($\mathbf{T}_8^P$)
test statistics:

$$\mathbf{T}_7^P = \frac{\bar{\mathbf{x}}_{1t(1)} - \bar{\mathbf{x}}_{2t(1)}}{\sqrt{\frac{(n_1-2)s_{1w(1)}^2 + (n_2-2)s_{2w(1)}^2}{n_1+n_2-2}}\sqrt{\frac{1}{n_1}+\frac{1}{n_2}}} \cdot \frac{n_1+n_2-4}{n_1+n_2-2}$$

$$\mathbf{T}_8^P = \frac{\bar{\mathbf{x}}_{1w(1)} - \bar{\mathbf{x}}_{2w(1)}}{\sqrt{\frac{(n_1-2)s_{1w(1)}^2 + (n_2-2)s_{2w(1)}^2}{n_1+n_2-2}}\sqrt{\frac{1}{n_1}+\frac{1}{n_2}}} \cdot \frac{n_1+n_2-4}{n_1+n_2-2}$$

with (i = 1, 2)

$$\bar{\mathbf{x}}_{it(1)} = \begin{cases} \bar{\mathbf{x}}_{it(1,0)}, & \text{if } \left|\mathbf{x}_i^{(1)} - \bar{\mathbf{x}}_i\right| > \left|\mathbf{x}_i^{(n_i)} - \bar{\mathbf{x}}_i\right| \\ \bar{\mathbf{x}}_{it(0,1)}, & \text{if } \left|\mathbf{x}_i^{(1)} - \bar{\mathbf{x}}_i\right| < \left|\mathbf{x}_i^{(n_i)} - \bar{\mathbf{x}}_i\right|, \end{cases}$$

$$\bar{\mathbf{x}}_{iw(1)} = \begin{cases} \bar{\mathbf{x}}_{iw(1,0)}, & \text{if } \left|\mathbf{x}_i^{(1)} - \bar{\mathbf{x}}_i\right| > \left|\mathbf{x}_i^{(n_i)} - \bar{\mathbf{x}}_i\right| \\ \bar{\mathbf{x}}_{iw(0,1)}, & \text{if } \left|\mathbf{x}_i^{(1)} - \bar{\mathbf{x}}_i\right| < \left|\mathbf{x}_i^{(n_i)} - \bar{\mathbf{x}}_i\right|, \end{cases}$$

and

$$s_{iw(1)}^2 = \begin{cases} s_{iw(1,0)}^2, & \text{if } \left|\mathbf{x}_i^{(1)} - \bar{\mathbf{x}}_i\right| > \left|\mathbf{x}_i^{(n_i)} - \bar{\mathbf{x}}_i\right| \\ s_{iw(0,1)}^2, & \text{if } \left|\mathbf{x}_i^{(1)} - \bar{\mathbf{x}}_i\right| < \left|\mathbf{x}_i^{(n_i)} - \bar{\mathbf{x}}_i\right|, \end{cases}$$

with

$$\bar{\mathbf{x}}_{it(1,0)} = \frac{1}{n_1-1}\sum_{j=2}^{n_i} \mathbf{x}_i^{(j)}, \quad \bar{\mathbf{x}}_{it(0,1)} = \frac{1}{n_1-1}\sum_{j=1}^{n_i-1} \mathbf{x}_i^{(j)},$$

$$\bar{\mathbf{x}}_{iw(1,0)} = \frac{1}{n_i}\left(2\mathbf{x}_i^{(2)} + \sum_{j=3}^{n_i} \mathbf{x}_i^{(j)}\right).$$

$$\bar{\mathbf{x}}_{iw(0,1)} = \frac{1}{n_i}\left(2\mathbf{x}_i^{(n_i-1)} + \sum_{j=1}^{n_i-2} \mathbf{x}_1^{(j)}\right)$$

$$s_{iw(1,0)}^2 = \frac{1}{n_i-2}\left[2\left(\mathbf{x}_i^{(2)} - \bar{\mathbf{x}}_{iw(1,0)}\right)^2 + \sum_{j=3}^{n_i}\left(\mathbf{x}_i^{(j)} - \bar{\mathbf{x}}_{iw(1,0)}\right)^2\right]$$

and

$$s_{iw(0,1)}^2 = \frac{1}{n_i-2}\left[2\left(\mathbf{x}_i^{(n_i-1)} - \bar{\mathbf{x}}_{iw(0,1)}\right)^2 + \sum_{j=1}^{n_i-2}\left(\mathbf{x}_i^{(j)} - \bar{\mathbf{x}}_{iw(0,1)}\right)^2\right]$$

(i = 1,2)

critical regions: (for T_7^P and T_8^P)

$$K^{(1)} = \{t,\ t \in \mathbf{R}:\ t > t(n_1+n_2-4;\ 1-\alpha)\}$$

$$K^{(2)} = \{t,\ t \in \mathbf{R}:\ t < -t(n_1+n_2-4;\ 1-\alpha)\}$$

$$K^{(3)} = \{t,\ t \in \mathbf{R}: |t| > t(n_1+n_2-4;\ 1-\alpha/2)\}$$

– The one sided trimmed Welch test $(\mathbf{T}_9^P)$ and the one-sided Winsorised Welch test $(\mathbf{T}_{10}^P)$

test statistic:

$$\mathbf{T}_9^P = \frac{\bar{\mathbf{x}}_{1t(1)} - \bar{\mathbf{x}}_{2t(1)}}{\sqrt{\frac{s^2_{1w(1)}}{n_1-1} + \frac{s^2_{2w(1)}}{n_2-1}}}$$

$$\mathbf{T}_{10}^P = \frac{\bar{\mathbf{x}}_{1w(1)} - \bar{\mathbf{x}}_{2w(1)}}{\sqrt{\frac{s^2_{1w(1)}}{n_1-1} + \frac{s^2_{2w(1)}}{n_2-1}}}$$

critical regions: (for $\mathbf{T}_9^P$ and $\mathbf{T}_{10}^P$)

$$K^{(1)} = \{t,\ t \in \mathbf{R}:\ t > t(n^*;\ 1-\alpha)\}$$

$$K^{(2)} = \{t,\ t \in \mathbf{R}:\ t < -t(n^*;\ 1-\alpha)\}$$

$$K^{(3)} = \{t,\ t \in \mathbf{R}:\ |t| > t(n^*;\ 1-\alpha/2)\}$$

with

$$\frac{1}{n} = \frac{c^2}{n_1-2} + \frac{(1-c)^2}{n_2-2} \quad \text{and} \quad c = \frac{s^2_{1w(1)}/(n_1-1)}{s^2_{1w(1)}/(n_1-1) + s^2_{2w(1)}/(n_2-1)}$$

– Use of a preliminary test of variance homogeneity

test statistic:

$$\mathbf{T}_{11}^P = \begin{cases} \mathbf{T}_1^P, & \text{if } \mathbf{F} \notin K^* \\ \mathbf{T}_2^P, & \text{if } \mathbf{F} \in K^* \end{cases}$$

with the "robust" Box-Andersen test (Nürnberg (1982))

$$\mathbf{F} = \frac{\max\{s_1^2, s_2^2\}}{\min\{s_1^2, s_2^2\}},$$

where

$$K^* = \begin{cases} \left\{t,\ t \in \mathbf{R}: t > F\left((n_1-1)\left(1+\frac{g_2'}{2}\right)^{-1}, (n_2-1)\left(1-\frac{g_2'}{2}\right)^{-1}; 1-\alpha/2\right)\right., & \text{if } s_1^2 > s_2^2 \\ \left\{t,\ t \in \mathbf{R}: t > F\left((n_2-1)\left(1+\frac{g_2'}{2}\right)^{-1}, (n_1-1)\left(1-\frac{g_2'}{2}\right)^{-1}; 1-\alpha/2\right)\right., & \text{if } s_1^2 < s_2^2. \end{cases}$$

$$g_2' = \frac{2n_1n_2\left(n_2\sum_{j=1}^{n_1}(x_{1j}-\bar{x}_1)^4 + n_1\sum_{j=1}^{n_2}(x_{2j}-\bar{x}_2)^4\right)}{\left[n_2\sum_{j=1}^{n_1}(x_{1j}-\bar{x}_1)^2 + n_1\sum_{j=1}^{n_2}(x_{2j}-\bar{x}_2)^2\right]^2} - 3$$

critical regions:

F ∉ K* → critical regions of the t-test

F ∈ K* → critical regions of the Welch-test

– Tiku's T-test

test statistic:

$$\mathbf{T}_{12}^P = \frac{\hat{\mu}_1 - \hat{\mu}_2}{\hat{\sigma}(1/m_1 + 1/m_2)^{1/2}}$$

$(i = 1,2)$

with Tiku's (1980) modified maximum likelihod (MML) estimators

$$\mu_i = \frac{1}{m_i}\left(\sum_{j=r_i+1}^{n_i-r_i} \mathbf{x}_i^{(j)} + r_i\beta_i\left(\mathbf{x}_i^{(r_i+1)} + \mathbf{x}_i^{(n_i-r_i)}\right)\right)$$

$$\hat{\sigma}_i = \frac{\mathbf{B}_i + \sqrt{(\mathbf{B}_i^2 + 4A_i\mathbf{C}_i)}}{2\sqrt{A_i(A_i-1)}}$$

$$\hat{\sigma}^2 = \frac{1}{A_1+A_2-2}\left((A_1-1)\hat{\sigma}_1^2 + (A_2-1)\hat{\sigma}_2^2\right)$$

with

$r_i = [0.5 + 0.1\,n_i]$ ([g] denotes the integer part of g)

$$A_i = n_i - 2r_i$$

$$m_i = n_i - 2r_i + 2r_i\beta_i$$

$$B_i = r_i\alpha_i\left(\mathbf{x}_i^{(n_i-r_i)} - \mathbf{x}_i^{(r_i+1)}\right)$$

$$C_i = \sum_{j=r_j+1}^{n_i-r_i}\left(\mathbf{x}_i^{(j)}\right)^2 + r_i\beta_i\left(\left(\mathbf{x}_i^{(r_i+1)}\right)^2 + \left(\mathbf{x}_i^{(n_i-r_i)}\right)^2\right) - m_i\mu_i^2$$

The constants α_i and β_i are obtained from the following equations

$$\beta_i = -f(t_i)\left(t_i - f(t_i)/q_i\right)/q_i;\ q_i = r_i/n_i$$

$$\alpha_i = \left(f(t_i)/q_i\right) - \beta_i t_i$$

t_i is determined by the equation $Q(t_i) = c_i$ where

$$Q(t_i) = 1 - \int_{-\infty}^{t_i} f(z)\,dz,$$

$$f(z) = \frac{1}{\sqrt{2\pi}}\exp(-z^2/2), \quad -\infty \le z \le \infty,$$

For several values of n_i table 2.1 contains the values of α_i and β_i:

Table 2.1

Some values of α_i and β_i

n_i	α_i	β_i
5	0.7410	0.7851
10	0.6737	0.8611
20	0.6873	0.8395
40	0.6896	0.8313

critical regions:

$$K^{(1)} = \{t,\ t \in \mathbf{R}:\ t > t(A_1+A_2-2;\ 1-\alpha)\}$$

$$K^{(2)} = \{t,\ t \in \mathbf{R}:\ t < -t(A_1+A_2-2;\ 1-\alpha)\}$$

$$K^{(3)} = \{t,\ t \in \mathbf{R}: |t| > t(A_1+A_2-2;\ 1-\alpha/2)\}$$

– Tiku's T_c-test

test statistic:

$$\mathbf{T}_{13}^P = \frac{\hat{\mu}_1 - \hat{\mu}_2}{\hat{\sigma}_1^2/m_1 + \hat{\sigma}_2^2/m_2}$$

critical regions:

$$K^{(1)} = \{t,\ t \in \mathbf{R}: t > t(n^*; 1-\alpha)\}$$

$$K^{(2)} = \{t,\ t \in \mathbf{R}: t < -t(n^*; 1-\alpha)\}$$

$$K^{(3)} = \{t,\ t \in \mathbf{R}: t > t(n^*; 1-\alpha/2)\}$$

with

$$n^* \text{from } \frac{1}{n^*} = \frac{c^2}{A_1 - 1} + \frac{(1-c)^2}{A_2 - 1}; \quad c = \frac{\hat{\sigma}_1^2/m_1}{\hat{\sigma}_1^2/m_1 + \hat{\sigma}_2^2/m_2}.$$

– The test of Lord (for small n_i)

test statistic: $\mathbf{T}_{14}^{P} = \dfrac{\bar{\mathbf{x}}_1 - \bar{\mathbf{x}}_2}{\mathbf{w}_1 + \mathbf{w}_2}$

where $\mathbf{w}_i = \mathbf{x}_i^{(n_i)} - \mathbf{x}^{(1)} \qquad (i=1,2)$

critical regions:

$$K^{(1)} = \{t,\ t \in \mathbf{R}:\ t > \tau(n_1, n_2; 1-\alpha)\}$$

$$K^{(2)} = \{t,\ t \in \mathbf{R}:\ t < -\tau(n_1, n_2; 1-\alpha)\}$$

$$K^{(3)} = \{t,\ t \in \mathbf{R}: |t| > \tau(n_1, n_2; 1-\alpha/2)\}$$

τ (FG, p) denotes the critical values of this test.

– The test of Bartlett and Scheffé

Let $n \leq n_2$.

test statistic:

$$\mathbf{T}_{15}^{P} = \frac{\sqrt{n_1(n_1-1)}(\bar{\mathbf{x}}_1 - \bar{\mathbf{x}}_2)}{\sqrt{\sum_{n=1}^{n_i} \mathbf{u}_j^2}}$$

with

$$\mathbf{u}_j = \mathbf{x}_{1j} - \bar{\mathbf{x}}_1 - \left(\frac{n_1}{n_2}\right)^{1/2} \mathbf{x}_{2j} + (n_1 n_2)^{-1/2} \sum_{j=1}^{n_1} \mathbf{x}_{2j}$$

critical regions:

$$K^{(1)} = \{t,\ t \in \mathbf{R}:\ t > t(n_1 - 1;\ 1-\alpha)\}$$

$$K^{(2)} = \{t,\ t \in \mathbf{R}:\ t < -t(n_1 - 1;\ 1-\alpha)\}$$

$$K^{(3)} = \{t,\ t \in \mathbf{R}:\ |t| > t(n_1 - 1;\ 1-\alpha/2)\}$$

– The approximative test of Chochran (1964)

test statistic:

$$\mathbf{T}_{16}^{P} = \frac{\bar{\mathbf{x}}_1 - \bar{\mathbf{x}}_2}{\sqrt{\frac{\mathbf{s}_1^2}{n_1} + \frac{\mathbf{s}_2^2}{n_2}}}$$

critical regions:

$$K^{(1)} = \{t,\ t \in \mathbf{R}:\ t > t'(n_1 + n_2 - 2;\ 1-\alpha)\}$$

$$K^{(2)} = \{t,\ t \in \mathbf{R}:\ t < -t'(n_1 + n_2 - 2;\ 1-\alpha)\}$$

$$K^{(3)} = \{t,\ t \in \mathbf{R}:\ |t| > t'(n_1 + n_2 - 2;\ 1-\alpha/2)\}$$

Whith the approximative critical values

$$t'(n_1 - n_2 - 2; p) = \frac{w_1' t(n_1 - 1; p) + w_2' t(n_2 - 1; p)}{w_1' + w_2'}$$

where $w_i' = \dfrac{s_i^2}{n_i} \qquad (i=1,2)$

– The approximative BM-test of Banerjee (1960) and McCullough, Gurland and Rosenberg (1960)

test statistic:

$$\mathbf{T}_{17}^{P} = \frac{\bar{\mathbf{x}}_1 - \bar{\mathbf{x}}_2}{\sqrt{\frac{\mathbf{s}_1^2}{n_1} + \frac{\mathbf{s}_2^2}{n_2}}}$$

critical regions:

$$K^{(1)} = \{t,\ t \in \mathbf{R}:\ t > t''(n_1 + n_2 - 2;\ 1-\alpha)\}$$

$$K^{(2)} = \{t,\ t \in \mathbf{R}:\ t < -t''(n_1 + n_2 - 2;\ 1-\alpha)\}$$

$$K^{(3)} = \{t,\ t \in \mathbf{R}: |t| > t''(n_1 + n_2 - 2;\ 1-\alpha/2)\}$$

with the approximative critical values

$$t''(n_1 + n_2 - 2; p) = \frac{\left[w_1' t^2(n_1 - 1; p) + w_2' t_2\ (n_2 - 1; p)\right]^{1/2}}{w_1' + w_2'}$$

where $w_i' = \dfrac{s_i^2}{n_i} \qquad (i=1,2).$

2.2. Nonparametric procedures

— The X-test of van der Waerden

test statistic:

$$\mathbf{T}_1^{NP} = \sum_{j=1}^{n_i} \psi\left(\frac{r(\mathbf{x}_{1j})}{n_1 + n_2 + 1}\right)$$

$r(x_{1j})$ denotes the rank of x_{1j} in the common sample $\mathfrak{X}_1 \cup \mathfrak{X}_2$, which is ordered according to size, $\psi(\cdot) = \Phi^{-1}(\cdot)$ and $\Phi(\cdot)$ is the N (0,1)-distribution function

critical regions:

$$K^{(1)} = \{t,\ t \in \mathbf{R}:\ t > t_x(n_1, n_2;\ \alpha)\}$$

$$K^{(2)} = \{t,\ t \in \mathbf{R}:\ t < -t_x(n_1, n_2;\ \alpha)\}$$

$$K^{(3)} = \{t,\ t \in \mathbf{R}:\ |t| > t_x(n_1, n_2;\ 1-\alpha/2)\}$$

$t_x(n_1, n_2; p)$ denotes the critical values of the X-test. For $n_i \to \infty$ we get the asymptotically critical regions:

$$K_{as}^{(1)} = \left\{t \mid t \in \mathbf{R}:\ t > \sqrt{\frac{n_1 n_2}{n}}\ Q_n u(1-\alpha)\right\}$$

$$K_{as}^{(2)} = \left\{t \mid t \in \mathbf{R}:\ t < -\sqrt{\frac{n_1 n_2}{n}}\ Q_n u(1-\alpha)\right\}$$

$$K_{as}^{(3)} = \left\{t \mid t \in \mathbf{R}: |t| > \sqrt{\frac{n_1 n_2}{n}}\ Q_n u(1-\alpha/2)\right\}$$

where $Q_n = \dfrac{1}{n} \sum_{j=1}^{n} \psi^2\left(\dfrac{j}{n+1}\right)$, $n = n_2 + n_2$,u(p) p-quantile of the N (0,1)-distribution.

– The U-test (Wilcoxon or Mann-Whitney test)

Let

$$\mathbf{U}_{jj'} = \begin{cases} 1, & \text{if } \mathbf{x}_{1j} > \mathbf{x}_{2j'}, \quad j = 1, \ldots, n_1, \\ 0, & \text{else} \qquad\qquad j' = 1, \ldots, n_2 \end{cases}$$

in the common sample $\mathfrak{X}_1 \cup \mathfrak{X}_2$.

test statistic: (one sided test):

$$\mathbf{T}_2^{NP} = \sum_{j=1}^{n_1} \sum_{j'=1}^{n_2} \mathbf{U}_{jj'}$$

Remark: T_2^{NP} is easier to compute by

$$\mathbf{T}_2^{NP} = \sum_{j=1}^{n_1} r(\mathbf{x}_{1j}) - \frac{n_1(n_1+1)}{2}$$

where $r(x_{1j})$ is the rank of x_{1j} in the ordered common sample.

For the one-sided test one can also make use of the test statistic

$$\mathbf{T}_2^{NP'} = \sum_{j=1}^{n_2} r(\mathbf{x}_{2j}) - \frac{n_2(n_2+1)}{2}$$

In the two sided case $T_2^{NP''} = \min \{\mathbf{T}_2^{NP}, T_2^{NP'}\}$ is to be used.

critical regions:

$$K^{(1)} = \{t, t \in \mathbf{R}: t < t'_u(n_1, n_2, \alpha)\}$$

$$K^{(2)} = \{t, t \in \mathbf{R}: t > n_1 n_2 - t'_u(n_1, n_2, \alpha)\}$$

$$K^{(3)} = \{t, t \in \mathbf{R}: t < t_u(n_1, n_2, \alpha)\}$$

t'_u $(n_1, n_2; \alpha)$ and $t_u(n_1, n_2, \alpha)$ denote the critical values of the U-test with

t'_u $(n_1, n_2; \alpha) = t_u$ $(n_1, n_2; 2\alpha)$.

For $n_i \to \infty$ we get the asymtotical critical regions:

$$K_{as}^{(1)} = \left\{t, t \in \mathbf{R}: t < \frac{n_1 n_2}{2} - u(1-\alpha) \sqrt{\frac{1}{12} n_1 n_2 (n_1 + n_2 + 1)}\right\}$$

$$K_{as}^{(2)} = \left\{t, t \in \mathbf{R}: t > \frac{n_1 n_2}{2} + u(1-\alpha) \sqrt{\frac{1}{12} n_1 n_2 (n_1 + n_2 + 1)}\right\}$$

$$K_{as}^{(3)} = \left\{t, t \in \mathbf{R}: t < \frac{n_1 n_2}{2} - u(1-\alpha/2) \sqrt{\frac{1}{12} n_1 n_2 (n_1 + n_2 + 1)}\right\}$$

– A quick test of Tukey

test statistic:

$$\mathbf{T}_3^{NP} = \begin{cases} 0, \text{ if } Z_1 = Z_{n_1+n_2} = 1 \vee Z_1 = Z_{n_1+n_2} = 0 \\ \text{Sum of the lengths of the first and last runs,} \\ \text{one being of 0's and the other of 1's.} \end{cases}$$

critical region:

$$K^{(3)} = \{t, t \in \mathbf{N}: t \geq t_Q(n_1, n_2; \alpha)\}.$$

3. Simulation experiment

According to Nürnberg (1983) and Rasch et al. (1978), the sample size necessary for the simulation is given by

$$N_S \approx \frac{u^2(1-\alpha^*/2)(\alpha+\varepsilon)(1-\alpha-\varepsilon)}{d^{*2}}$$

where $\varepsilon = \alpha/5$

$$d^* = \frac{\varepsilon}{2}$$

$$\alpha^* = 0.05$$

and u (P) is the P-quantile of the N (0, 1)-normal distribution. Table 3.1 contains the necessary numbers N_S of the simulations for a few values of α. For reasons of economy we take $N_S = 10\,000$ (caution when interpreting the results).

Table 3.1

Necessary numbers N_S of the simulations

α	N_S
0.005	114 552
0.010	45 546
0.025	17 904
0.050	8 667
0.100	4 057

Since the sample sizes n_1 and n_2 play a major role in studies on the robustness of tests used for comparing two means, rather extensive simulation studies using both equal and unequal sample sizes should be performed.

Table 3.2

Sample sizes

n_1	5	5	5	10	10	20	40
n_2	5	10	20	20	40	20	40

The pseudorandom numbers which are distributed with a given skewness γ_1 and kurtosis γ_2 (c.f. table 3.3)

Table 3.3

Values of γ_1 (skewness) and γ_2 (kurtosis) for the simulations

γ_1	0	0	0	0	1.0	1.5	2.0
γ_2	0	1.5	3.75	7.0	1.5	3.75	7.0

required for simulation studies on robustness should be generated in accordance with the following scheme:

Step 1:

Generation of (0, 1)-uniformly distributed pseudorandom numbers by means of the generator ZZGD

Step 2:

Transformation of the (0, 1)-uniformly distributed random rumbers (Ode/Evans (1974)) into N (0, 1) random numbers

Step 3:

Application to traisform the N (0, 1) random numbers into random numbers with a specified skewness and kurtosis.

For several distributions in the γ_1-γ_2-plane that cannot be reached by the polynomial transformation, pseudorandom numbers should be generated by the system of truncated normal distributions (Rasch, Teuscher (1982)).

To be able to perform corresponding robustness studies, it is necessary to estimate the first kind risks, α (n_1, n_2, g). and second kind risks, $1-\beta$ (n_1, n_2, g, δ)

$$\text{where } \delta = \frac{|\mu_1 - \mu_2|}{\left(\frac{\sigma_1^2}{n_1} + \frac{\sigma_2^2}{n_2}\right)^{1/2}}$$

The parameter δ is required to ensure that the power functions of the tests being investigated are comparable. Apart for the estimation at $\delta = 0$ (the first kind risk is estimated in this case), the power function should be estimated at three further points.

In order to choose three suitable δ-points, we first consider the power function of the t-test under the normal distribution for $\sigma_1^2 = \sigma_2^2 = 1$, $\alpha = 0.025$ for $H_A^{(1)}$.

Table 3.4

δ-values for given power of the t-test for $n_1 = n_2 = n$

	δ		
$1-\beta$	n = 5	n = 20	n = 40
0.10	0.910	0.699	0.698
0.50	2.306	2.021	1.990
0.90	3.703	3.351	3.282

The following values of δ are thus suitable for the simulation experiment:

$\delta_1 = 0$
$\delta_2 = 0.7$
$\delta_3 = 2.0$
$\delta_4 = 3.5.$

For each experimental design (n_1, n_2), each quotient

$$\frac{\sigma_1^2}{\sigma_2^2} \in \left\{\frac{1}{2}, 1, 2\right\}$$

and each point (γ_1, γ_2) in the (γ_1, γ_2)-plane to be investigated, we obtain, after 10 000 simulations, the estimation

$$\widehat{\alpha(n_1, n_2, g)} = \frac{N_A^{(1)}(T_k | \delta_1)}{N_S}$$

for the first kind risk and

$$\widehat{1 - \beta(n_1, n_2, g, \delta_r)} = \frac{N_A^{(1)}(T_k | \delta_r)}{N_S}; \quad r = 2, 3, 4$$

for the second kind risk.

$N_A^{(l)}(T_k \mid \delta_r)$ with $r = 1, 2, 3, 4$; $l = 1, 2, 3$; $T_k \in \mathfrak{T}_P \cup \mathfrak{T}_{NP}$ denotes the number of rejections of the null hypothesis in favour of the alternative hypotheses $H_A^{(l)}$ for the test T_k under the condition δ_r.

The simulation experiments will be performed mainly on a EC 1055 computer.

4. References

AFIFI, A. A., KIM, P. J.
Comparison of some Two-Samples Location Test for Nonnormal Alternatives.
J. R. S. S. Ser. B, 34 (1972), 448–455.

CAMS, D. J.
Corrected and Extended Tables for Tuckey's Quick Test.
Technometrics 23 (1981), 193–195.

GUIARD, V. (ed.)
Robustheit II.
Probleme der angewandten Statistik – FZ für Tierproduktion Dummerstorf-Rostock, Heft 5, (1981).

HERRENDÖRFER, G. (ed.)
Robustheit I.
Probleme der angewandten Statistik – FZ für Tierproduktion Dummerstorf-Rostock, Heft 4, (1980).

LEE, E. T., DESU, M. M., GEHAN, E. A.
A Monte Carlo study of the power of some two-sample tests.
Biometrika 62 (1975) 2, 425–432.

MURPHY, P. B.
Some two-sample tests when the variances are unequal: a simulation study.
Biometrika 54 (1967) 4, 679–683.

NEAVE, H. R., GRANGER, C. W. J.
A Monte Carlo Study Comparing Various two sample tests.
Technometrics 10 (1968) 3, 509–522.

NÜRNBERG, G.
Beiträge zur Versuchsplanung für die Schätzung von Varianzkomponenten und Robustheitsuntersuchungen zum Vergleich zweier Varianzen.
Probleme der angewandten Statistik – FZ für Tierproduktion Dummerstorf-Rostock (1982) 6.

POSTEN, H. O.
The robustness of the two-sample t-test over the Pearson system.
J. Stat. Comp. Simul. 6 (1978), 295–311.

PRATT, J. W.
Robustness of some Procedure for the Two-sample Location Problem.
JASA 59 (1964), 665–680.

RASCH, D., HERRENDÖRFER, G. (ed.)
Robustheit III.
Probleme der angewandten Statistik – FZ für Tierproduktion Dummerstorf-Rostock, Heft 7, (1982).

RASCH, D., HERRENDÖRFER, G., BOCK, J., BUSCH, K.
Verfahrensbibliothek Versuchsplanung und -auswertung.
Berlin: VEB Deutscher Landwirtschaftsverlag Bd. I, II: 1978; Bd. III: 1981.

RASCH, D., TEUSCHER, F.
Das System der gestutzten Normalverteilungen.
7. Sitzungsbericht der IG Mathematische Statistik Berlin 1982, 51–60.

TIKU, M. L.
Robust two-sample tests and robust regression and analysis of variance with MML-estimators.
Statistical Report Nr. 2 (1960) McMaster University Hamilton, Canada.

WELCH, B. L.
The significance of the difference between two means when the population variances are unequal.
Biometrika 29 (1937), 250–262.

YUEN, K. K.
The two-sample trimmed t for unequal population variances.
Biometrika 61 (1974) 1, 165–170.

Institute of Econometrics and Statistics, University of Łódź, Poland

On the Measurement of the Tests Robustness

ZBIGNIEW WASILEWSKI

Abstract

In the paper some possible evaluation criteria of robustness of statistical tests are presented. The criteria have been used in evaluating robustness of the test for significance of the parameters of linear regression model in the case of outliers existence.

I. Introduction

Great interest in robustness as an approach to make advances in statistics was evoked by the observed fact that we never have exact information on the true form of distribution generating the real data, and that the behaviour of some classical statistical procedures (tests, estimators) is very sensitive with respect to even small changes in the form of distribution generating the data. The essence of robustness consists in protection against the influence of some possible departures from the assumed statistical model (likely to occur in practice) on the properties of statistics derived under this model. The concept of robustness has been introduced to statistics by Box and Anderson (1955) and concerned the robustness of statistical test. They postulated that the tests should be:

(i) sensitive to the changes of verified quantities,

(ii) non-sensitive (robust) to the changes of non-verified quantities that play the role of external factors in the verified hypothesis.

The new concept was not, however, defined from the start formally enough which led to many different attempts to define it more precisely. To get an idea of diversity of interpretations of robustness it is enough to look into paper of Huber (1964, 1972), Hampel (1971), Bickel (1976), Dutter (1979), Bartoszyński-Pleszczyńska (1980), Zieliński (1980, 1981). In the case of general statistical decision problems we can state that a given solution method derived under $\mathfrak{M}_0$ changes a little if we replace $\mathfrak{M}_0$ by its approximation. The approximation error influences of course both the robust methods as well as non-robust ones. In the case of robust methods this influence is however, small. In order to characterize this "smallness" it is necessary to define natural extensions (neighbourhoods) of the standard model $\mathfrak{M}_0$, and then to formulate criteria (measures) of evaluation of robustness of a given statistic "T" with respect to some of the properties (such as bias, variance, mean square error, power function of the test etc.) which we are specially interested in. Some interesting ideas on a formal description of this problem were given by Zieliński (1981).

In this paper we restrict our attention to the problem of the evaluation criteria of robustness of statistical tests. We propose some measures of the robustness of the tests' power function and use them to analyze the robustness of the t-Student test for significance of parameters of the linear regression model when some of the non-verified assumptions taken into account by the construction of the test are broken.

2. Evaluation Criteria of Robustness of Significance Tests

Let $\mathfrak{M}_0$ be a standard statistical model describing a real system. We can distinguish stochastic and non-stochastic elements of this structure. The only tested elements are stochastic structure' elements. We shall call them parameters of stochastic structure of the model $\mathfrak{M}_0$. By a model's parameter we understand both a probability measure on the sample space (a distribution function of this measure) as well as traditionally understood parameters of a probability distribution (e. g. location and scale parameters of a given probability function). By values of these parameters we mean a concrete functional shape of distribution function (e. g. normal distribution function) in the first case, and concrete numerical values of the probability function parameters (e. g. mean of univariate distribution function is equal 2). Within the statistical decision problem $\mathfrak{D}$ concerning stochastic structure of the model $\mathfrak{M}_0$ its parameters would be (in accordance to Box-Anderson postulates) separated into two groups:

a) verified by a given test,

b) non-verified, external factors in the verified hypothesis.

More formally, let $\mathfrak{Z}$, Z_j, $j = 0, \ldots, l$ denote the set of the non-verified (in $\mathfrak{D}$) parameters of $\mathfrak{M}_0$ and the j-th possible set of the externally (to $\mathfrak{D}$) assumed values of these parameters, respectively. $\mathfrak{W}$, W_i, $i = 0, \ldots, m$ denote respectively the set of tested (in $\mathfrak{D}$) parameters of $\mathfrak{M}_0$ and the i-th verified set of their values. The choice of the (i, j)-form of the model $\mathfrak{M}_0 = \mathfrak{M}_0(W_i, T_j)$ is equivalent to the assumed (i, j) stochastic structure of the analyzed system.

For the model $\mathfrak{M}_0$ we can define two groups of statistical hypotheses concerning their parameters. These are:

I $\quad H_{oo}: (\mathfrak{W} = W_o) \,|\, (\mathfrak{Z} = Z_o)$

$\quad\; H_{io}: (\mathfrak{W} = W_i \neq W_o) \,|\, (\mathfrak{Z} = Z_o) \qquad i = 1, \ldots, m$

and

II $\quad \bar{H}_{oj}: (\mathfrak{W} = W_o) \,|\, (\mathfrak{Z} = Z_j \neq Z_o) \qquad j = 1, \ldots, l$

$\quad\;\; \bar{H}_{ij}: (\mathfrak{W} = W_i \neq W_o) \,|\, (\mathfrak{Z} = Z_j \neq Z_o) \qquad i = 1, \ldots, m,$

where

H_{00}, $\bar{H}_{0j}$ — denote given "zero" hypotheses and H_{i0}, $\bar{H}_{ij}$ — alternative (to H_{00}, $\bar{H}_{0j}$) hypotheses concerning the form of $\mathfrak{W}$

W_0 — denotes the set of verified parameters values within H_{00} and $\bar{H}_{0j}$

"|" — denotes "under condition"

Z_0 — the set of non-verified parameters' values, which are fixed externally to $\mathfrak{D}$

W_i ; $i = \overline{1, m}$ — denotes the i-th (alternative to W_0) set of tested parameters' values for the model $\mathfrak{M}_0$

$Z_j \neq Z_0$; $j = \overline{1, l}$ — denotes the j-th alternative to Z_0 set of non-tested values of parameters for the model $\mathfrak{M}_0$.

Group I of hypotheses matches the classical testing situation when $\mathfrak{Z}$ is fixed externally, its form does not change

and the testifiable parameters set is $\mathfrak{W}$. In this situation we evaluate test performance by analyzing the power of a given test under the significance point equal to α.
Group II of hypotheses has been distinguished with respect to the change of form of set $\mathfrak{Z}$. Thus the hypothesis $\overline{H}_{ij}$ is an alternative for both $\mathfrak{W} = W_0$ and $\mathfrak{Z} = Z_0$. In this situation we evaluate test performance by analyzing a given significance test.
Using the above notation we can formulate

Definition 1

A significance level of test T will be called robust with respect to changes of values of non-verified (by the use of the significance test T) model $\mathfrak{M}_0$ parameters iff

$$\forall \varepsilon>0\ \exists \delta>0\ \forall Z_j \in \mathfrak{Z}\ \forall W_0 \in \mathfrak{W}:$$

$$\varrho_Z(Z_0, Z_j) < \delta \Rightarrow \varrho_S\Big(\{P[\mathrm{val}(T) \in \varphi \mid H_{00}]\},\ \{P[\mathrm{val}(T) \in \varphi \mid \overline{H}_{0j}]\}\Big) < \varepsilon,$$

where:

Z — a measurable space of external parameter values,
S — a probability space (e. g. $S = (\mathfrak{U}, \mathfrak{F}, P)$, $\mathfrak{U}$ — the set of elementary events, $\mathfrak{F}$ is the σ-borel field of $\mathfrak{U}$ subsets, P — a probability measure defined on $\mathfrak{F}$.
φ — is the critical region of T, T — a test statistic,
ϱ_Z, ϱ_S — are some metrics in space Z and S.

Definition 2

Power function of a significance level α of test T will be called robust with respect to changes of values of non-verified (by the use of T) parameters of model $\mathfrak{M}_{0i}$ iff

$$\forall \varepsilon>0\ \exists \delta_{(\varepsilon)}>0\ \forall Z_j \in \mathfrak{Z}\ \forall W_i \in \mathfrak{W}:$$

$$\varrho_Z(Z_0, Z_j) < \delta \Rightarrow \varrho_S\Big(\{P[\mathrm{val}(T) \in \varphi \mid H_{i0}]\},\ \{P[\mathrm{val}(T) \in \varphi \mid \overline{H}_{ij}]\}\Big) < \varepsilon.$$

Due to lack of appropriate numerical methods of calculating the distances between probability distributions a direct study of robustness behaviour on the ground of Definition 1 and 2 is very difficult. There is, however, a possibility to formulate and use indirect criteria of test robustness evaluation when the level α is fixed. Some of these criteria are as follows:
C1: Test's conventional power as a function of parameters belonging to $\mathfrak{W}$:

$$\pi_1(\alpha) = P[\mathrm{val}(T) \in \varphi_\alpha \mid H_{i0}],$$

where φ_α is such that $\mathfrak{P}$ [val (T) $\in \varphi_\alpha \mid H_{00}] = \alpha$, α is fixed;
C2: Relative change of a given test's significance level α corresponding to the change of $\mathfrak{Z} = Z_0$ into $\mathfrak{Z} = Z_j \mid j = \overline{1,1}$ as a function of Z_j. It measures the stability of α with respect to changes from Z_0 into Z_j and has the form

$$\pi_2(Z_j) = \pi_2(Z_j, \alpha) = \left(P[\mathrm{val}(T) \in \varphi_\alpha \mid \overline{H}_{0j}] - \alpha\right)/\alpha;$$

C3: Test's guaranteed significance level

$$\pi_3(Z_j^{max}) = \pi_3(Z_j^{max}, \alpha) = \max_{\substack{Z_j \\ j=\overline{1,1}}} \{P[\mathrm{val}(T) \in \varphi_\alpha \mid \overline{H}_{0j}]\};$$

C4: Relative change of a given test's power corresponding to the change from Z_0 into Z_j, $j = \overline{1,1}$ as a function of Z_j. The criterion function measures the stability of test's power with respect to the changes of non-verified parameters' values Z_j, $j = \overline{0,1}$, i. e.

$$\pi_4(Z_j, \alpha) = \left(P[\mathrm{val}(T) \in \varphi_\alpha \mid \overline{H}_{ij}] - \pi_1(\alpha)\right)/\pi_1(\alpha);$$

C5: Test's guaranteed power for the joint alternative hypothesis $\overline{H}_i = \overline{H}_{i1}, \overline{H}_{i2} \ldots \overline{H}_{i1}$, i. e.

$$\pi_5(Z_j^{min}) = \min_{\substack{Z_j \\ j=\overline{1,1}}} \{P[\mathrm{val}(T) \in \varphi_\alpha \mid \overline{H}_{ij}]\}$$

Criteria π_2, π_4 enable us to carry out robustness evaluation of test size and power on changes from Z_0 into Z_i, $j = \overline{1,1}$.
Criteria π_3 and π_5 enable us to determine the extreme values of test size and power for the specified Z_j, $j = 1, 1$. They are, in some sense, the criteria of choice of the most robust test. Zieliński (1980) proposes two other robustness measures for the given extension of $\mathfrak{M}_0$. We use this proposition in the case of the distinguished set family Z_j, $j = \overline{0,1}$, taking a difference between sup and inf values of test significance levels and powers over the sets Z_j, $j = \overline{0,1}$.
The successive two criteria for evaluation of test robustness are

C6: Range of significance levels on the neighbourhoods of H_{00}:

$$\pi_6(\alpha) = \sup_{\substack{Z_j, Z_0 \\ j=\overline{1,1}}} \{P[\mathrm{val}(T) \in \varphi_\alpha \mid \overline{H}_{0j} \cup H_{00}]\} - \inf_{\substack{Z_j, Z_0 \\ j=\overline{0,1}}} \{P[\mathrm{val}(T) \in \varphi_\alpha \mid \overline{H}_{0j} \cup H_{00}]\}$$

C7: Range of powers on the neighbourhoods of H_{i0}:

$$\pi_7(\alpha, W_i) = \sup_{\substack{Z_j, Z_0 \\ j=\overline{1,1}}} \{P[\mathrm{val}(T) \in \varphi_\alpha \mid H_{i0} \cup \overline{H}_{ij}]\} - \inf_{\substack{Z_j, Z_0 \\ j=\overline{1,1}}} \{P[\mathrm{val}(T) \in \varphi_\alpha \mid H_{i0} \cup \overline{H}_{ij}]\}.$$

Small values of π_6, π_7 indicate great robustness of test on values changes of non-verified parameter Z_j, $j = \overline{0,1}$. All the presented criteria of robustness evaluation have been chosen having in mind their direct usefulness in Monte Carlo studies of tests robustness. An illustration how to use some of the measures $\pi_1, \ldots, \pi_7$ will be given in the next section.

3. Some Results of Measuring the Robustness of Chosen Tests

The measures of robustness proposed above have been used in evaluating the robustness of t-Student test for significance of parameters in a linear regression model.
We have considered the linear regression model with contamination:

$$\mathfrak{M}_0 := \left(R^{n \times k}, S, Y = X\beta + \Xi + \sum_{j=1}^{n} \mathring{\mu}_j j_j,\ k_0 = k,\ n_0 = n,\ P_\Xi \sim \mathfrak{N}_\Xi(\mu, \Omega),\ \mu = 0,\ \Omega = \sigma^2 I\right)$$

where $R^{n \times k}$ — set of $(n \times k)$ real matrices
$S = (\mathfrak{U}, \mathfrak{F}, P)$ — where $\mathfrak{U}$ denotes the sample space, $\mathfrak{F}$ is a borel σ-field of $\mathfrak{U}$ subsets, P is a measure satisfying the condition $P(\mathfrak{U}) = 1$;

$Y, \Xi \in S$

$\beta \in R^{k \times 1}$

$X \in R^{n \times k}$

$k_0 =)k$, $n_0 = n$ denote ranks of matrices X and Ω respecticely

$P_{\Xi} \sim N_{\Xi}(\mu, \Omega)$ denotes that the probability distribution function of Ξ belongs to the class of normal distribution functions with expected value $\mathfrak{E}(\Xi) = \mu$ and variance-covariance matrix $\mathfrak{D}(\Xi) = \Omega$.

$\mathring{\mu}_i$ — denotes the contamination constant added to the i-th component of vector Y, and

j_i — denotes $(n \times 1)$ vector for which i-component is equal to 1 and others are equal to 0.

For model $\mathfrak{M}_0$ we have formulated two groups of hypotheses concerning the verification of the significance of the parameter β_1 by means of t-Student test.

$$\text{I} \quad H_{oo}: \left(\mathfrak{W} = W_o = \{\beta_1 = 0\}\right) \Big| \left(\mathfrak{Z} = Z_o = \{P_{\Xi} \sim \mathfrak{N}_{\Xi}(\mu, \Omega), \; \mu = 0, \Omega = \sigma^2 I, \forall_i: \mathring{\mu}_i = 0\}\right)$$

$$H_{io}: \left(\mathfrak{W} = W_i = \{\beta_1 = a_i\}\right) \Big| \left(\mathfrak{Z} = Z_o\right) \qquad i = 1, \ldots, 7$$

$$\text{II} \quad \bar{H}_{oj}: \left(\mathfrak{W} = W_o\right) \Big| \left(\mathfrak{Z} = Z_j = \{P_{\Xi} \sim \mathfrak{N}_{\Xi}(\mu, \Omega), \; \mu = 0, \Omega = \sigma^2 I, \mathring{\mu}_i = 0, \; i \neq q, p, \mathring{\mu}_p = c_j, \mathring{\mu}_q = d_j\}\right)$$

$$\bar{H}_{ij}: \left(\mathfrak{W} = W_i\right) \Big| \left(\mathfrak{Z} = Z_j\right) \qquad i = 1, \ldots, 7$$

In the case of the first group of hypotheses the test statistic $t = b_1 / S(b_1)$, where b_1 is the 1-component of the least squares (1—s) estimator, $b = (X^T X)^{-1} X^T Y$, $S(b_1) = \frac{E^T E}{n-k} \operatorname{diag} (X^T X)^{-1}_1$ and $\operatorname{diag} (X^T X)^{-1}_1$ denotes the 1-th diagonal element of matrix $(X^T X)^{-1}$, has by zero hypothesis, t-Student distribution with $n - k$ degrees of freedom. In the case of the second group of hypotheses, however, (which assume the existence of outliers in the model) we do not know this distribution. It is because of the biasedness of the estimate b_1

$$\mathfrak{E}(b_1) = \beta_1 + \sum_{i=1}^{n} \left(X^T X\right)^{-1} X_i^T, \mathring{\mu}_i j_i$$

and the non-centrality of the $\sigma^2 \chi^2_{n-k}$ distribution of residual sum of squares. The value of the parameter of non-centrality is equal $\frac{1}{2} \mathring{\mu}^T M \mathring{\mu}$ where

$$M = (I - X (X^T X)^{-1} X^T) \text{ and } \mathring{\mu} = \sum_{i=1}^{n} \mathring{\mu} j_i$$

and the same as biasedness of b_1 depends on the values of $\mathring{\mu}_i$. Similar difficulty arises when we use different than $1 - s$ method of estimation of the vector β, for instance, the method of minimization of absolute values of residuals (msae). To answer the question to what degree such factors as the above mentioned ones, influenced the power function of the t-Student test, a Monte Carlo experiment has been carried out. We limited our attention to the standardized form of the model $\mathfrak{M}_0$, generating sample data with uncorrelated explanatory variables from standardized normal distribution and var (Y) = 1. By these assumptions the parameters of the model defined the value of the correlation bexween the explained variable Y and respective explanatory variables (i. e. $\beta_i = \operatorname{corr}(Y, X_i)$ $i = 1, \ldots, k$). The multiple correlation coefficient was given by $\varrho = \sqrt{\sum_{i=1}^{k} \beta_i^2}$ and equal 0.07, and var $(\Xi) = \sigma = 0.174$. The values of sample size and number of parameters were equal $n = 20$ and $k = 3$, respectively. The power of the test was calculated for 8 different from zero values of β_2 equal —0.3, —0.1, —0.05, 0.05, 0.1, 0.15, 0.2, 0.3, respectively, and 7 versions of contamination of the first and last component of the vector Y. Each combination of the β value and values of outliers charakterized one of the hypotheses $\bar{H}_{ij}$ $(i = \overline{1.8}, j = \overline{1.7})$.

Table 1.
Values of contaminated constants given as the multiplication of the standard deviation of the error term in the model

j	0	1	2	3	4	5	6	7
level of contamination								
μ^1	0	0	0	5σ	10σ	-5σ	-5σ	-5σ
μ^2_0	0	5σ	10σ	5σ	10σ	0	5σ	10σ

For each form of sets $\mathfrak{W}$ and $\mathfrak{Z}$ 500 samples were generated and the values of the estimate of power of the test were computed. The critical values of test were fixed for the significance level equal 0.1. This was done both in the case when parameters of the model were estimated by 1 — s method and by minimization of the sum of absolut errors (msae). Then the measures of the robustness of the test were evaluated. Below in Tables 2 and 3 some of the results are given.

Figure 1
The shape of the power function fo $t_{0.1}$ test in the case of $1 - s$ and msae estimation
$P[\text{val}(t) \in \varphi_{0.1} \mid \bar{H}_{ij})$ $\quad i = 0.9 \quad j = 0.7$

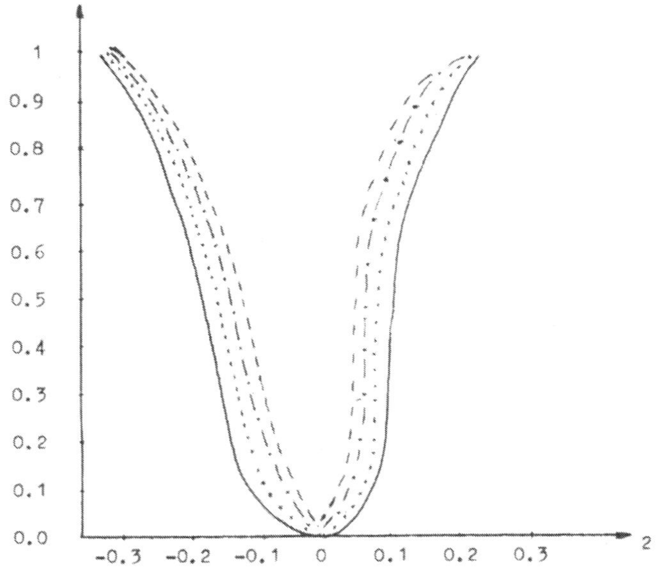

- - - - $1 - s$ estimation without outliers
· · · · $1 - s$ estimation with one outlier equal 5
——— $1 - s$ estimation with one outlier equal 10
- · - · msae estimation without outliers

Table 2

Values of the power function of the test in the case of 1 — s estimates

		Variant of contamination							
i	Value of parameter	j = 0	1	2	3	4	5	6	7
1	—0.3	1.0	1.0	0.980	1.0	0.910	1.0	1.0	0.85
2	—0.1	0.717	0.333	0.0	0.167	0.0	0.25	0.083	0.0
3	—0.05	0.350	0.033	0.0	0.067	0.0	0.05	0.0	0.0
0	0.0	0.1	0.017	0.0	0.017	0.0	0.017	0.01	0.0
4	0.05	0.283	0.167	0.0	0.0	0.0	0.133	0.067	0.0
5	0.1	0.617	0.450	0.017	0.1	0.0	0.417	0.15	0.0
6	0.15	0.933	0.733	0.183	0.317	0.17	0.867	0.7	0.185
7	0.2	0.983	0.967	0.567	0.633	0.083	1.0	0.85	0.35
8	0.3	1.0	1.0	0.940	0.990	0.510	1.0	1.0	0.99

Table 3

Values of $\pi_4(Z_j, W_i) = \dfrac{\left(P\left[val(t) \in \varphi_{0.1} \middle| \bar{H}_{ij}\right] - P\left[val(t) \in \varphi_{0.1} \middle| H_{io}\right]\right)}{P\left[val(t) \in \varphi_{0.1} \middle| H_{io}\right]}$

for $j = \overline{0.7}$ and estimates given by least squares and minimization of sum of absolute errors

1 — s estimates

Value of β_2	Variants of contamination j = 0	1	2	3	4	5	6	7
—0.3	0.0	0.0	0.0	—0.02	—0.09	0.0	0.0	—0.15
—0.1	0.0	—0.534	—1.0	—0.767	—1.00	—0.651	—0.882	—1.0
—0.05	0.0	—0.906	—1.0	—0.809	—1.00	—0.857	—1.0	—1.0
0.0	0.0	—0.83	—1.0	—0.83	—1.00	—0.83	—1.0	—1.0
0.05	0.0	—0.413	—1.0	—1.0	—1.00	—0.565	—0.767	—1.0
0.1	0.0	—0.27	—0.97	—0.838	—1.00	—0.324	—0.757	—1.0
0.15	0.0	—0.214	—0.804	—0.661	—0.983	—0.072	—0.250	—0.801
0.2	0.0	—0.017	—0.424	—0.350	—0.916	—0.017	—0.135	—0.644
0.3	0.0	0.0	0.0	—0.06	—0.49	0.0	0.0	0.01

msae estimates

Value of β_2	Variants of contamination j = 0	1	2	3	4	5	6	7
—0.3	0.0	0.0	—0.06	—0.0	—0.28	0.0	—0.01	—0.16
—0.1	—0.046	—0.512	—0.907	—0.604	—1.00	—0.488	—0.697	—1.0
—0.05	—0.143	—0.714	—1.00	—0.666	—1.00	—0.571	—1.00	—1.0
0.0	—0.33	—0.50	—1.0	—0.83	—1.0	—0.5	—1.0	—1.0
0.05	—0.177	—0.53	—1.0	—0.943	—1.0	—0.53	—0.803	—1.0
0.1	—0.134	—0.297	—0.919	—0.753	—0.972	—0.514	—0.730	—0.972
0.15	—0.089	—0.268	—0.804	—0.482	—0.987	—0.146	—0.429	—0.911
0.2	—0.05	—0.05	—0.458	—0.237	—0.88	—0.017	—0.186	—0.712
0.3	0.0	—0.01	—0.13	—0.02	—0.43	0.0	—0.02	—0.13

Table 4

Sensitivity of the power of the test measured by criterion π_7

1 — s estimates

β_2	—0.3	—0.1	—0.05	0.05	0.1	0.15	0.2	0.3
π_7	0.15	0.717	0.350	0.283	0.617	0.817	0.9	0.490
	msae estimates							
π_7	0.280	0.683	0.367	0.267	0.6	0.817	0.866	0.430

From Tables 2 and 3 it can be easily seen that within the alternative hypothesis $\bar{H}_{ij}$ $(i = \overline{1, 7})$ the power of the t-Student test rapidly decreases with the increase of the contamination of the sample, and the guaranteed level of power is greater than 0.9 just for $|\beta| \geq 0.3$. In our experiments the values of parameters define the level of correlation between Y and explanatory variables, so we can state that the existence of outliers in a sample makes it difficult to test the significance of parameters when a suitable explanatory variable is low correlated with the explained variable (corr (Y, X) $<$ 0.3).

In the case of msae estimates the values of P[val (t) $\in \varphi_{0.1} \mid H_{i0}$] were taken for 1 — s estimates.

The use of the t-Student test for testing significance of parameters in the case when they were obtained by minimization of absolute sum of residuals instead of squared residuals seems to be quite reasonable in spite of the fact that the real distribution of the test's statistics is unknown. In the case of lack of contamination in a sample the power characteristics of the test are only a bit worse for small values of parameter, and the existence of outliers makes these characteristics even better than in the case of the least squares. It is connected with smaller values of biasedness of msae estimates in the case of outliers existence. The guaranteed level of the test significance was nearly all the time equal to its assumed value 0.1 or even less than this value.

The comparison of values of the sensitivity of the power of the test shows its high sensitivity to outliers existence, their level and number in a sample.

4. Final Remarks

The presented analysis seems to confirm the necessity of a careful examination of test roubstness. Despite the fact that the list of alternatives used in our experiments was rather short, we can formulate some remarks concerning the robustness of the t-Student test as well ts the possibilities of using it in the case of estimates obtained by the msae method. Especially important seems to be a quite good performance of the test in the case of estimating the parameters by means of the msae method which is the most natural way of estimation and which leads to estimates which are more robust than the l — s estimates. On the other hand, the obtained results show rather low robustness of the t-Student test and by this fact the necessity of a careful examination of the tested sample.

5. References

BARNET, V., LEWIS, T.
Outliers in Statistical Data.
J. Wiley and Sons Inc., New York, (1978).

BARTOSZYŃSKI-PLESZCZYŃSKA
Reducibility of Statistical Structures and Decisions Problems.
Math. Stat. Banach Center Publications. Vol. 6, Warsaw PWN, (1980), 29–38.

BICKEL, P.
Another Look at Robustness.
Scandin. J. of Statistics (1976) 145–168.

BOX, G., ANDERSON, S.
Permutation Theory in the Derivation of Robust Criteria and the Study of Departures from Assumptions.
J. Roy. Statist. Soc. Ser. B, B, (1955) 1–34.

DUTTER, R.
Robuste Regression.
Bericht Nr. 135, Institut für Statistik, Techn. Univ. Graz, (1979) 111.

GREEN, R. F.
Outlier Prone and Outlier Resistant Distributions.
IASA 71 (1976) 502–505.

HAMPEL, F.
A General Qualitative Definition of Robustness.
Ann. Math. Statist. (1971), 1887–1896.

HUBER, P.
Robust Estimation of a Location Parameters.
Ann. Math. Statist. (1964), 73–101.

HUBER, P.
Robust Statistics.
Ann. Math. Statist. (1972), 1041–1067.

MILO, W., WASILEWSKI, Z.
Comparison of Methods for Detecting Outliers.
Publications Ecconométriques, 2 (1979) 43–53.

MILO, W., WASILEWSKI, Z.
The Analysis of Robustness of Chosen Tests for Outliers.
Paper within the Contract R. III. 9., 1981.

NEYMAN, J., SCOTT, E. L.
Outlier Proneness of Phenomena and of Related Distributions.
n: Jagolish S. Rustay ed.: Optimizing Methods in Statistics, Academic Press Inc., New York, 1971, 413–430.

ZIELIŃSKI, R.
Robustness: a Quantitative Approach.
Banach Center Publications Vol. 6: Mathematical Statistics PWN, Warsaw, 1980, 353–354.

ZIELIŃSKI, R.
Robustness: Discussion of an Approach.
Invited Lecture presented to the 14th European Meeting of Statisticians, Wrocław, August 31–September 4, 1981.

A Robust Estimate of Variance in a Linear Model

RYSZARD ZIELIŃSKI
Inst. Math. Polish Acad. Sci. Warsaw

WOJCIECH ZIELIŃSKI
Dept. Statistics, Univ. of Agriculture SGGW-AR, Warsaw

(Summary)

Let us consider a standard linear model $y = X\beta + \varepsilon$, where $\varepsilon = (\varepsilon_1, \ldots, \varepsilon_n)^T$, ε_i being iid random variables such that $E\varepsilon_i = 0$, Var $\varepsilon_i = \sigma^2$ and kurtosis γ_2 of ε_i equals a fixed number γ_2^0. Suppose that the model is violated in such a way, that γ_2 runs over an interval $(\underline{\gamma_2}, \overline{\gamma_2})$.

Lat $\mathfrak{A}$ be the class of matrices A such that y^TAy is the unbiased and shift-invariant estimate of the variance σ^2. We are looking for an estimate of variance the variance of which depends on kurtosis as little as possible, i.e. for a matrix $A \in \mathfrak{A}$ which minimizes

$$\sup_{\gamma_2} \mathrm{Var}\, y^T Ay - \inf_{\gamma_2} \mathrm{Var}\, y^T Ay.$$

It appears that in many practical situations there exist infinitely many such matrices A. From among them we choose that with minimal variance; such an estimate is uniquely determined. It is interesting that if $X = (1, \ldots, 1)^T$, i.e. if y is a vector of iid observations then the robust estimate with minimal variance coincides with the standard one.

The full text of this paper will appear in "Matematyka Stosowana" (in Polish).

Institute of Mathematics, Polish Academy of Sciences, Warsaw

Minimax Versus Robust Experimental Design: Two Simple Examples

RYSZARD ZIELIŃSKI

Abstract

There exist many approaches to assessing robustness of statistical procedures. We discuss two of them: one connected with stability of the performance of a given statistic when passing to a supermodel, and the second connected with constructing the procedure which is minimax in the supermodel. Both approaches are compared in simple practical situations.

1. Introduction

Consider a given statistical model $M_0 = (\mathfrak{X}, \mathfrak{B}, \mathfrak{P}_0)$ where $(\mathfrak{X}, \mathfrak{B})$ is a sample space and $\mathfrak{P}_0 \subset \mathfrak{P}$ is a family of distributions on $(\mathfrak{X}, \mathfrak{B})$, $\mathfrak{P}$ being the family of all distributions on the sample space. Usually, the adapted statistical model M_0 describes a real situation under consideration only approximately so that the statistician should take into account a supermodel $M_1 = (\mathfrak{X}, \mathfrak{B}, \mathfrak{P}_1)$, $\mathfrak{P}_0 \subset \mathfrak{P}_1 \subset \mathfrak{P}$. There are many ways of doing this (see e.g. Pearson 1931, Hoeffding 1955, Box and Andersen 1955, Huber 1981, Hampel 1971, Zieliński 1977, Zolotariev 1977) but in the problems which we consider below there are apparently only two competitive approaches: one in the spirit of the minimax, as in Chapter 9 of the well known Huber's book (1981), and the other in the spirit of stability of the solution when passing from M_0 to M_1, as in Zieliński (1977). The former can be described as a minimax solution for the statistical model M_1. To explain the idea of the latter, consider a statistical procedure T (for example a test or an estimate) and a property ϱ (for example the size or the bias). Any nontrivial property of the procedure T depends on the distribution P of the observations so we write $\varrho = \varrho_T(P)$. The procedure T is robust if $\varrho_T(P)$ does not change significantly when P leaves $\mathfrak{P}_0$ and runs through $\mathfrak{P}_1$.

2. A Regression Problem: Estimation

Consider the simple linear model $y = ax + \varepsilon$, $0 \le x \le 1$, $E\varepsilon = 0$, $\operatorname{Var}\varepsilon = \sigma^2$, with an unknown a. Suppose that we are interested in estimating a so that we could be able to predict y at a given point $t \in (0, 1]$, and that to this end we are allowed to perform only one experiment consisting in observation of y at a freely chosen point $x \in (0, 1]$. The standard estimate of a is then $\hat{a}_x = y/x$ and the standard prediction of y at t is $\hat{a}_x \cdot t$. How to choose x? The mean square error of the prediction $E(\hat{a}_x t - at)^2 = \sigma^2 t^2/x^2$ and hence the answer is: whatever t, take $x = 1$. Now suppose that, due to measurement errors or due to some other extraneous factors, a small quadratic term may appear, and consider the supermodel $y = ax + \beta_0 + \beta_2 x^2 + \varepsilon$ with x, a and ε as before and with β_0 and β_2 such that $\beta_0^2 + \beta_2^2 \le \eta^2$ for a given small positive number η. How should we choose x now?

For the mean square error of prediction of y at t, when the observation of y is taken at x, we have

$$\operatorname{MSE}(\hat{y}(t); x, \beta_0, \beta_2) = \left[\left(\frac{t}{x} - 1\right)\beta_0 + t(x-t)\beta_2\right]^2 + \left(\frac{t}{x}\right)^2 \sigma^2.$$

The minimax solution x_{minmax} is that which minimizes

$$\sup_{\beta_0^2+\beta_2^2 \le \eta^2} \operatorname{MSE}(\hat{y}(t); x, \beta_0, \beta_2) = \eta^2\left[\left(\frac{t}{x} - 1\right)^2 + (x-t)^2 t^2\right] + \left(\frac{t}{x}\right)^2 \sigma^2. \quad (1)$$

The robust solution x_{rob} is that for which $\operatorname{MSE}(\hat{y}(t); x, \beta_0, \beta_2)$ is stable and changes as little as possible when (β_0, β_2) runs over the set $\{(\beta_0, \beta_2) : \beta_0^2 + \beta_2^2 \le \eta^2\}$. That means that x_{rob} minimizes

$$\sup_{\beta_0^2+\beta_2^2 \le \eta^2} \operatorname{MSE}(\hat{y}(t); x, \beta_0, \beta_2) - \inf_{\beta_0^2+\beta_2^2 \le \eta^2} \operatorname{MSE}(\hat{y}(t); x, \beta_0, \beta_2) = \eta^2\left[\left(\frac{t}{x} - 1\right)^2 + (x-t)^2 t^2\right]. \quad (2)$$

The obvious solution is $x_{rob} = t$.

To compare the both solutions numerically suppose that $t = \frac{1}{2}$, $\eta = \sigma = 1$. Then $x_{minmax} = 0.875$, $x_{rob} = 0.5$, and (1), (2) and

$$\inf_{\beta_0^2+\beta_2^2 \le \eta^2} \operatorname{MSE}(\hat{y}(t); x, \beta_0, \beta_2) = \left(\frac{t}{x}\right)^2 \sigma^2 \quad (3)$$

are as in Fig. 1.

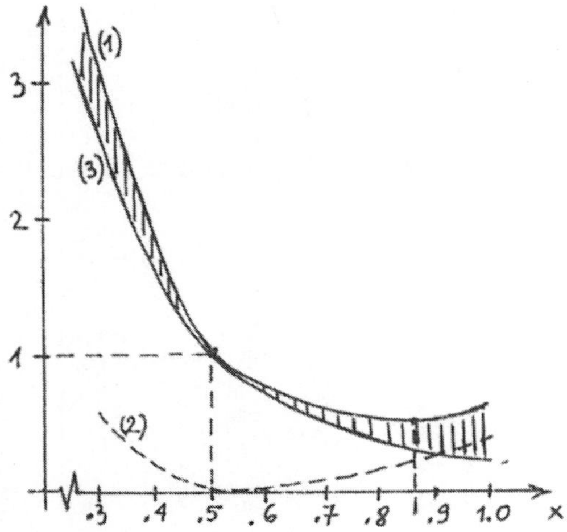

Figure 1

The minimax solution seems better in that its maximal MSE of the prediction is smaller for every regression function in the supermodel. We may also expect a shorter confidence interval although the construction of that interval might be somewhat troublesome (e.g. if ε is normally distributed, the confidence interval is based on noncentral t). On the other hand, the robust solution seems better in that its MSE of prediction is constant over the supermodel; the underlying estimate $\hat{y}(t)$ of $y(t)$ is unbiased so that MSE equals to the variance ; and the confidence interval, possibly longer, does not depend on the regression function.

The following more general situation was considered by Huber (1981, Chapter 9). Suppose we want to estimate the parameters a and b of the standard linear model M_0: $Ey(x) = a + bx$, $-\frac{1}{2} \le x \le \frac{1}{2}$, $\operatorname{Var} y(x) = \sigma^2$, by $\hat{a}$ and $\hat{b}$, respectively, so to minimize the integrated mean square error of prediction

$$E\int_{-1/2}^{1/2}\left[(a+bx)-(\hat{a}+\hat{b}x)\right]^2 dx.$$

Given a continous experimental design ξ (a probability measure on $\left[-\frac{1}{2},\frac{1}{2}\right]$), the well known least squares estimates are

$$\hat{a} = \int_{-1/2}^{1/2} y(x)\xi(dx), \quad \hat{b} = \frac{1}{\gamma}\int_{-1/2}^{1/2} x\cdot y(x)\xi(dx).$$

$$\text{where } \gamma = \int_{-1/2}^{1/2} x^2\xi(dx).$$

Now suppose that the original model M_0 is violated in such a way that $Ey(x) = f(x)$, f being an „almost linear" function which means that $f \in \mathfrak{F}_\eta$ with

$$\mathfrak{F}_\eta = \left\{ f \in L_2\left[-\frac{1}{2},\frac{1}{2}\right]: \ \|f(x)-(a_f+b_f x)\|^2 \le \eta \right\}$$

where $a_f = \int_{-1/2}^{1/2} f(x)dx$, $b_f = \int_{-1/2}^{1/2} x\cdot f(x)dx/12$, so that $a_f + b_f x$ is the best linear approximation of f. The symbol $\|\cdot\|$ states for the norm in the space L_2 of all square integrable functions on $\left[-\frac{1}{2},\frac{1}{2}\right]$ and η is a small positive constant.

If "true" regression function is $f \in \mathfrak{F}_\eta$, then the mean square error of prediction equals

$$Q(f,\xi) = E\left\|f(x)-(\hat{a}+\hat{b}x)\right\|^2$$

The minimax experimental design ζ_{minmax} is that minimizing

$$\sup_{f\in\mathfrak{F}_\eta} Q(f,\xi)$$

and the robust experimental design ζ_{rob} minimizes the variation of $Q(f,\xi)$ over the supermodel:

$$\sup_{f\in\mathfrak{F}_\eta} Q(f,\xi) - \inf_{f\in\mathfrak{F}_\eta} Q(f,\xi).$$

The Huber's solution is the density m_{minmax} minimizing

$$\sup_{f\in\mathfrak{F}_\eta} Q(f,\xi) = \eta\left(1+\|m-1\|^2\right) + \left(1+\frac{1}{12\gamma}\right)\frac{\sigma^2}{n}$$

where m is the density of the experimental design ξ (so that $\xi(dx) = m(x)dx$) and n is the number of adopted experimental points. The density m_{minmax} has a rather complicated form so we do not give it here (see Huber (1981), p. 246). It is easy to observe that

$$\inf_{f\in\mathfrak{F}_\eta} Q(f,\xi) = \left(1+\frac{1}{12\gamma}\right)\frac{\sigma^2}{n}$$

which is attained by any linear function f. It follows that

$$\sup_{f\in\mathfrak{F}_\eta} Q(f,\xi) - \inf_{f\in\mathfrak{F}_\eta} Q(f,\xi) = \eta\left(1+\|m-1\|^2\right),$$

and hence the most robust experimental design ξ_{rob}, which minimizes this quantity, is the uniform experimental design with the density $m_{rob}(x) \equiv 1$ for $-\frac{1}{2} \le x \le \frac{1}{2}$. Some numerical results for certain η_0 such that $\eta = \eta_0\delta^2/n$, are presented in Fig. 2.

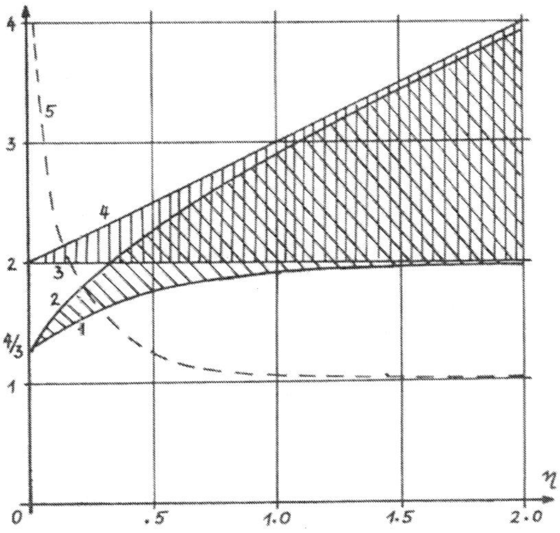

1) $\frac{n}{\sigma^2}\inf_f Q(f,\xi_{minmax})$; 2) $\frac{n}{\sigma^2}\sup_f Q(f,\xi_{minmax})$

3) $\frac{n}{\sigma^2}\inf_f Q(f,\xi_{rob})$; 4) $\frac{n}{\sigma^2}\sup_f Q(f,\xi_{rob})$

5) $\left[\sup_f Q(f,\xi_{minmax}) - \inf_f Q(f,\xi_{minmax})\right] \Big/ \left[\sup_f Q(f,\xi_{rob}) - \inf_f Q(f,\xi_{rob})\right]$

Figure 2

The minimax solution seems more acceptable than the robust one, especially for small violations of the linear model (for small values of η_0) although the ratio of the variation of $Q(f,\xi_{minmax})$ in the supermodel to that of $Q(f,\xi_{rob})$ tends to infinity when η_0 tends to zero.

3. A Regression Problem: Hypotheses Testing

Consider the linear model $M_0: y = ax + \varepsilon$, $0 \le x \le 1$, where ε is a random variable distributed as N (0, 1), and suppose that the problem is to verify the hypothesis $H: a = 0$ against $K: a > 0$, at a given significance level $\alpha \in (0, 1)$. To this end we are allowed to perform a series of experiments which consist in observing y at $n \ge 1$ freely chosen points $x_1, x_2, \ldots, x_n \in [0, 1]$; the choice of these points should intend to achieve as powerful test as possible. The observed values of y will be denoted by $y_1, y_2, \ldots, y_n$ and $X = \{x_1, x_2, \ldots, x_n\}$ will be refered to as an experimental design.

Suppose that the linear model under consideration is slightly violated and consider the supermodel M_1: $y = ax + g(x) + \varepsilon$, where g (x) and x are linearly independent

$$\int_0^1 x\,g(x)\,dx = 0 \qquad (4)$$

and g (x) represents a "small violation"

$$\sup_{0\le x\le 1} |g(x)| \le \eta \qquad (5)$$

for a fixed small positive number η. The class of functions g satisfying (4) and (5) will be denoted by G_η.

The well known minimax (the more adequate term here is "maximin") procedure is as follows: from among all tests satisfying

$$\sup_{g \in G_\eta} P\{\text{rejecting H when } y = g(x) + \varepsilon\} \le \alpha \tag{6}$$

choose that maximazing (uniformly in $a > 0$, if possible)

$$\inf_{g \in G_\eta} P\{\text{rejecting H when } y = ax + g(x) + \varepsilon\}. \tag{7}$$

In our case, for every experimental design X we can construct the uniformly most powerful α-test $\xi_X(y_1, y_2, \ldots, y_n)$ and then choose the experimental design X so to maximize (7). Given X, the test $\xi_X(y_1, y_2, \ldots, y_n)$ has the form

$$\varphi_X(y_1, y_2, \ldots, y_n) = \begin{cases} 1, \text{ if } \sum_{i=1}^{n} x_i y_i \ge K, \\ 0, \text{ otherwise,} \end{cases} \tag{8}$$

where $K = \Phi^{-1}(1-\alpha) \cdot \left(\sum_{i=1}^{n} x_i^2\right)^{1/2} + \eta \cdot \sum_{i=1}^{n} |x_i|$, Φ being the cumulative distribution function of the standard normal variable N (0, 1).

Denote by $\mathfrak{M}_X(a)$ the probability of rejecting H when $Ey = ax + g(x)$ and the test ξ_X is applied. When g runs over G_η, we have

$$1 - \Phi\left(\Phi^{-1}(1-\alpha) - a\sqrt{\sum x_i^2} + 2\eta \frac{\sum |x_i|}{\sqrt{\sum x_i^2}}\right) \le \mathfrak{M}(a) \le$$

$$\le 1 - \Phi\left(\Phi^{-1}(1-\alpha) - a\sqrt{\sum x_i^2}\right)$$

so that the minimax design X is that maximazing the left-hand side of this inequality.

Now consider the robust design. Given X, the uniformly most powerful α-test $\xi_X(y_1, y_2, \ldots, y_n)$ in the original model M_0 has the form (8) with

$$K = \Phi^{-1}(1-\alpha) \cdot \left(\sum_{i=1}^{n} x_i^2\right)^{1/2}.$$

If g runs over G_η then the power function changes in the interval

$$1 - \Phi\left(\Phi^{-1}(1-\alpha) - a\sqrt{\sum x_i^2} + \eta \frac{\sum |x_i|}{\sqrt{\sum x_i^2}}\right) \le \mathfrak{M}(a) \le$$

$$\le 1 - \Phi\left(\Phi^{-1}(1-\alpha) - a\sqrt{\sum x_i^2} - \eta \frac{\sum |x_i|}{\sqrt{\sum x_i^2}}\right)$$

To construct the test with the power function which is as stable as possible under the violation $g \in G_\eta$ we have to choose X for which the above interval is as short as possible. This amounts to minimizing $\sum x_i / \sqrt{\sum x_i^2}$.

but now

$$1 - \Phi\left(\Phi^{-1}(1-\alpha) - a + \eta\right) \le \mathfrak{M}(a) \le 1 - \Phi\left(\Phi^{-1}(1-\alpha) - a - \eta\right)$$

The results are presented in Fig. 3. An advantage of the minimax solution is that the size of the test never exceeds the adapted level α but the price is a loss in power of the test in the original model M_0 (see the dotted line in Fig. 3). On the other hand, the robust solution gives us the better power function (the best one in the original model) but the price in an augmentation of the size of the test under violations of the original model.

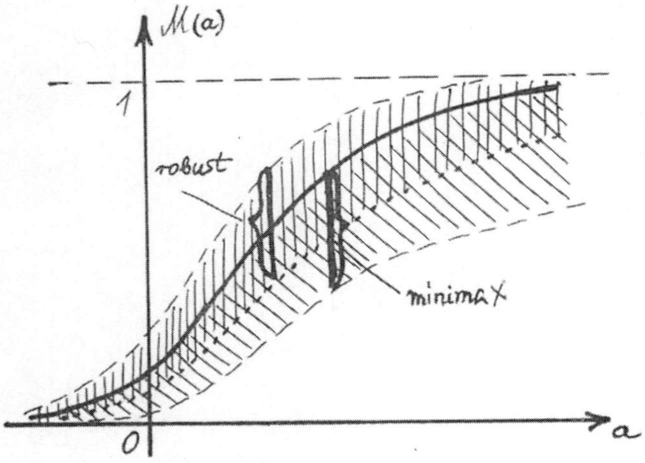

Fig. 3

4. A Final Remark

The aim of my talk was not to advocate the robust methods versus minimax ones or vice versa, but to visualize some differencies and some consequences of using ones instead of others. In my opinion, in robustness the emphasis lays rather on stability than on efficiency and in the minimax approach the main problem is efficiency in an adequate supermodel.

5. References

BOX, G. E. P., ANDERSEN, S. L.
Permutation theory in the derivation of robust criteria and the study of departures from assumptions.
J. Roy. Statist. Soc., Ser. B, 17 (1955), 1–34.

HAMPEL, F. R.
A general qualitative definition of robustness.
Ann. Math. Statist. 42 (1971), 1887–1896.

HOEFFDING, W.
The role of assumptions in statistical decisions.
Proc. Third Berkeley Symp. Math. Statist. and Probability. Vol. I (1955), 105–114.

HUBER, P. J.
Robust statistics.
John Wiley and Sons. New York – Chichester – Brisbane – Toronto 1981.

PEARSON, E. S.
The analysis of variance in case of non-normal variation.
Biometrika 23 (1931), 114–133.

ZIELIŃSKI, R.
Robustness: a quantitative approach.
Bull. Acad. Polonaise Sci. Ser. math., astr. phys. Vol. XXV, (1977), No. 12, 1281–1286.

ZOLOTARIEV, V. M.
General problems of the stability of mathematical models.
Proc. 41st Session of the International Statistical Institute, New Delhi 1977, 382–401.